Latent Class Analysis of Survey Error

WILEY SERIES IN SURVEY METHODOLOGY
Established in Part by WALTER A. SHEWHART AND SAMUEL S. WILKS

A complete list of the titles in this series appears at the end of this volume.

Latent Class Analysis of Survey Error

PAUL P. BIEMER

RTI International
University of North Carolina—Chapel Hill

Published by John Wiley & Sons, Inc., Hoboken, New Jersey
Published simultaneously in Canada

For general information on our other products and services or for technical support, please contact our Customer Care Department within the United States at (800) 762-2974, outside the United States at (317) 572-3993 or fax (317) 572-4002.

Wiley also publishes its books in a variety of electronic formats. Some content that appears in print may not be available in electronic formats. For more information about Wiley products, visit our web site at www.wiley.com.

Library of Congress Cataloging-in-Publication Data:

Biemer, Paul P.
Latent class analysis of survey error / Paul P. Biemer
p. cm.
Includes index.
ISBN 978-0-470-28907-5 (cloth)

oBook ISBN: 978-0-470-89115-5
ePDF ISBN: 978-0-470-89114-8

10 9 8 7 6 5 4 3 2 1

To Judy, Erika, and Lianne

Contents

Preface

Survey estimates are based on a sample and are therefore subject to sampling error. However, there are many other sources of error in surveys. These errors are referred to collectively as *nonsampling errors*. Nonsampling errors arise from interviewers, respondents, data processors, other survey personnel, and operations. Evaluating sampling error is easily accomplished by estimating the standard error of an estimate that any commercial analysis software package can do. Evaluating nonsampling errors is quite difficult, often requiring data not normally collected in the typical survey. This book discusses methods for evaluating the nonsampling error in survey data focusing primarily on data that are categorical and errors that result in misclassifications. The book concentrates on a general set of models and techniques referred to as collectively as latent class analysis.

Latent class analysis (LCA) encompasses a wide range of techniques and models that can be used for numerous applications, and this book covers many of those. The general methods literature often views LCA as the categorical data analog of factor analysis. That is not the approach taken in this book. Rather, this book treats LCA as a generalization of classical test theory and the traditional survey error modeling approaches. Readers who wish to apply LCA in factor analytic applications can benefit from the techniques considered here, but may wish to supplement their study of LCA with other literature cited in Chapter 4.

This book was written because there is currently no comprehensive resource for survey researchers and practitioners to learn the methods needed in the modeling and estimation of classification errors, particularly LCA techniques. Survey error evaluation methodology emerged in the 1960s primarily at the US Bureau of the Census. Over the years, a number of journal articles have been published that describe various models for estimating classification error in survey results; however, the work has been relatively sparse and scattered. This book collects the many years of experience of the author and other authors in the field of survey misclassification to provide a guide for the practitioner as well as a text for the student of survey error evaluation. It combines the theoretical, methodological, and practical aspects of estimating classification error and interpreting the results for the purposes of survey data quality improvement.

The audience for the book is individuals in government, universities, businesses, and other organizations who are involved in the development, implementation, or evaluation of surveys. Students of statistics or survey methodology who wish to learn how to evaluate the measurement error in survey data could also use this book as a resource for their studies. The book's content should be accessible to anyone having an intermediate background in statistics and sampling methods. An ideal reader is anyone possessing a working understanding of elementary probability theory, expectation, bias, variance, multinomial distribution, hypothesis testing, and model goodness of fit. Knowledge of categorical data analysis is helpful but is not required to grasp the essential concepts of the book. A primer on categorical data analysis containing the essential concepts used in the book is provided in Appendix B.

The book provides a very general statistical framework for modeling and estimating classification error in surveys. Following a general discussion of surveys and nonsampling errors in Chapter 1, Chapter 2 examines some of the early models for survey measurement error including the Census Bureau model and the classical test theory model. In this chapter, the similarities and differences, strengths, and weaknesses of the approaches are described. This background serves to introduce the basic concepts of measurement error modeling for categorical data in Chapter 3, beginning with the very elementary model proposed by Bross. A latent class model for two measurements (Hui–Walter model) is introduced that also serves to establish the essential principles of LCA. This chapter introduces the reader to the concept of the true value of a variable as an unobserved (latent) variable and the survey response as a single indicator of this latent variable. The early models are shown to be special cases of this general latent class model. Chapter 3 also describes the expectation–maximization (EM) algorithm and its pivotal role in latent class model parameter estimation.

The general latent class model for three or more indicators is introduced in Chapter 4. This chapter provides all the details regarding how to estimate the model parameters, how to build good models and test their fit, and other issues related to the interpretation of the model parameter estimates. Chapter 4 also introduces the LCA software package, ℓEM, written by Dr. Jeroen Vermunt, which can be downloaded free from Vermunt's Website. ℓEM input statements are provided for most of the examples thoughout the book, and readers are encouraged to replicate the analysis with this software. Chapter 5 contains a number of advance topics in LCA, including how one deals with sparse data, boundary values, unidentifiability, and local maxima. Chapter 5 also provides an extensive discussion of local dependence in LCA and how to model its effects. The chapter ends with a discussion of latent class modeling with complex survey data. A number of advanced applications of LCA are included in Chapter 6.

Chapter 7 discusses models and analysis techniques that are appropriate for evaluating the measurement error in panel survey data referred to as *Markov latent class analysis* (MLCA). MLCA is an important area for survey evaluation because it provides a means for estimating classification error directly from the data collected from the survey without the need for special reinterview or response replication studies. Essential in the application of these models is some evaluation or assessment of the extent to which the model assumption holds. The chapter concludes with a discussion of these issues and the primary approaches for model validation.

Finally, Chapter 8 provides an overview of LCA and MLCA, tracing the history of the methodology and summarizing the current state of the art. The chapter considers some of the criticisms and pitfalls of LCA, when such criticism may be justified, and how to avoid the pitfalls and criticisms by an appropriate analysis of the data. Glimpsing into the future, the chapter concludes with a discussion of areas where further research is needed to advance the field.

Acknowledgments

A number of individuals and organizations deserve recognition and my appreciation for their support and encouragement throughout the preparation of this book. First, much appreciation is due to RTI International, who, through the RTI Fellows Program, supported much of my time to write this book. I am also indebted to Wayne Holden, my supervisor, for his encouragement throughout this project. This book has benefited substantially from my associations with Marcus Berzofsky, who wrote his Ph.D. thesis in this area, and Bill Kalsbeek, who codirected Berzofsky's dissertation with me. Berzofsky also read a draft of the manuscript and offered valuable suggestions for improvement. In that regard, many thanks also to Lars Lyberg, Frauke Kreuter, Juliana Werneburg, and Adam Carle, who also read drafts of the manuscript and offered important ideas for improving the material.

I am very grateful for the many contributions of Clyde Tucker and Brian Meekins at BLS to this work. Our collaboration on applying MLCA to the Consumer Expenditure Survey provided considerable insights into the many complexities associated with real-world applications. John Bushery, Chris Wiesen, Gordon Brown, and Ken Bollen coauthored papers with me in this area, and I am sincerely grateful for the substantial body of knowledge I gained from these associations. Thanks also to the many students to whom I have taught this material. Their questions, comments, and critiques helped me clarify the concepts and helped to reveal areas of the methodology that are more difficult for the novice. Jeroen Vermunt deserves my sincere gratitude for his advice over the years. He has taught me much about LCA through our discussions, email exchanges, and other writings. The field of LCA also owes him a great debt of gratitude for writing the ℓEM software and making it available to anyone for free.

Finally, this book would not have been possible without the support, sacrifice, and encouragement of my loving wife, Judy. To her I express love and sincere appreciation.

Raleigh, NC
November 2010

PAUL P. BIEMER

Abbreviations

AIC	Akaike information criterion
ALVS	Agriculture Land Values Study
ANES	American National Election Studies
ANOVA	analysis of covariance (ANACOVA—analysis of covariance)
ARL	administrative records list
BIC	Bayesian information criterion
BLS	Bureau of Labor Statistics (US)
BVR	bivariate residual
CATI	computer-assisted telephone interview(ing)
CASI	computer-assisted self-interview(ing) (ACASI—audioCASI)
CBCL	child behavior checklist
CDAS	Categorical Data Analysis System (proprietary software)
CEIS	Consumer Expenditure Interview Survey
CPS	Current Population Survey (US)
CR	Cressie–Read
DA	data acquisition
deff	design effect
df	degree(s) of freedom
DSF	delivery service file
ECM*	expectation–conditional maximization
EE	erroneously enumerated
EFU	Evaluation Followup (US Census Bureau)
EI	evaluation interview
EM	expectation–maximizatin (IEM–LCA software package for EM; proprietary to J. Vermunt)
EMP	employed
EPSEM	equal-probability selection method
ERS	Economic Research Service
FR	field representative (SFR—supervisory FR)
GLLAMM	generalized linear latent and mixed models

HT	Horvitz–Thompson
ICC	intracluster correlation
ICE	independent classification error
ICM	integrated coverage measurement
i.i.d.	independent and identically distributed
IPF	iterative proportional fitting
IRT	item response theory
KD20	Kuder–Richardson Formula 20
LC	latent class
LCA	latent class analysis (MLCA—Markov LCA)
LCMS	latent class mover–stayer (model)
LD	local dependence
LF	labor force
LISREL*	*li*near *s*tructural *rel*ations
LL	loglinear–logit (model; loglinear and logit combined)
LOR	log-odds ratio (LORC—LOR check)
LTA	latent transition analysis
MAR	missing at random (MCAR—missing completely at random; NMAR—not missing at random)
ML	maximum likelihood (MLE—ML estimation)
MLC	Markov latent class
MM	manifest Markov (model)
MMS	manifest mover–stayer (model)
MSA	metropolitan statistical area
MSE	mean-squared error
NASS	National Agricultural Statistics Service (US)
NDR	net difference rate
NHSDA	National Household Survey on Drug Abuse (US)
NHIS	National Health Interview Survey (US)
NLF	not (in) labor force
npar	number of (free unconstrained π or u) parameters
NR	nonresponse
NSDUH	National Survey on Drug Use and Health (US)
PES	postenumeration survey
PML	pseudo-maximum-likelihood
PPS	probabilities proportional to size
Pr	probability (in equations; Prob in tables)
PS	poststratification
PSU	primary-stage sampling unit (SSU—secondary-stage SU)
RB	relative bias
RDD	random-digit dialing
SAS*	statistical analysis system
s.e.	standard error (in equations; Std Err in tables)
SEM	structural equation model
SIPP	Survey of Income Program Participation (US)

SPSS*	statistical package for social sciences; statistical product and service solutions
SRS	simple random sampling
SRV	simple response variance
SSM	scale score measure(ment)
SV	sampling variance
UNE	unemployed
UWE	unequal weighting effect

*ECM, LISREL, SAS, and SPSS are proprietary names; these acronyms are commonly used in industry and elsewhere (original unabbreviated meanings given here are rarely used).

CHAPTER 1

Survey Error Evaluation

1.1 SURVEY ERROR

1.1.1 An Overview of Surveys

This book focuses primarily on the errors in data collected in sample surveys and how to evaluate them. A natural place to start is to define the term "survey." The American Statistical Association's Section on Survey Research Methods has produced a series of 10 short pamphlets under the rubric *What Is a Survey?* (Scheuren 1999). That series defines a survey as a *method* of gathering information from a *sample* of objects (or *units*) that constitute a *population*. Typically, a survey involves a questionnaire of some type that is completed by either an informant (referred to as the *respondent*), an interviewer, an observer, or other agent acting on behalf of the survey organization or sponsor. The population of units can be individuals such as householders, teachers, physicians, or laborers or other entities such as businesses, schools, farms, or institutions. In some cases, the units can even be events such as hospitalizations, accidents, investigations, or incarcerations. Essentially any object of interest can form the population.

In a broad sense, surveys also include censuses because the primary distinction is just the fraction of the sample to be surveyed. A survey is confined to only a sample or subset of the population. Usually, only a small fraction of the population members is selected for the sample. In a *census*, every unit in the population is selected. Therefore, much of what we say about surveys also applies to censuses.

If the survey sample is selected randomly (i.e., by a probability mechanism giving known, nonzero probabilities of selection to each population member), valid statistical statements regarding the parameters of the population can be made. For example, suppose that a government agency wants to estimate the proportion of 2-year-old children in the country that has been vaccinated against infectious diseases (polio, diphtheria, etc.). A randomly selected sample

Latent Class Analysis of Survey Error By Paul P. Biemer

of 1000 children in this age group is drawn and their caregivers are interviewed. From these data, it is possible to determine the proportion of children who are vaccinated within some specified *margin of error* for the estimate. Sampling theory [see, e.g., Cochran (1977) or more recently, Levy and Lemeshow (2008)] provides specific methods for estimating margins of error and testing hypotheses about the population parameters.

Surveys may be cross-sectional or longitudinal. *Cross-sectional* surveys provide a "snapshot" of the population at one point in time. The products of cross-sectional surveys are typically descriptive statistics that capture distributions of the population for characteristics of interest, including health, education criminal justice, economics, and environmental variables. Cross-sectional surveys may occur only once or may be repeated at some regular interval (e.g., annually). As an example, the National Heath Interview Survey (Centers for Disease Control and Prevention 2009) is conducted monthly and collects important data on the health characteristics of the US population.

Longitudinal or *panel* surveys are repeating surveys where at least some of the same sample units are interviewed at different points in time. By taking similar measurements on the same units at different points in time, investigators can more precisely estimated changes in population parameters as well as individual characteristics. A *fixed panel* (or *cohort*) survey interviews the entire sample repeatedly usually over some significant period of time such as 2 or more years. As an example, the Panel Study of Income Dynamics (Hill 1991) has been collecting income data on the same 4800 families (as well as families spawned from these) since 1968.

A *rotating panel survey* is a type of longitudinal survey where part of the sample is replaced at regular intervals while the remainder of the sample is carried forward for additional interviewing. This design retains many of the advantages of a fixed panel design for estimating change while reducing the burden and possible conditioning effects on sample units caused by repeatedly interviewing them many times. An example is the US Current Population Survey (CPS) (US Census Bureau 2006), which is a monthly household survey for measuring the month-to-month and year-to-year changes in labor force participation rates. The CPS uses a somewhat complex rotating panel design, where each month about one-eighth of the sample is replaced by new households. In this way, households are interviewed a maximum of 8 times before they are *rotated* out of the sample.

Finally, a *split-panel* survey is a type of longitudinal survey that combines the features of a repeated cross-sectional survey with a fixed panel survey design. The sample is divided into two subsamples: one that is treated as a repeated cross-sectional survey and the other that follows a rotating panel design. An example of a split-sample design is the American National Election Studies [American National Election Studies (ANES), 2008]. Figure 1.1 compares these four survey designs.

As this book will explain, methods for evaluating the error in surveys may differ depending on the type of survey. Many of the methods discussed can be

	Time 1	Time 2	Time 3	Time 4
		Repeated Cross-Sectional		
Sample 1	X			
Sample 2		X		
Sample 3			X	

		Fixed Panel		
Sample 1	X	X	X	X

		Rotating Panel		
Sample 1	X	X		
Sample 2		X	X	
Sample 3			X	X

		Split-Panel		
Sample 1-A	X			
Sample 1-B	X	X		
Sample 2-A		X		
Sample 2-B		X	X	
Sample 3-A			X	
Sample 3-B			X	X

Figure 1.1 Reinterview patterns for four survey types.

applied to any survey while others are appropriate only for longitudinal surveys. The next section provides some background on the problem of survey error and its effects on survey quality.

1.1.2 Survey Quality and Accuracy and Total Survey Error

The terms *survey quality*, *survey data quality*, *accuracy*, *bias*, *variance*, *total survey error*, *measurement validity*, and *reliability* are encountered quite often in the survey error literature. Unfortunately, their definitions are often unspecified or inconsistent from study to study, which has led to some confusion in the field. In this section, we provide definitions of these terms that are reasonably consistent with conventional use, beginning with perhaps the most ambiguous term: survey quality.

Because of its subjective nature, *survey quality* is a vague concept. To some data producers, survey quality might mean *data* quality: large sample size, a high response rate, error-free responses, and very little missing data. Statisticians, in particular, might rate such a survey highly on some quality scale. Data users, on the other hand, might still complain that the data were not timely or accessible, that the documentation of the data files is confusing and incomplete, or that the questionnaire omitted many relevant areas of inquiry that are essential for research in their chosen field. From the user's perspective, the survey exhibits very poor quality.

These different points of view suggest that survey quality is a very complex, multidimensional concept. Juran and Gryna (1980) proposed a simple definition of quality that can be appropriately applied to surveys, namely, the quality of a product is its "fitness for use." But, as Juran and Gryna explain, this definition is deceptively simple because there are really two facets of quality: (1) freedom from deficiencies and (2) responsiveness to customers' needs. For survey work, facet 1 might be translated as error-free data, data accuracy, or high data quality, while facet 2 might be translated as providing product features that result in high user satisfaction. The latter might include data accessibility and clarity, timely data delivery, collection of relevant information, and use of coherent and conventional concepts.

When applied to statistical products, the definition "fitness for use" has another limitation in that it implies a single use or purpose. Surveys are usually designed for multiple objectives among many data users. A variable in a survey may be used in many different ways, depending on the goals of the data analyst. For some uses, timeliness may be paramount. For other uses, timeliness is desirable, but *comparability* (i.e., ensuring that the results can be compared unambiguously to prior data releases from the same survey) may be more critical.

In the mid-1970s, a few government statistical offices began to develop definitions for survey quality that explicitly took into account the multidimensionality of the concept [see, e.g., Lyberg et al. (1977) or, more recently, Fellegi (1996)]. This set of definitions has been referred to as a *survey quality framework*. As an example, the quality framework used by Statistics Canada includes these seven quality dimensions: relevance, accuracy, timeliness, accessibility, interpretability, comparability, and coherence. Formal and accepted definitions of these concepts can be found at Statistics Canada (2006). Eurostat has also adopted a similar quality framework [see, e.g., Eurostat (2003)].

Given this multidimensional conceptulization of quality, a natural question is quality to be maximized in a survey? One might conceptualize a one-dimensional indicator that combines these seven dimensions into an overall survey quality indicator. Then the indicator could be evaluated for various designs and the survey design maximizing this quantity could be selected. However, this approach oversimplifies the complexity of the problem since there is no appropriate way for combining the diverse dimensions of survey quality. Rather, quality reports or quality declarations providing information on each dimension have been used to summarize survey quality. A quality report might include a description of the strengths and weaknesses of a survey organized by quality dimension, with emphasis on sampling errors, nonsampling errors,[1] key release dates for user data files, forms of dissemination,

[1]Nonsampling errors are discussed in more detail in Section 1.1.3. Nonsampling errors, which are inevitable in survey data collection, arise from many sources, including interviewers, respondents, data processors, other survey personnel, and operations. As the term implies, they are essentially all errors in a survey apart from sampling errors.

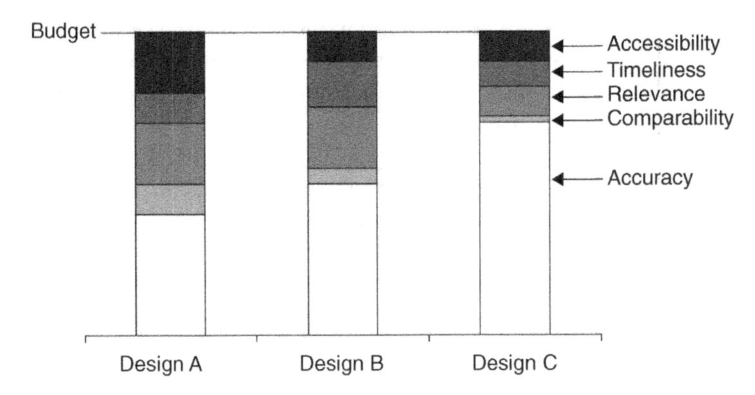

Figure 1.2 Comparisons of three cost-equivalent survey designs. (*Source*: Biemer 2010).

availability, and contents of documentation, as well as special features of the survey approach that may be of importance to most users. A number of surveys have produced extended versions of such reports, called *quality profiles*. A quality profile is a document that provides a comprehensive picture of the quality of a survey, addressing each potential source of error. Quality profiles have been developed for a number of US surveys, including the Current Population Survey (CPS) (Brooks and Bailar 1978), the Survey of Income and Program Participation (Jabine et al. 1990), US Schools and Staffing Survey (Kalton et al. 2000), American Housing Survey (Chakrabarty and Torres 1996), and the US Residential Energy Consumption Survey (Energy Information Administration 1996). Kasprzyk and Kalton (2001) review the use of quality profiles in US statistical agencies and discuss their strengths and weaknesses for survey improvement and quality declaration purposes.

Note that *data quality* or *accuracy* is not synonymous with *survey quality*. Good survey quality is the result of optimally balancing all quality dimensions to suit the specific needs of the primary data users. As an example, if producing timely data is of paramount importance, accuracy may have to be compromised to some extent. Likewise, if a high level of accuracy is needed, temporal comparability may have to be sacrificed to take advantage of the latest and much improved methodologies and technologies. On the other hand, *data quality* refers to the amount of error in the data. As such, it focuses on just one quality dimension—accuracy.

To illustrate this balancing process, Figure 1.2 shows three cost-equivalent survey designs, each with a different mix of five quality dimensions: accessibility, timeliness, relevance, comparability, and accuracy. The shading of the bars in the graph represents the proportion of the survey budget that is to be allocated for each quality dimension. For example, design C allocates about two-thirds of the budget to achieve data accuracy while design A allocates less to accuracy so that more resources can be devoted to the other four dimensions. Design B represents somewhat of a compromise between designs A and C.

Determining the best design allocation depends on the purpose of the survey and how the data will ultimately be used. If a very high level of accuracy is required (e.g., a larger sample or a reduction of nonsampling errors), design C is preferred. However, if users are willing to sacrifice data quality for the sake of greater accessibility, relevance, and comparability, then design A may be preferred. Since each design has its own strengths and weaknesses, the best one will have a mix of quality attributes that is most appropriate for the most important purposes or the majority of data users.

Total survey error refers to the totality of error that can arise in the design, collection, processing, and analysis of survey data. The concept dates back to the early 1940s, although it has been revised and refined by a many authors over the years. Deming (1944), in one of the earliest works, describes "13 factors that affect the usefulness of surveys." These factors include sampling errors as well as nonsampling errors: the other factors that will cause an estimate to differ from the population parameter it is intended to estimate. Prior to Deming's work, not much attention was being paid to nonsampling errors, and, in fact, textbooks on survey sampling seldom mentioned them. Indeed, classical sampling theory (Neyman 1934) assumes that survey data are error-free except for sampling error. The term *total survey error* originated with an edited volume of the same name (Andersen et al. 1979).

Optimal survey design is the process of minimizing the total survey error subject to cost constraints [see, e.g., Groves (1989) and Fellegi and Sunter (1974)]. Biemer and Lyberg (2003) extended this idea to include other quality dimensions (timeliness, accessibility, comparability, etc.) in addition to accuracy. They advocate an approach that treats the other quality dimensions as additional constraints to be met as total survey error is minimized (or equivalently, accuracy is maximized). For example, if the appropriate balance of the quality dimensions is as depicted by design B in Figure 1.2, then the optimal survey design is one that minimizes total survey error within that fraction of the budget allocated to achieving high data accuracy represented by the unshaded area of the bar. As an example, in the case of design B, the budget available for optimizing accuracy is approximately 50% of the total survey budget. The optimal design is one that maximizes accuracy within this budget allocation while satisfying the requirements established for the other quality dimensions shown in the figure.

Mean-Squared Error (MSE)

The prior discussion can be summarized by stating that the central goal of survey design should be to minimize total survey error subject to constraints on costs while accommodating other user-specified quality dimensions. *Survey methodology*, as a field of study, aims to accomplish this goal. General textbooks on survey methodology include those by Groves (1989), Biemer and Lyberg (2003), Groves et al. (2009), and Dillman et al. (2008), as well as a number of edited volumes. The current book focuses on one important facet of survey methodology—the evaluation of survey error, particularly measurement error. The book focuses on the current best methods for assessing the accuracy of survey estimates subject to classification error. A key concept in the survey methods literature is the *mean squared error* (MSE), which is a measure of the accuracy of an estimate. The next few paragraphs describe this concept.

Let $\hat{\mu}$ denote an estimate of the population parameter μ based on sample survey data. *Survey error* may be defined as the difference between the estimate and the parameter that it is intended to estimate:

$$\text{Survey error} = \hat{\mu} - \mu \tag{1.1}$$

There are many reasons why $\hat{\mu}$ and μ may disagree and, consequently, the survey error will not be zero. One obvious reason is that the estimator of μ is based upon a sample and, depending on the specific sample selected, $\hat{\mu}$ will deviate from μ, sometimes considerably so, especially for small samples. However, even in very large samples, the difference can be considerable due to *nonsampling errors*, meaning errors in an estimate that arise from all sources other than sampling error. The survey responses themselves may be in error because of ambiguous question wording, respondent errors, interviewer influences, and other sources. In addition, there may be missing data due to nonresponding sample members (referred to as *unit nonresponse*) or when respondents do not answer certain questions (referred to as *item nonresponse*). Data processing can also introduce errors. All these errors can cause $\hat{\mu}$ and μ to differ even when there is no sampling error as in a complete census.

In the survey methods literature, the preferred measure of *total survey error of an estimate* (i.e., the combination of sampling and nonsampling error sources) is the MSE, defined as

$$\text{MSE}(\hat{\mu}) = E(\hat{\mu} - \mu)^2 \tag{1.2}$$

which can be rewritten as

$$\text{MSE}(\hat{\mu}) = B^2 + Var(\hat{\mu}) \tag{1.3}$$

where $B = E(\hat{\mu} - \mu)$ is the bias of the estimator and $Var(\hat{\mu})$ is the variance. In these expressions, *expected value* is broadly defined with respect to the sample design as well as the various random processes that generate nonsampling errors. Optimal survey design attempts to minimize (1.3) given the budget, schedule, and other constraints specified by the survey design. This is a very challenging task. It is facilitated, first and foremost, by some assessment of the magnitude of the MSE for at least a few key survey characteristics.

The preferred approach to evaluating survey error is to first decompose the total error into components associated with the various sources of error in a survey. Then each error source can be evaluated separately in a "divide and conquer" fashion. Some sources may be ignored while others may be targeted in special evaluation studies. Biemer and Lyberg (2003, Chapter 2) suggest a mutually exclusive and exhaustive list of error sources applicable to most surveys. The list includes sampling error and five nonsampling error sources: specification error, measurement error, nonresponse error, frame error, and data processing error. These errors sources are briefly described next.

1.1.3 Nonsampling Error

The evaluation of sampling error is considered a best practice in survey research. Nonsampling errors are rarely fully evaluated in surveys, although many examples of evaluations focus on one or perhaps two error sources. In this section, we consider the five sources of nonsampling error in more detail and then discuss some methods that have been used in their evaluation.

Specification Error

A *specification error* arises when the concept implied by the survey question and the concept that should have been measured in the survey differ.[2] When this occurs, the wrong parameter is estimated by the survey and, thus, inferences based on the estimate are likely to be erroneous. Specification error is often caused by poor communication between the researcher (or subject matter expert) and the questionnaire designer. This concept is closely related to the concept of *construct validity* in psychometric literature [see, e.g., Nunnally and Bernstein (1994)] and *relevance* in official statistics [see, e.g., Dalenius (1985)].

Specification errors are particularly common in surveys of business establishments and organizations where many terms that have precise meanings to accountants are misspecified or defined incorrectly by the questionnaire designers. Examples are terms such as *revenue*, *asset*, *liability*, *gross service fees*, and *information services*, which have different meanings in different contexts. Such specialized terms should be clearly defined in surveys to avoid specification error.

As an example, consider the measurement of unemployment in the US Current Population Survey (CPS). The US Bureau of Labor Statistics (BLS) considers the unemployed population as comprising two types of persons: those who are "looking for work" and those who are "on layoff." Persons on layoff are defined as those who are separated from a job and await a recall to return to that job. Persons who are "looking for work" are the unemployed who are not on layoff and who are pursuing certain specified activities to find employment. Distinguishing between these two groups is important for labor economists. Prior to 1994, the CPS questionnaire did not consider or collect information as to whether a person classified as "laid off" expected to be

[2]This usage of the term should not be confused with the econometric term *model specification error* or *misspecification*. The latter error arises through omission of important variables from a statistical model [see, e.g., Ramsey (1969)] or failures of model assumptions to hold.

recalled to work at some point in the future. Rather, respondents were simply asked "Were you on layoff from a job?" This question was later determined to be problematic because, to many people, a "layoff" could mean *permanent termination* from the job rather than the temporary loss of work as the BLS economists defined the term.

In 1994, the BLS redesigned the labor force status questions, as part of that redesign, attempted to clarify the concept of layoff in the questionnaire. The revised questions now ask, "Has your employer given you a date to return to work?" and "Could you have returned to work if you had been recalled?" These questions brought the concept of "on layoff" in line with the specification being used by BLS economists. Specification errors can be quite difficult to detect without the help of subject matter experts who are intimately familiar with the survey concepts and how they will ultimately be used in data analyses, because questions may be well-worded while still completely missing essential elements of the variable to be measured.

Biemer and Lyberg (2003, p. 39) provide another example of specification error from the Agriculture Land Values Survey (ALVS) conducted by the US National Agricultural Statistics Service. The ALVS asked farm operators to provide the market value for a specific tract of land that was randomly selected within the boundaries of the farm. Unfortunately, the concepts that were essential for the valid valuation of agricultural land were not accurately stated in the survey—a problem that came to light only after economists at the Economic Research Service (ERS) were consulted regarding the true purpose of the questions. These subject matter experts pointed out that their models required a value that did not include capital improvements such as irrigation equipment, storage facilities, and dwellings. Because the survey question did not exclude capital improvements, the survey specification of agricultural land value was inconsistent with the way the ERS economists were using the data.

Measurement Error

Whereas *total survey error* is defined for a statistic (or estimator), *measurement error* is defined for an observation. Let μ_i denote the true value of some characteristic measured in a survey for a unit i, and let y_i denote the corresponding survey measurement of μ_i. Then

$$\text{Measurement error} = y_i - \mu_i \tag{1.4}$$

that is, the difference between the survey measurement and the true value of the characteristic. Measurement error has been studied extensively and is often reported in the survey methods literature [for an extensive review, see Biemer and Lyberg (2003, Chapters 4–6)]. For many surveys, measurement error can also be the most damaging source of error. It includes errors arising from respondents, interviewers, and survey questions. Respondents may either deliberately or otherwise provide incorrect information in response to questions. Interviewers can cause errors in a number of ways. They may, by their appearance or comments, influence responses; they may record responses incorrectly, or otherwise fail to comply with prescribed survey procedures; and, in some cases, they may deliberately falsify data. The questionnaire can be a major source of error if it is poorly designed. Ambiguous questions, confusing instructions, and easily misunderstood terms are examples of questionnaire problems that can lead to measurement error.

Measurement errors can also arise from the information systems that respondents may draw on to formulate their responses. For example, a farm operator or business owner may consult records that may be in error and, thus, cause an error in the reported data. It is also well known (Biemer and Lyberg 2003, Chapter 6) that the mode of data collection can affect measurement error. As an example, mode comparison studies (Biemer 1988; de Leeuw and van der Zouwen 1988; Groves 1989) have found that data collected by telephone interviewing are, in some cases, less accurate than the same information collected by face-to-face interviewing. Finally, the setting or environment within which the survey is conducted can also contribute to measurement error. When collecting data on sensitive topics such as drug use, sexual behavior, or fertility, the interviewer may find that a private setting is more conducive to obtaining accurate responses than one in which other members of the household are present. In establishment surveys, topics such as land use, financial loss and gain, environmental waste treatment, and resource allocation can also be sensitive. In these cases, assurances of confidentiality may reduce measurement errors that result from intentional misreporting. Biemer et al. (1991) provides a comprehensive review of measurement error in surveys.

Frame Error

Frame error arises in the process for constructing, maintaining, and using the sampling frame(s) for selecting the survey sample. The sampling frame is defined as a list of population members or some other mechanism used for drawing the sample. Ideally, the frame would contain every member of the population with no duplicates. Also, units that are not part of the population would not be on the frame. Likewise, information on the frame that is used in the sample selection process should be accurate and up to date. Unfortunately, sampling frames rarely satisfy these ideals, often resulting in various types of frame errors.

There are essentially three types of sampling frames: area frames, list frames, and implicit frames. *Area frames* are typically used for agricultural and household surveys. An area frame is constructed by first dividing an area to be sampled (say, a state) into smaller areas (such as counties, census tracts, or blocks). A random sample of these smaller areas is drawn and a *counting and listing* operation is implemented in the selected areas to enumerate all the ultimate sampling units. For household surveys, the counting and listing operation is intended to identify and list every dwelling unit in the sampled smaller areas. Following the listing process, dwelling units may be sampled according to any appropriate randomization scheme. The process is similar for agricultural surveys, except rather than a dwelling unit, the ultimate sampling unit may be a farm or land parcel.

The omission of eligible population units from the frame (referred to as *noncoverage error*) can be a problem with area samples, primarily as a result of errors made during the counting–listing phase. Enumerators in the field may miss some dwelling units that are hidden from view or are mistaken as part of other dwelling units (e.g., garages that have been converted to apartments). Boundary units may be erroneously excluded or included because of inaccurate maps or enumerator error. Boundary units can also be a source of duplication error if they are included for areal units on both sides of the boundary.

More recent research has considered the use of *list frames* for selecting household samples [see, e.g., O'Muircheartaigh et al., (2007), Dohrmann et al. (2007), and Iannacchione et al. (2007)]. One such list is the US Postal Service *delivery sequence file* (DSF). This frame contains all the delivery point addresses serviced by the US Postal Service. Because sampling proceeds directly from this list, a counting–listing operation is not needed, saving considerable cost. Noncoverage error may be an important issue in the use of the DSF, particularly in rural areas [see, e.g., Iannacchione et al. (2003)]. Methods for reducing the noncoverage errors, such as the *half-open interval method* [see, e.g., Groves et al. 2009)] have met with varying success (O'Muircheartaigh et al., 2007). List frames are also commonly used for sampling special populations such as teachers, physicians, and other professionals. Establishment surveys make extensive use of list frames drawn from establishment lists purchased from commercial vendors.

A sampling frame may not be a physical list, but rather an implicit list as in the case of random-digit dialing (RDD) sampling. For RDD sampling, the frame is implied by the mechanism generating the random numbers. Frame construction may begin by first identifying all telephone exchanges (e.g., in the United States and Canada, the area code plus the 3-digit prefix) that contain at least one residential number. The implied frame is then all 10-digit telephone numbers that can be formed using these exchanges, although the numbers in the sample are the only telephone numbers actually generated and eventually dialed. *Intercept sampling* may also use an implicit sampling frame. In intercept sampling, a systematic sample of units is selected as they are encountered during the interviewing process; examples where an explicit list of population units is not available include persons in a shopping mall or visitors to a website.

To ensure that samples represent the entire population, every person, farm operator, household, establishment, or other element in the population should be listed on the frame. Ineligible units should be identified and removed from the sample as they are selected. Further, to weight the responses using the appropriate probabilities of selection, the number of times that each element is listed on the frame should also be known, at least for the sampled units. To the extent that these requirements fail, frame errors occur.

Errors can occur when a frame is constructed. Population elements may be omitted or duplicated an unknown number of times. There may be elements on the frame that should not be included (e.g., in a farm survey, businesses that are not farms). Erroneous omissions often occur when the cost of creating a complete frame is too high. We may be well aware that the sampling frame for the survey is missing some units but the cost of completing the frame is quite high. If the number of missing population members is small, then it may not be worth the cost to provide a complete frame. Duplications on a frame are a common problem when the frame combines a number of lists. For the same reason, erroneous inclusions on the frame usually occur because the available information about each frame member is not adequate to determine which units are members of the population and which are not. Given these frame imperfections, the population represented by the frame does not always coincide with the population of interest in the survey. The former population is referred to as the *frame population* and the latter as the *target population.*

Nonresponse Error

Nonresponse error is a fairly general source of error encompassing both unit and item nonresponse. Unit nonresponse occurs when a sampled unit (e.g., a household, farm, school or establishment) does not respond to any part of a questionnaire, such as a household that refuses to participate in a face-to-face survey, a mail survey questionnaire that is never returned, or an eligible sample member who refuses or whose telephone is never answered. Item nonresponse error occurs when the questionnaire is only partially completed because an interview was prematurely terminated or some items that should have been answered were skipped or left blank. For example, income questions are typically subject to a high level of item nonresponse from respondent refusals.

For open-ended questions, even when a response is provided, nonresponse may occur if the response is unusable or inadequate. As an example, a common open-ended question in socioeconomic surveys is "What is your occupation?" A respondent may provide some information about his or her occupation, but perhaps not enough to allow an occupation and industry coder to assign an occupation code number during the data-processing stage.

Data-Processing Error

The final source of nonsampling error is data processing. Data-processing error includes errors in editing, data entering, coding, weighting, and tabulating of the survey data. As an example of editing error, suppose that a data editor is instructed to call the respondent back to verify the value of some budget line item whenever the value of the item exceeds a specified limit. In some cases, the editor may fail to apply this rule correctly, thus generating errors in the data.

For open-ended items that are subsequently coded, coding error is another type of data-processing error. The coders may make mistakes or deviate from prescribed procedures. The system for assigning the code numbers—for variables such as place of work, occupation, industry in which the respondent is employed, and field of study for college students—may itself be ambiguous and prone to error. As a result, code numbers may be inconsistently and inappropriately assigned, resulting in significant levels of coding error.

The survey weights that statistically compensate for unequal selection probabilities, nonresponse error, and frame coverage errors may be calculated erroneously, or there may be programming errors in the estimation software that computes the weights. Errors in the tabulation software may also affect the final data tables. For example, a spreadsheet used to compute the estimates may contain a cell-reference error that goes undetected. As a result, the weights are applied incorrectly and the survey estimates are in error.

Decomposing Total Survey Error

While each source of error can increase the total error, some pose a much greater risk for bias and/or variance than others. Biemer and Lyberg (2003) provide a rough assessment of the risk of *variable* and *systematic* errors for each error source based on a synthesis of the nonsampling error literature. They define variable errors as errors that are distributed with zero mean and nonzero variance. Systematic errors are defined as errors having nonzero mean and zero or trivially small variance. As an example, in a survey using acceptable probability sampling methods, the sampling error distribution has zero mean and there is no risk of bias. Of course, sampling variance is an unavoidable consequence of sampling.

Specification error contributes systematic errors because measuring the wrong concept in a survey causes the estimate to consistently differ from the true concept. Nonresponse error typically poses a greater risk to systematic error, although nonresponse adjustment methods such as imputation and weighting can contribute importantly to the variance, especially when the nonresponse rates are high. Frame errors, particularly noncoverage errors, are viewed primarily as a source of systematic error, although, as with nonresponse, coverage adjustments can also increase the variance. Measurement errors (emanating primarily from interviewers, respondents, and the questions themselves) can pose a serious risk for both systematic and variable errors, as detailed in the later chapters. Finally, data-processing operations involving human operators, such as coding and editing, can add both systematic and variable errors to the total error.

Using these risk classifications, Biemer and Lyberg posit an expanded version of the MSE that is applicable to estimates of means, totals, and proportions and that includes specific terms for each major source of error. In their formulation, the B^2 component is expanded to include bias components for all the sources of error having a high risk of systematic error: specification bias, B_{SPEC}; frame bias, B_{FR}; nonresponse bias, B_{NR}; measurement bias, B_{MEAS}; and data-processing bias, B_{DP}. These components sum together to produce the total bias component, B:

$$B^2 = (B_{\mathrm{SPEC}} + B_{\mathrm{FR}} + B_{\mathrm{NR}} + B_{\mathrm{MEAS}} + B_{\mathrm{DP}})^2 \tag{1.5}$$

Likewise, the variance of an estimator can be decomposed into components associated with the major sources of variable error: sampling variance (Var_{SAMP}), measurement error variance (Var_{MEAS}), and data-processing variance (Var_{DP}). Combining these components, an expanded version of the MSE formula showing components for all the major sources of bias and variance can be written as

$$\text{MSE}(\hat{\mu}) = (B_{\text{SPEC}} + B_{\text{FR}} + B_{\text{NR}} + B_{\text{MEAS}} + B_{\text{DP}})^2 + Var_{\text{SAMP}} + Var_{\text{MEAS}} + Var_{\text{DP}} \tag{1.6}$$

Although complex, this expression for the MSE is still oversimplified because it ignores some interaction terms that may be important for some applications. The article by Fellegi (1964) contains an excellent discussion of these interaction terms. The use of this "divide and conquer" approach greatly simplifies the task of evaluating total survey error because scarce evaluation resources can be targeted to the error source(s) with the greatest effect on survey accuracy, provided the effect can be assessed.

Another way to classify the nonsampling error sources is by errors associated with missing data, or *errors of nonobservation* (viz., nonresponse and frame errors) and those that affect the content of the observations (viz., specification, measurement, and data processing). The focus of this book is on the latter, which are sometimes referred to as *content* errors[3] but more often as *measurement errors*.

1.2 EVALUATING THE MEAN-SQUARED ERROR

As discussed in this section, estimating the MSE or any of its components, often requires the collection of additional data, complex methods of analysis, or both. These are costly and may consume resources that could otherwise be put toward the data collection effort to improve survey quality. Why then should survey organizations bear this expense and allocate resources for survey evaluation? This question is addressed in the next section.

1.2.1 Purposes of MSE Evaluation

Historically, total survey error components have been estimated for at least four purposes:

To compare the accuracy of data from alternative modes of data collection or estimation methods

[3]Groves (1989) refers to these errors as *errors of observation*. We prefer our terminology to reflect those errors (such as specification and data processing errors) that occur either before or after data are observed, yet still affect the content of the data.

To optimize the allocation of resources for the survey design
To reduce the nonsampling error contributed by specific survey processes
To provide information to data users regarding the quality of the data or the reported estimates

The first is one of the most common uses of total survey error evaluations. As an example, a survey methodologist may wish to compare the accuracy of health data collected by mail and by telephone. Because mail is usually the cheaper mode, a larger mail survey could be afforded that would reduce sampling error, but it is possible that the total survey error would increase because of nonsampling errors. To compare the two modes, a mode comparison study could be conducted using a *split-ballot design* in which half the sample is collected by telephone and the other half is collected by mail. But while the split-ballot approach is sufficient for determining whether the two modes give differing results, it is seldom sufficient for determining the more accurate mode. For evaluating accuracy, the MSEs of the estimates from each mode must be evaluated and compared (Biemer 1988).

In many cases, it is also important to determine whether the differences in MSEs stem from bias or variance components and which components contribute most to the differences. For example, if the response rates differ considerably for the two modes, it may be tempting to attribute the difference in accuracy for some key characteristics to nonresponse bias. For other characteristics, the real culprit may be measurement bias or variance. Knowledge of which MSE components are most responsible for the differences in MSE will provide clues as to how the differences can be minimized or eliminated.

Another purpose of survey error evaluation is design optimization. For this purpose, the survey designer would like to know how much of the error is contributed by each major error source. Then resources could be allocated for mitigating the sources of the largest errors. An alternative is to use expert judgment or intuition for allocating resources, but the risk with that approach is that the allocation may be far from optimal for minimizing total survey error.

An example of optimal survey design is Dillman's (2007) *tailored design method* for mail surveys. This methodology has been developed by combining the results of many experiments across many surveys and on a very wide range of topics. Dillman and others have used meta-analysis and other techniques for integrating this vast collection of research results. In the process, they have identified the most essential factors of optimal design and the best combinations of sample design, questionnaire design, and implementation techniques for minimizing nonresponse and measurement error and reducing survey costs.

Another important reason for estimating the MSE is that information on the magnitudes of nonsampling error components contributed by specific survey operations is also useful for identifying where improvements are needed for ongoing survey operations. As an example, a study of errors made by interviewers may determine that interviewers contribute considerably to the total variance as well as the bias of estimates. Additional testing and experimentation could be done to identify the root causes of the error focusing on the interviewer variance component. Further study may reveal that improvements are needed in interviewer training, the questionnaire, interviewer monitoring or supervision, or other areas. Over the years, studies of the components of total survey error have led to many improvements in survey accuracy. For example, studies of enumerator variance in the 1950 US census led to the decision to adopt a self-enumeration census methodology (Eckler 1972, p. 105).

Finally, the estimation of the total MSE and its individual components can provide data users with objective information on the relative importance of different errors, which can aid their understanding of the limitations of the data. Measures of nonsampling error indicating excellent or very good data quality create high user confidence in the quality of the data, while measures that imply only fair to poor data quality can serve as a warning to users to proceed with caution in interpretation of the data.

As an example, reports on survey quality often contain estimates of nonresponse bias for the key estimates produced from the survey data. This information is quite informative for assessing the accuracy of the estimates and whether nonsampling error should be a concern for interpreting the research findings. Likewise, an analysis of measurement error could be useful for explaining why an analysis failed to replicate findings in the literature or why unexpected and inexplicable relationships among the variables were found.

To understand the causes of nonsampling error and develop strategies for its prevention, the errors must be measured often. Continuous quality improvement requires current knowledge of which error sources are the most problematic so that scarce survey resources can be most effectively allocated. For some error components, this might involve interviewing a small representative sample of the target population using *cognitive interviewing methods* [see, e.g., Forsyth and Lessler (1991), Tourangeau et al. (2000), and Willis (2005)] rather than a large study aimed at estimating a bias component. However, small-scale laboratory investigations used in conjunction with large-scale error component evaluation studies may be ideal for most purposes. Evaluation studies aimed at describing the effect of alternate design choices on total survey error are also extremely important, because without them total survey design optimization is not possible.

1.2.2 Effects of Nonsampling Errors on Analysis

One of the primary motivations for wanting to minimize survey errors in surveys is their potentially devasting effects on all types of data analysis. Cochran (1968), Fuller (1987), and more recently Biemer and Trewin (1997) consider the effects of nonsampling errors on a range of estimators and analysis methods including the estimation of means, quantiles and their standard errors, correlations, regression coefficients [including analysis of covariance and analysis of variance (ANOVA)], goodness of fit, and association test statistics in contingency table analysis. As they show, the damaging effects of nonsampling errors on estimation and inference can be quite severe. Some evidence of this damage is summarized here; however, more details regarding these effects is provided in later chapters.

As noted above, variable errors may not be biasing for estimators of means, totals, and proportions. Variable errors will inflate the variance of these estimators. Biemer and Trewin show the inflation factor to be the reciprocal of the reliability *reliability ratio* for the measurements.[4] The reliability ratio, denoted by R, may be defined roughly as the proportion of the total variance of an observation that is not measurement error variance. They further show that the usual estimator of the standard error will be unbiased when nonsampling errors are variable only. In other words, although the standard error is inflated, the usual estimator or the standard error correctly reflects the increase. An exception is when the nonsampling errors are correlated in ways not reflected in the survey design.

Correlated errors occur when survey operators (interviewers are a prime example) contribute errors to the observations that vary in magnitude across the operators [see, e.g., Kish (1962), Fellegi (1964) Fellegi and Sunter (1974), Biemer and Stokes (1991), and Biemer and Lyberg (2003)]. For example, suppose that interviewers are asked to estimate the average value of housing in the neighborhoods where their sampling units are located. Some interviewers who are not current on the housing values may consistently undersestimate these values for the units in their work assignments. Others may overestimate these values for other reasons. Similarly, interviewers, by the way they dress or act in an interview, may influence responses in a systematic way. These systematic errors may be regarded as random effects that covary positively across the units in an interviewer's assignment. Thus, two observations on different units in the same interviewer's assignment may be positively correlated as a result of these *correlated interviewer errors*.

Similar to a "clustering effect" in two-stage sampling [see, e.g., Kish (1965, pp. 257–259)], correlated interviwer (or, more genereally, operator) errors increase the variance of estimators by a multiplicative factor that is a function of average caseload size and the intra-cluster correlation due to operators. For interviewing, the *interviewer clustering effect* is given by

$$\delta_{\text{int}} = [1 + (m-1)\rho_{\text{int}}] \tag{1.7}$$

[4]The reliability ratio is similar to the *signal-to-noise* ratio in physics. It will be discussed in detail in the Chapter 2.

where m is the average size (in number of units) of the interviewers' caseloads and ρ_{int} is the intraclass correlation of the interviewers.[5] Note the similarity of (1.7) to the design effect defined for cluster sampling (Kish 1965). As an example, suppose that the interviewers for a survey interviewed an average of 50 households each. Suppose further that ρ_{int} is estimated to be 0.01. [Biemer and Lyberg (2003) describe the methodology for estimating ρ_{int} in some detail.] Then, according to (1.7), the variance of an estimator of the mean will be increased by $1 + (49)0.01 = 1.49$ as a result of correlated interviewer error. This is essentially equivalent to a reduction of sample size by two-thirds for estimating the mean of the item. In fact, the sample size n divided by (1.7) is often referred to as the *effective sample size*.

A number of authors [see, e.g., Hansen et al. (1964), Cochran (1968), and Biemer and Trewin (1997)] have shown that, although standard errors may be inflated by variable errors, the traditional estimators of standard errors are still unbiased as long as the errors are uncorrelated. This is somewhat of a "silver lining" on the variable error "cloud." However, as Hansen et al. (1951) showed, correlated errors will cause the usual standard error estimators to be negatively biased. The extent of this bias is roughly the same as the magnitude of the increase in standard error from correlated errors. From the illustration above for interviewers, this bias can be substantial even when the intrainterviewer correlation is quite small (i.e., 0.01 or less).

For regression and correlation coefficients, variable errors tend to bias coefficients toward 0 referred to as *attenuation* toward 0. Fuller (1987) shows that, in ordinary regression analysis, the expected value of the estimated regression coefficient is $R \times \beta$, where R is the reliability ratio associated with the predictor variable and β is the regression coefficient assuming error-free measurements. In addition, the *coefficient of determination* (i.e., the usual regression R^2 measure), which measures the proportion of variance in the dependent variable explained by the independent variables, cannot exceed the reliability of the dependent variable in the regression (Fuller 1987).

Fuller further showed that the usual Pearson product moment correlation coefficient between two variables x and y is shown to have expected value $\sqrt{R_x R_y}\,\rho_{xy}$, where R_x and R_y are the reliability ratios associated with x and y, respectively, and ρ_{xy} is the correlation coefficient when x and y are measured without error. When nonsampling errors are correlated, the regression and correlation coefficients can be unpredictably biased in either a positive or a negative direction.

Biemer and Trewin (1997) show that estimators of population quantiles are biased whether the errors are variable (uncorrelated or correlated) or systematic. Likewise, parameter estimates from ANOVA and analysis of covariance (ANACOV), as well as chi-square (χ^2) tests for goodness of fit and association are biased even in the benign case of uncorrelated variable errors.

[5]The formula is discussed further in Section 8.3.

These results underscore the importance of collecting survey data that are nearly error-free. Essentially all statistical inferences are biased in the presence of nonsampling error, except in the case of linear point estimation and uncorrelated variable errors. Methods are now available for correcting data analysis for the effects of uncorrelated nonsampling errors. For example, structural equation models (SEMs) [see, e.g., Bollen (1989)], errors-in-variables regression analysis [see, e.g., Fuller (1987)], and latent class analysis [see, e.g., Hagenaars and McCutcheon (2002) as well as this book] are techniques that have been developed to deal with uncorrelated error in statistical inference. Software packages that employ these techniques are widely available. However, methods for handling correlated variable errors and systematic errors in statistical analysis are not so common. Exceptions include methods for compensating for the effects of item nonresponse and frame undercoverage through the application of postsurvey weight adjustments (Biemer and Christ 2008). For item nonresponse, methods for imputing missing data have been developed as a way of reducing the bias in data analysis [see de Waal and Haziza (2009) for a recent overview of the literature]. However, these methods are only partially effective at best. Notwithstanding these advances, eliminating nonsampling error during the survey process is still the best approach for avoiding nonsampling error bias in estimation and data analysis.

1.2.3 Survey Error Evaluation Methods

Survey data quality evaluation is a critical branch of the field of survey research because without it, the process of survey quality improvement is essentially guesswork. The best practices for conducting surveys are based on the results of survey quality evaluations.

A number of methods are available for evaluating survey error. Some methods can be applied during the design and pretesting stages to guide the designers as they attempt to optimize survey design. Several methods can be applied concurrently with data collection and data processing to monitor the quality of the data and to alert survey managers when important errors enter the survey process. Other methods can be applied on completion of the survey to describe the magnitudes of the error components and provide valuable information to survey designers for improving future surveys. Thus, quality evaluation can be seen as a continuous process that is carried from the design stage to the postsurvey analysis stage.

Table 1.1 summarizes of some of the most frequently used methods for survey error evaluation. These methods are organized in the order in which they might be used in the survey process: the survey design stage, the pretesting stage, the survey data collection stage, or the postsurvey stage. For a description of these methods, see Biemer and Lyberg (2003). As shown in Table 1.1, cognitive methods and interviewer debriefings tend to be used in the design and pretesting stages to identify questionnaire problems. Supervisor observations of interviewers, quality control methods, and the methods for

Table 1.1 Some Evaluation Methods and Their Purposes

Stage of the Survey Process	Evaluation Method	Purpose

Design/pretest	Expert review
	Cognitive methods
	  Behavior coding
	  Cognitive interviewing
	Debriefings
	  Interviewer group discussions
	  Respondent focus groups
		Identify problems with questionnaire layout, format, question wording, question order, and instructions
		Evaluate one or more stages of the response process
		Evaluate data collection and processing procedures

Pretest/survey	Observation
	  Supervisor observation
	  Telephone monitoring
	  Recorded interviews
	Data quality control
	Nonresponse reduction analysis
		Evaluate interviewer performance
		Identify questionnaire problems
		Refine data collection procedures
		Refine nonresponse followup procedures

Postsurvey	Postsurvey analysis
	  Experimental design
	  Embedded repeated measures
	  Internal consistency
	  External validation
	Postsurvey data collection
	  Reinterview surveys
	  Record check studies
	  Nonresponse followup studies
		Compare alternative methods of data collection
		Estimate one or more MSE components
		Validate survey estimates
		Identify problem questions

reducing nonresponse tend to be used during the survey. Postsurvey methods might include experimental design, reinterviews, and other methods aimed at estimating one or more components of the MSE. These categorizations are not definitive. For example, experimental design methods can also be used at the design and pretest stages to choose among competing methods. Likewise, interviewer and respondent debriefings can be used postsurvey to evaluate the effectiveness of various survey approaches.

For many surveys, data quality evaluation is limited to just a few activities, for instance, pretesting the questionnaire and data collection quality control monitoring procedures such as response rate monitoring. There may be additional quality control procedures for data processing operations such as data capture and coding. As household response rates to surveys continue to fall, evaluations of the nonresponse bias in the survey estimates are becoming more common. These evaluations may involve postsurvey interviews of nonrespondents to determine why they were not interviewed and how their characteristics might differ from those of respondents. More often descriptive studies of the nonrespondents are conducted using whatever data are available on the nonrespondents, possibly from the sampling frame. In rare instances, embedded experiments and postsurvey evaluations might be conducted to estimate other components of the MSE such as reliability, measurement biases, or interviewer and other operator variances.

For surveys conducted on a continuing basis or repeated periodically, survey evaluation plays a critical role. It is important not only for continuous quality improvement but also as a vehicle for communicating the usefulness and limitations of the data series. For this reason, a number of US federal surveys have developed quality profiles (Biemer and Lyberg 2003) as a way of summarizing all that is known, as well as pointing out the gaps in knowledge, about the sampling and nonsampling errors in the surveys.

1.2.4 Latent Class Analysis

Latent class analysis (LCA), the main topic of this book, is a powerful method for estimating one or more parameters of a survey error model. These estimates can be used to evaluate the MSE of an estimate and the error probabilities associated with a survey question, as well as many other purposes. Application of latent class models is appropriate when the data to be analyzed are categorical (either nominal or ordinal categories). In that situation, the measurement errors in the observations are referred to collectively as *classification errors* or simply *misclassification*.

Evaluations of classification error are important because many survey variables are categorical. Demographic questions may ask individuals to classify themselves by gender, age, race, education, income level, tenure, marital status, and so on. Many survey questions are closed-ended—meaning that a small number of response options are presented from which the respondent must choose. The closed-ended response format is popular because it is simple for respondents to answer and responses are self-coded, making them easy to key as well. In addition, the information contained in the response options can indicate the type of responses the designer expects, which can often clarify the meaning of the question (Tourangeau et al. 2000). Responses to open-ended questions are seldom useful in data analysis unless they are coded to a reasonably small number of discrete categories prior to analysis. For example, a person's occupation is often collected as an open-ended response and then coded to one of a number of occupation categories. LCA is applicable in all of these situations.

As an example, consider the case where we want to classify respondents as either smokers or nonsmokers but our methods of questioning do not accurately convey what we mean by smoking or some respondents may be unwilling to give us accurate information on their smoking habits. As a result, responses to a question on smoking habits are subject to misclassification. LCA is a statistical method for predicting the true classifications of individuals according to their observed classifications. In addition, LCA will provide estimates of the probabilities of respondents being misclassified by the question.

To apply LCA, we might ask two or three questions about smoking that are somewhat different but are all aimed at determining whether an individual smokes. LCA uses a statistical model that incorporates characteristics of the individual along with his or her responses to estimate a probability that the individual smokes. Under this model, an individual that answers "no" to all three questions still has a positive probability of being a smoker. LCA attempts to estimate the probability that individuals with varying patterns of responses to the smoking questions truly smoke. Each response pattern is associated with a probability of smoking that may vary depending on the other characteristics of the individual (age, race, sex, income, etc.). LCA will provide estimates of the true proportion of the population who smokes as well as error rates associated with each question about smoking.

However, as this book shows, many applications of LCA or its counterpart for panel surveys, Markov latent class analysis (MLCA), go beyond the estimation of classification error. Although this book focuses on measurement error, we also provide examples of applications of LCA for estimating census coverage error and for adjusting for nonresponse bias. New applications of LCA are still being discovered for this powerful statistical method.

1.3 ABOUT THIS BOOK

As implied by the title, the focus of this book is on the use of LCA as a tool for evaluating survey error. Survey error evaluations serve a number of purposes as noted in Section 1.2.1. LCA can be a valuable tool for all of these purposes. Although LCA finds application is evaluating all the error sources described in Section 1.1.3, our focus is primarily on measurement errors or, more broadly, content errors. A few applications are considered in Chapters 6 and 7 that aim to evaluate nonresponse and frame errors. These are included primarily to show the wide applicability of LCA. The book also focuses on error in cateogorical measurements (i.e., classification errors) rather than the errors in continuous measurements. The latter types of error are briefly treated in the next chapter to show the linkages with continuous and classification error models. This is not an important restriction because many of the variables used in survey data analysis are categorical. In fact, a common practice among survey analysts is to discretize continuous variables and treat them as either ordinal or nominal categorical variables in the analysis. This often leads to results that are more easily interpreted.

This book differs fundamentally from other books on LCA in that it focuses exclusively on survey error evaluation rather than more factor analytic purposes of LCA such as typological analysis, cluster analysis, and stage-sequential dynamics. Unlike more general uses of LCA, our framework assumes that the variable to be measured is well defined and is directly observable. The true value of the variable exists and can be measured accurately under ideal conditions. However, in the survey setting, it is measured with error. In this context, the primary objective of LCA is to estimate the magnitudes of the errors in measurements. These estimates can be used to identify problematic questionnaire items, to study the causes of the errors, and ultimately for survey improvement. A secondary objective is to estimate the prevalence of population characteristics that are corrected for classification error. The differences between LCA for survey error evaluation and general LCA methodology are further discussed and illustrated in Section 4.1.1.

The next three chapters trace the evolution of survey error evaluation from the early models of Hansen and his colleagues at the US Census Bureau to the modern and more sophisticated loglinear models with latent variables, of which the latent class model is a special case. In these chapters the linkages between the early models to the modern models is emphasized because a thorough understanding of LCA for survey error evaluation requires an understanding of its roots. It should be clear by Chapter 4 that LCA for survey evaluation is an extension of classical test theory concepts and ideas and the measurement error models of the 1950s and 1960s.

Chapter 2 describes a very general model for measurement error, one that has its roots in the early days of survey research. In that chapter, some of the basic concepts associated with the study of survey response and measurement error are described. A general model is described that can be applied to virtually any type of data. More useful models can be obtained by considering special cases of the model based on assumptions that are specific to either continuous or discrete data types. This chapter ends with a very simple model for binary data, where simple random sampling is assumed. The basic concepts form the foundations for the analysis of polytomous categorical variables under complex survey sample designs.

Chapter 3 picks up this thread and carries the ideas further into the realm of probability modeling when only two measurements of the true characteristic (i.e., the latent variable) are available. This situation is probably the most common in survey evaluations because obtaining more than two measurements of the same characteristic often requires additional efforts and resources. An interview–reinterview study, especially the *test–retest* reinterview (defined in Chapter 2), is one vehicle for obtaining remeasurements. Embedding remeasurements within the same questionnaire is also common; however, it must be used sparingly (say, for only a few characteristics) to avoid overburdening the respondent and unduly lengthening the interview. An advantage is that three or more replicate measurements of the same characteristics can be obtained in one interview. The basic ideas of LCA are also developed in this chapter.

Chapter 4 extends the latent class modeling concepts and specifications to $K \geq 2$ repeated measurements. As discussed in this chapter, an important advantage of adopting a latent class modeling framework for evaluating survey misclassification is the availability of general software for estimating the error components. Chapter 4 describes the standard latent class (LC) model and shows why its assumptions are unnecessarily restrictive for most applications. It introduces a more general modeling framework based on a loglinear model with latent variables and shows how the standard LC model is just a special case of the more general model. Within this general modeling framework, a wide range of error structures and error evaluation designs is discussed and analyzed using familiar loglinear modeling notation, methods and techniques. Chapter 5 considers some advanced topics in the study of LCA, including models for ordinal data and methods for complex surveys. Chapter 6 discusses some more advanced models and applications, including models for ordinal data and their use, the agreement model, and the capture–recapture latent class model.

Chapter 7 extends the ideas of LC models to panel survey data by considering Markov latent class (MLC) models. MLC models are important because they provide a means for estimating classification error directly from the data collected in the panel survey without special reinterview or response replication studies. Essential in the application of these models is some evaluation or assessment of the extent to which the model assumptions hold. These issues are discussed and the primary approaches for model validation are described.

Throughout these chapters, the emphasis is on the application of the models to survey error evaluations. In this regard, some topics that are relevant to general-purpose uses of latent class modeling (e.g., population typological analysis and latent transition analysis) are not treated in much detail. At the same time, topics often deemphasized in other discussions of LCA (such as error structure modeling) are emphasized in this book. Numerous examples and illustrations are presented to demonstrate the estimation methods as well as how to interpret the modeling results. The utility of the models for evaluating and improving survey data quality are discussed and demonstrated. The primary software package used for data analysis is the ℓEM software for fitting a wide range of LC models. This software is quite intuitive and can be downloaded from the Internet at no charge.

Finally, Chapter 8 summarizes the major innovations in the field of LCA and discusses the current state of the art. This discussion concludes by reviewing several issues where further research is needed and other matters that have been neglected in the previous chapters. Most of these issues are spawned from criticisms of the methodology over modeling assumptions and a general lack of rigor in prior applications of the methodology. These criticisms and how they can be dealt with in future applications are also discussed.

CHAPTER 2

A General Model for Measurement Error

Some measurement errors are easily recognized: the 4-year-old mother of two, the corner gas station with 1500 employees, a 50-item set of questions using a 5-point scale where every response is "3," and the list goes on. Responses to questions that are impossible or highly unlikely are indicative of measurement errors. But more often, errors in the data can look as plausible as error-free values. How can these errors be identified in an evaluation? This is one of the major challenges in the analysis of measurement errors. Except for the most egregious errors resulting in highly unlikely values, measurement errors are often hidden in the data. Occasionally they may be discovered by their inconsistencies with other responses. As an example, suppose that a person reported to be 3 years old is also reported to have graduated from high school. Obviously, both responses cannot be correct and, thus, at least one error has been identified. Inconsistent data can indeed identify some errors. However, in the age of computer-assisted surveys, such inconsistencies are quite rare since they can easily be detected and repaired during the interview.

Indeed, most errors remain hidden in the data unless an evaluation study is conducted that is designed especially to identify the errors. Usually, this involves embedding items in the survey specifically for this purpose of checking consistency or by using external data of some type, for example, remeasurements of the same characteristics for the same individuals. However, these methods are not perfect and measurements errors can remain hidden despite our best efforts to identify them.

To illustrate the problem, consider the data in Table 2.1 from a study conducted by Kreuter et al. (2008) of individuals who received undergraduate degrees from the University of Maryland from 1989 to 2002. A total of

---

Latent Class Analysis of Survey Error By Paul P. Biemer

Table 2.1 Cross-Classification of Questions 1 and 2

	Question Q2	
Question Q1	Yes	No
Yes	231	203
No	11	509

954 students responded to two questions regarding their undergraduate grades:

Q1. Did you ever receive a grade of "D" or "F" for a class?

Q2. Did you ever receive an unsatisfactory or failing grade?

Table 2.1 shows the total the number of students for each possible combination of the responses "yes" and "no" to the two questions. In the study, each student's college transcripts were also reviewed for any evidence of a "D" or "F" grade to evaluate the accuracy of the question responses. Those data will be discussed subsequently. The diagonal cells of Table 2.1 show the number of students whose Q1 and Q2 responses agreed. The off-diagonal cells represent the disagreements.

Although the two questions are worded slightly differently, they should generate essentially the same responses. But, in fact, the responses disagree for about 22% of respondents (203 + 11 = 214 divided by 954). A total of 11 students who replied "no" to question Q1 responded "yes" to question Q2 and even more students (203) who responded "yes" to question Q1 said "no" to question Q2. Because the questions are intended to measure the same characteristic for the same students, the disagreements are evidence of classification errors in the data.

A *classification error* occurs when respondents who really have a poor or failing grade respond that they do not (i.e., a *false-negative* response) or when respondents who do not have a poor or failing grade respond that they do (i.e., a *false-positive* response). Note that, although the table provides ample evidence of error, it alone is not sufficient to determine which of the two responses for a student is the correct response. For that purpose, data that can be regarded as highly accurate or truthful are required.

An important question in survey research is—which question, Q1 or Q2, is better for collecting these data? That is, which of the two questions elicits the truth more often? For that, Kreuter et al. obtained the official transcripts for all 954 students and determined which students had ever received a "D" or "F" grade. Students with a grade of "D" or "F" on their transcripts should have responded "yes" to both questions. Otherwise, they should have responded "no." Although they are not perfect, the data from the transcripts may be regarded the most accurate information available and truthful enough for the purposes of evaluating Q1 and Q2. The term *gold standard*, which will be

Table 2.2 Cross-classification of Q1 and Q2 by the Transcript Classification

	Transcript	
	Yes	No
	Q1	
Yes	423	11
No	150	370
	Q2	
Yes	235	7
No	338	374

discussed in greater detail later, is often used to describe such data. The question eliciting responses that agree most often with the gold standard is the better question.

Table 2.2 compares the responses for each question with the official transcript. Clearly, the table shows that neither question is error-free. Q1 misclassified 150 students who, according to their transcripts, have at least one poor or failing grade. By contrast, Q2 misclassified more than twice as many of these students, namely, 338 students. For students who have a "D" or "F" grade on their transcripts, Q1 misclassified 11 and Q2 misclassified 7. It appears from these data that, for classifying students who truly have a poor or failing grade, Q1 is vastly superior to Q2. But for classifying students who do *not* have poor or failing grades, there is not much difference between the two questions. This example illustrates the usefulness of data quality evaluations for identifying questions having lower levels of misclassification.

This type of analysis may be unavailable for some characteristics. There may be no reasonable gold standard values for highly malleable personal characteristics, for example, an individual's psychological or emotional states. It may not even be plausible to assume that true values exist for some characteristics. But when they do exist, it may be virtually impossible to measure them in a survey. In such cases, evaluating the measurement error in the data poses quite a challenge.

This chapter develops the basic theory for evaluating measurement errors in situations where the underlying true values, if they exist, are unknown. Data inconsistencies, such as those apparent in Table 2.1, are often sufficient to uncover important information about measurement errors in a survey process. Although they may not reveal much about response bias, they can still tell us a lot about response variability. At the end of the chapter, situations where true values are known are considered as well as the additional information on measurement errors that they can provide. As an example, we have already seen that comparisons such as those in Table 2.2 can reveal much information about both misclassification variance and bias. This chapter provides the basic concepts of measurement error theory that will be used throughout the book.

2.1 THE RESPONSE DISTRIBUTION

To describe the potential effects of measurement errors on the observations and estimators used in survey research, a statistical model (either explicit or tacit) of the response process is required. The available models are gross oversimplifications of highly complex processes that underlie survey error, yet they are often surprisingly useful for understanding and estimating the effects of error on survey measurement. The model described in this section combines two streams of research that progressed virtually independently in developmental years of this field: one approach arising from survey statistics literature and the other from the psychometrics literature. The survey statistics approach is embodied in the work of Morris Hansen and his colleagues at the US Bureau of the Census (Hansen et al. 1961, 1964). A similar approach using a very different terminology and with somewhat different goals is embodied in the so-called *classical test theory* approach that evolved in the field of psychometric testing and psychological research. Key references for this work are Lord and Novick (1968) and, more recently, Nunnally and Bernstein (1994). In the next section, only a few essential concepts are reviewed, borrowing extensively from both literatures. For a more extensive discussion of the similarities and contrasts between these two approaches, see Groves (1989, Chapter 1) and Biemer and Stokes (1991).

2.1.1 A Simple Model of the Response Process

First, let us consider a simple statistical model for describing the process of obtaining a survey response for a single unit[1] in the population subject to measurement or, more generally, content errors.[2] In the psychometric literature, the model is known as the *classical test theory model*, harkening back to its roots in psychological and educational testing. Let $\mathcal{Y}$ denote the survey variable of interest or characteristic to be measured. To simplify the discussion, suppose that $\mathcal{Y}$ is measured by a single survey question or item. For instance, $\mathcal{Y}$ could represent the number of alcoholic drinks a that person consumed in the last 3 days or a 0–1 indicator variable for whether a respondent ever received a failing grade. Even more generally, $\mathcal{Y}$ could define a charac-

[1]Recall from Chapter 1 that a unit can be a person, business, a school, or any other object for which the characteristics of interest will be observed.

[2]As discussed in Chapter 1, *content errors* include measurement errors as well as data processing errors and other error sources than may change the value of a response to a question.

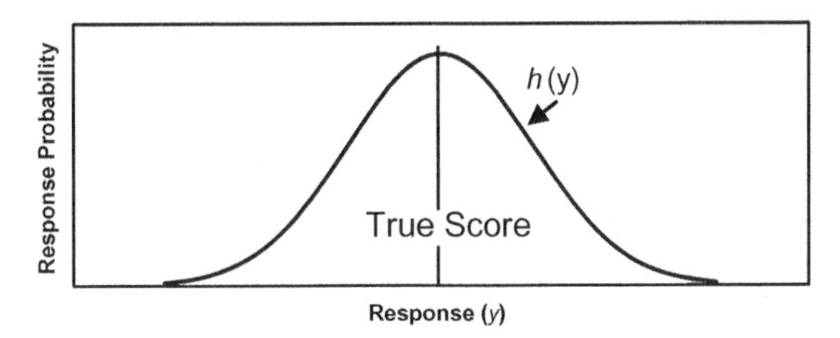

Figure 2.1 Hypothetical response distribution for continuous data.

teristic measured by a series of questions whose responses are recoded[3] to form a single variable. For example, a common practice in labor force surveys is to form a variable denoting the employment status of an individual by combining the responses of a series of questions about employment, including type of work, loss of work and efforts to find work. Later in this chapter we will consider situations were $\mathcal{Y}$ is defined as a composite of multiple questionnaire items. Such multiple item responses can be summarized to form a new variable called a *scale score measurement* (SSM). Essentially, $\mathcal{Y}$ can be any uniquely defined characteristic of the units in the population measured on either a continuous or a discrete scale by one or more questions. For the present discussion, however, let us conceive of $\mathcal{Y}$ as being measurable by a single survey question.

To obtain a realization of $\mathcal{Y}$, a question on the questionnaire is delivered to which a response, y, is recorded. A response obtained for a specific unit, i, will be denoted by y_i. Let us assume that the recorded response is just one of many (potentially infinite) responses that may be recorded for the ith unit. That is, if the question were to be repeated many times, each time under the very same conditions and each time inducing amnesia among all parties involved in the interview, many independent responses could be recorded. Each potential response has an unknown, positive probability of being the response that would be offered by the respondent in response to the question. Therefore, we can associate a probability distribution of responses denoted by $h_i(\mathcal{Y})$ for the ith unit called the *response distribution* for unit i. The subscript i will be dropped when the unit identification is arbitrary or implied by the context.

To aid the discussion, the graphs in Figures 2.1 and 2.2 show the potential shapes for $h(\mathcal{Y})$ for continuous and dichotomous variables, respectively. The shape of the graphs reflects the cumulative effects of the many content error

[3]*Recoding* means to replace the responses to a series of questions on a topic with a single set of codes or an index that is interpreted as a new characteristic. For example, the US consumer price index (`http://www.bls.gov/cpi/`) is a recoded variable.

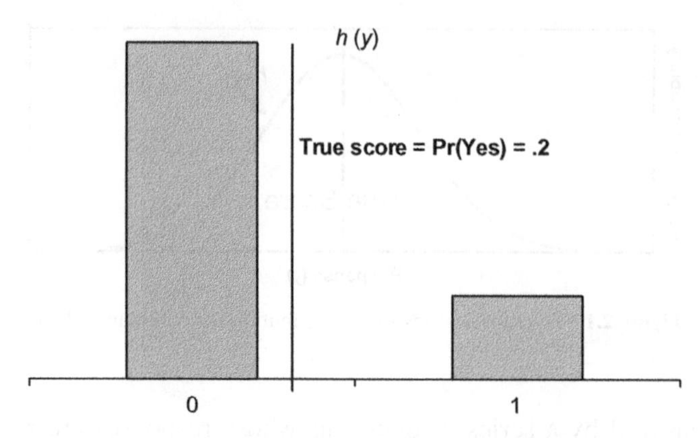

Figure 2.2 Hypothetical response distribution for dichotomous data.

sources described in Chapter 1. It is largely defined by the survey process and the survey environment but may vary across units in the sample by respondent characteristics, traits, and attitudes. Because $h(\mathcal{Y})$ depends on many external factors, it is not unique to the individual but may change as the external factors that influence individual responses, sometimes referred to as *general* or *essential survey conditions* (Hansen et al. 1961) change. Fixing the essential survey conditions as well as the unit i also fixes the distribution $h_i(\mathcal{Y})$.

Thus, the process of measuring $\mathcal{Y}$ is assumed to be equivalent to random selection from an individual's response distribution $h_i(\mathcal{Y})$. Let y_i denote a random draw from $h_i(\mathcal{Y})$. The conditional expectation of y_i with respect to $h_i(\mathcal{Y})$, is denoted by $E(y_i|i)$. This conditional expected value, denoted by τ_i, is called the *true score* for the ith unit in classical test theory. Thus, the true score for an individual is equal to the mean of the individual's response distribution $h_i(\mathcal{Y})$.

As an example, suppose that $\mathcal{Y}$ is a question regarding annual income. Suppose that we ask an individual the same question about income many times, holding all the conditions that may affect the response fixed (same survey setting, same type of interviewer, same question wording, essentially the same point in time, etc.). Fixing all these conditions is necessary so responses are drawn from the same $h(\mathcal{Y})$. Further suppose that any memory of being asked the question previously could be completely erased with each new repetition (i.e., induced amnesia). This condition assures that each repetition elicits an independent, random draw (i.e., response) from $h(\mathcal{Y})$. Under this scenario, the repeated responses may be regarded as independent and identically distributed (i.i.d.) random draws from $h(\mathcal{Y})$. In the psychometric literature [see, e.g., Nunnally and Bernstein (1994)], such i.i.d. measurements are said to be *parallel*. Taken to the limit, the resulting distribution of responses from this hypothetical experiment would empirically reproduce $h(\mathcal{Y})$. The average of these random draws or responses provides an estimate of the unit's true score.

Figure 2.1 shows a typical graph of $h(\mathcal{Y})$ for continuous data. It suggests a bell-shaped or normal distribution of responses for an individual in response to $\mathcal{Y}$. On most occasions, the individual provides responses that are similar. This is suggested by the heaping of the distribution around the middle of the graph. Extreme answers (shown at the tails of the distribution) are possible; however, they are relatively rare, as indicated by the very small response probabilities at the extremes of the distribution. For example, for annual income, an individual's report may vary slightly over independent repeated observations, but most reports are fairly close to each other. Figure 2.2 shows a response distribution $h(\mathcal{Y})$ that is more appropriate for a dichotomous variable. Here, most of the time the respondent responds negatively to the question but on some occasions may offer a positive response. When these positive responses occur depends on random factors that intervene during the interview process.

The true score for an individual is distinct from the individual's *true value*[4] or, as it is known in classical test theory, the *platonic true score*. Suppose an individual's true score for an income question is τ_i = \$32,000. Repeated draws from this person's response distribution could vary considerably from \$32,000 if $h(\mathcal{Y})$ has a large variance. Further, the individual's true income value could have little or no relationship with τ_i. For example, it could be \$45,000 or even \$110,000. The difference between τ_i and the true income is the result of systematic errors in the survey process.

Neither Figure 2.1 nor Figure 2.2 tells us anything about the individual's true value (if it exists), although, for a well-designed set of essential survey conditions, the true value should be close to the mean (true score) for Figure 2.1 and hopefully 0 for Figure 2.2. Methods for obtaining true values and using them in survey evaluations are discussed further in Section 2.5.

The previous discussion leads naturally to the following model for y_i

$$\begin{aligned} y_i &= E(y_i \mid i) + [y_i - E(y_i \mid i)] \\ &= \tau_i + e_i \end{aligned} \tag{2.1}$$

where e_i is the *random measurement error*[5] representing the deviation of a single measurement on unit *i* from its true score τ_i. Equation (2.1) states simply that an observation is equal to the true score plus a measurement error. By taking expectation of both sides of (2.1), we find that $E(e_i \mid i) = 0$; that is, the measurement error has 0 mean. Denote the variance of e_i by σ_{ei}^2. Because, conditional on $h_i(\mathcal{Y})$, τ_i is constant, it follows that

[4]Our usage of *true value* is the same as the *platonic true score* in Lord and Novick's (1968) terminology.

[5]The term *random* is used here to distinguish this type of measurement error from the total measurement error defined in Chapter 1 as the difference between the response and the *true value* of the characteristic.

$$E(\tau_i e_i \mid i) = \tau_i E(e_i \mid i) = 0 \tag{2.2}$$

and, thus, $Cov(\tau_i, e_i) = 0$. By the same reasoning, the conditional variance of y_i with respect to $h_i(\mathcal{Y})$ is

$$Var(y_i \mid i) = Var(\tau_i + e_i \mid i) = Var(e_i \mid i) = \sigma_{ei}^2 \tag{2.3}$$

because τ_i is fixed given unit i. In other words, the variance of an individual's response is also σ_{ei}^2. It is not necessary to assume any particular form of the response distribution at this state, and, in that sense, the theory is nonparametric. The next section introduces a key concept in classical test theory: the *reliability ratio*. This parameter summarizes the quality of a measurement process in terms of the magnitudes of the true score and measurement error variances.

2.1.2 The Reliability Ratio

In the above discussion, the selection of unit i was arbitrary. Suppose that a single unit is selected randomly with equal probability from a population consisting of N units, each having a distinct and independent response distribution $h_i(\mathcal{Y})$ for $i = 1,\ldots, N$. Let $E(\cdot)$ now denote expectation with respect to both $h_i(\mathcal{Y})$ and all possible random draws of a single unit i from the finite population. Note that $E(\cdot)$ can be decomposed into a condition expectation, $E(\cdot \mid i)$, which is with respect to $h_i(\mathcal{Y})$ holding the unit i constant and an unconditional expectation $E_i(\cdot)$ that is with respect to all possible samples of size 1 from the populaton. It follows that

$$E(y_i) = E_i E(y_i \mid i) = E_i(\tau_i) = \tau \tag{2.4}$$

where $\tau = N^{-1} \sum_{i=1}^{N} \tau_i$ is the population mean of the true scores. This essentially says that a single observation from on an individual is unbiased for that individual's true score.

Using a well-known result from sampling theory [see, e.g., Cochran (1977, p. 276)], the unconditional variance of y_i can be decomposed into two quantities representing *between-unit* (interunit) *variance* $[Var_i E(\cdot \mid i)]$ and *within-unit* (intraunit) *variance* $[E_i Var(\cdot \mid i)]$ components. Here $E(\cdot \mid i)$ is as defined above and $Var(\cdot \mid i)$ is a conditional variance with respect to $h_i(\mathcal{Y})$ that holds the unit i constant. Likewise, $Var_i(\cdot)$ is the variance over all samples of size 1 from the population. Thus, it follows that

$$Var(y_i) = Var_i E(y_i \mid i) + E_i Var(y_i \mid i) \tag{2.5}$$

In the survey literature, Hansen et al. (1961) refer to the between unit variance component $[Var_i E(y_i \mid i)]$ as the *sampling variance* (SV) and the within-unit variance component $[E_i Var(y_i|i)]$ as the *simple response variance* (SRV). They define the *inconsistency ratio* as

$$I = \frac{E_i\, Var(y_i \mid i)}{Var(y_i)} = \frac{\text{SRV}}{\text{SV} + \text{SRV}} \tag{2.6}$$

which is the proportion of the total variance of a single response that is measurement (or content) error variance in the survey process. Often, this definition is formulated as *the inconsistency ratio is the proportion of total variance that is simple response variance.* In general, the complement of the inconsistency ratio (i.e., $1 - I$) is referred to as the *reliability ratio*, R. Thus

$$R = \frac{Var_i\, E(y_i \mid i)}{Var(y_i)} = \frac{\text{SV}}{\text{SV} + \text{SRV}} \tag{2.7}$$

(Lord and Novick 1968, p. 208). Thus, *the reliability ratio is the proportion of total variance that is sampling variance.* Note that under our measurement error model, (2.5) can be rewritten as

$$\begin{aligned} Var(y_i) &= Var(\tau_i) + E_i(\sigma_{ei}^2) \\ &= \sigma_\tau^2 + \sigma_e^2 \end{aligned} \tag{2.8}$$

where

$$\sigma_\tau^2 = \frac{\sum_{i=1}^N (\tau_i - \tau)^2}{N} \quad \text{and} \quad \sigma_e^2 = \frac{\sum_{i=1}^N \sigma_{ei}^2}{N} \tag{2.9}$$

Classical test theory refers to σ_τ^2 as the *true score variance* and σ_e^2 as the *measurement error variance* [see, e.g., Nunnally and Bernstein (1994)]. Thus, under the classical test theory model, the SV component is the true score variance and the SRV component is the measurement error variance. Now reliability takes the following form:

$$R = \frac{\sigma_\tau^2}{\sigma_\tau^2 + \sigma_e^2} \tag{2.10}$$

Thus the reliability ratio is the proportion of the total variance of a single observation attributable to true score variance. The inconsistency ratio is then the proportion of total variance attributable to measurement error variance.

Note that R is always in the interval [0,1]. If $R = 1$ (corresponding to $I = 0$), then $\sigma_e^2 = 0$ and all the $h(\mathcal{Y})$ distributions are degenerate with probability mass 1 at $y = \tau$, for $i = 1,\dots, N$. In other words, the population units can be expected to always provide the same response to independent repetitions of the survey question. The population variance is solely a function of the true score variance (σ_τ^2) and the survey measurement process is said to be *completely reliable*. If $R = 0$ (corresponding to $I = 1$), then the variation in the observations is solely a function of measurement error variance (σ_e^2) because then the true score variance is 0. The survey process is said to be *completely unreliable*. This occurs when there is essentially no variance in true scores (e.g., all population members have the same true score, τ), yet respondents respond inconsistently to independent repetitions of the survey question. In that case, we might say that any variation in the observations is purely the result of measurement error or *random noise*. The information content, which gives rise to σ_τ^2, is nil.

In the next section, we put the results of this section to use in describing the effects of measurement error on survey inference, in particular, inference to the population mean.

2.1.3 Effects of Response Variance on Statistical Inference

Suppose one goal of a survey is to estimate the population mean for the characteristic $\mathcal{Y}$.[6] Let τ_i for $i = 1,\dots, N$ denote the true score values for N population units, and let $y_1,\dots, y_n$ denote the observed values of $\mathcal{Y}$ for the n units selected for the sample. To simplify the discussion, we assume simple random sampling, deferring the complications arising from more complex sample designs to Chapter 5. As another simplification, we consider only estimators that are linear functions of the observations such as means, totals, and proportions. The results are still widely applicable as much of survey data analysis involves descriptive statistics that tend to be linear estimators. The population (true score) mean or proportion can be written as

$$\tau = \frac{\sum_{i=1}^{N} \tau_i}{N} \tag{2.11}$$

and the population (true score) total, denoted by T, is given by $T = N\tau$.

A typical course in survey sampling would focus on determining unbiased estimators of these population parameters under various sampling designs. The usual starting point for the discussion of complex designs is simple random sampling. Although it is seldom used for practical survey work, simple random sampling provides the basic building blocks and key concepts on which all other, more complex designs are based. In this book, a similar

[6]The results that follow also apply to population totals and proportions.

strategy is taken. Initially, the effects of measurement errors on descriptive statistics is considered under simple random sampling. Subsequent chapters will build on these elementary concepts.

Traditional sampling theory assumes the characteristic $\mathcal{Y}$ is observed without error for all units in the sample: $\Pr(y_i = \tau_i) = 1$ or, equivalently, $\sigma_{ei}^2 = 0$ for all i. Under this assumption, the sample mean

$$\bar{y} = \frac{1}{n}\sum_{i=1}^{n} y_i \tag{2.12}$$

is an unbiased estimator of the true population mean $\bar{Y} = N^{-1}\sum_{i=1}^{N} y_i$; that is, $E(\bar{y}) = \bar{Y} = \tau$, where expectation is with respect to the sampling distribution of the sample mean. The traditional formulas for the variance of $\bar{y}$ and its unbiased estimator are

$$Var(\bar{y}) = \left(1 - \frac{n}{N}\right)\frac{S^2}{n} \quad \text{and} \quad v(\bar{y}) = \left(1 - \frac{n}{N}\right)\frac{s^2}{n} \tag{2.13}$$

respectively, where

$$S^2 = \frac{\sum_{i=1}^{N}(y_i - \bar{Y})^2}{N-1} \quad \text{and} \quad s^2 = \frac{\sum_{i=1}^{n}(y_i - \bar{y})^2}{n-1} \tag{2.14}$$

In the presence of content errors, the true values of the characteristics are *latent* or unobserved variables. As noted in Chapter 1, the usual estimators of population parameters may be biased and the variance formulas developed under classical sampling theory no longer hold. In the remainder of this chapter, we develop new formulas for inference to the population mean that are applicable when characteristics are observed with error and the error follows the model developed in the last section.

Recall that in obtaining expressions for the means and variances in the last section, it was useful to apply the expected value operator in two stages: one stage for the process of randomly selecting an individual from the finite population of individuals and the other for sampling a single observation from the infinite population of potential responses, namely, the response distribution, $h(\mathcal{Y})$. This two-stage sampling process can be viewed as a special case of *two-stage finite-population sampling*, a common sample design in field surveys. In traditional two-stage sampling, the first-stage units (called *primaries*) may be counties or provinces and the second-stage units (called *secondaries*) may be persons or households within the primaries. The two-stage sampling process is analogous to our conceptual model for the response process. For the response process, the primary units are individuals that are assumed to be selected by

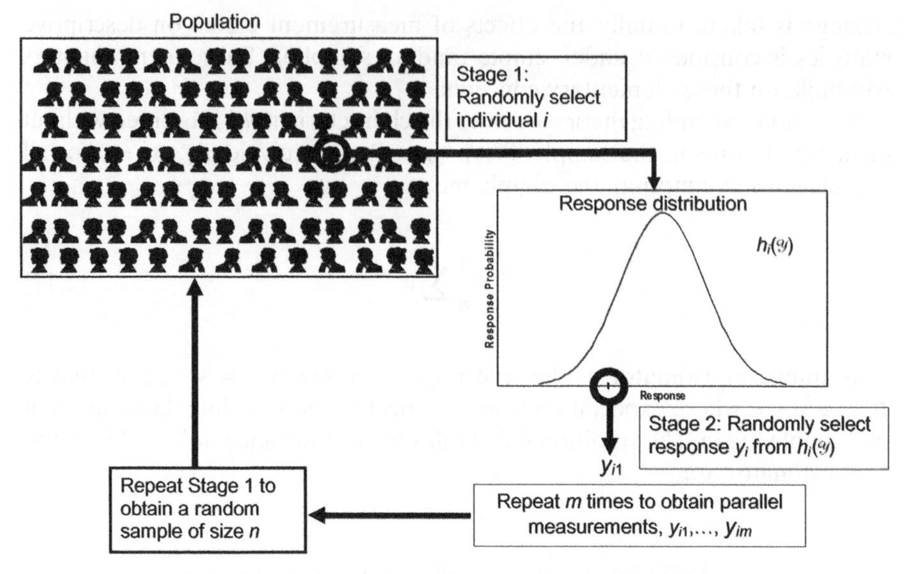

Figure 2.3 Survey response as a two-stage sampling process.

simple random sampling from the population. At the "secondary" stage (i.e., the response stage), a response is selected from an individual's response distribution (see Figure 2.3). This "secondary" selection occurs when an individual responds to the survey question; in other words, the act of responding is equivalent to a random selection from the individual' response distribution. One important difference is that, in classical sampling theory, the primary units such as counties, provinces, and school districts are finite in size. For the response process, the primary unit is an individual having (theoretically) an infinite number of possible responses (secondaries). Thus, the primaries can be considered infinite because the response distribution $h(\mathcal{Y})$ is infinite.

When the response process is viewed this way, formulas for unbiased estimation of the population mean and its variance can be readily obtained by simply applying the two-stage sampling formulas from traditional sampling theory [see, e.g., Cochran (1977)]. A short review of the essential two-stage sampling formulas can be found in Appendix A. In the following discussion, we apply the traditional two-stage sampling formulas to derive formulas for inference to the population mean, total, or proportion under the measurement error model of the previous section.

Because the sample mean under simple random sampling is unbiased for the population mean, $\bar{y}$ as defined in (2.12) is unbiased for $E(\bar{y}) = \tau$. Applying (A.3) in Appendix A, it follows that

$$Var(\bar{y}) = (1-f)\frac{S_1^2}{n} + \frac{S_2^2}{n} \tag{2.15}$$

where, in the notation of the measurement error model is

$$S_1^2 = \sum_{i=1}^{N} \frac{(\tau_i - \tau)^2}{N-1} = \frac{N}{N-1}\sigma_\tau^2 \tag{2.16}$$

and

$$S_2^2 = \frac{\sum_{i=1}^{N}\sigma_{ei}^2}{N} = \sigma_e^2 \tag{2.17}$$

Comparing the two expressions of $Var(\bar{y})$ in (2.15) and (2.13), we see that when errors are introduced into the observations, the usual expression for the variance of the sample mean no longer holds primarily because of the addition of the term S_2^2/n in (2.15). We see from (2.17) that this term reflects the average measurement error variance in the population. Note that (2.15) can be rewritten as

$$\begin{aligned} Var(\bar{y}) &= \left(1 - \frac{n-1}{N-1}\right)\frac{\sigma_\tau^2}{n} + \frac{\sigma_e^2}{n} \\ &\doteq \frac{1}{R}\frac{\sigma_\tau^2}{n} \end{aligned} \tag{2.18}$$

when N is large relative to n. In this formula, σ_τ^2/n, which is the usual large population variance of $\bar{y}$ *without* measurement error, is multiplied by the inverse of the reliability ratio, R. This result suggests that the consequences of measurement error (i.e., $R < 1$) is to increase the variance of the sample mean by the factor $1/R$. Or, stated another way, the *effective sample size*, defined as nR, decreases as reliability departs from unity. To illustrate, suppose that the sample size for a survey is $n = 5000$ and further that the reliability for the characteristic of interest is $R = 0.7$. Then the precision of the usual estimator of the population mean is equivalent to the precision that would be realized from a sample of $n = 3500$ and where the characteristic is measured without error. In this case, one could say that the *cost of poor quality* is the difference between 5000 and 3500 observations (or its equivalent in monetary terms). If steps could be taken to increase the reliability to, say, 0.8, the effective sample size would increase to 4000. The question the survey designer must ask is whether the cost associated with increasing reliability is offset by the increased effective sample size.

Unreliability not only increases the variance of the estimates but also tends to attenuate the correlations between variables as noted in Chapter 1. To see this, suppose that x and y are two variables subject to measurement errors. Using the measurement error model (2.1), denote their true by scores τ_x, τ_y, their true score variances by $\sigma^2_{\tau_x}, \sigma^2_{\tau_y}$, and their reliabilities by R_x, R_y, respectively. Since random measurement errors are uncorrelated with both the true scores and each other, the covariance between x and y, written $Cov(x,y)$, is equivalent to the covariance of their true scores, $Cov(\tau_x, \tau_y)$. Further, by (2.18), the variances of x and y are $\sigma^2_{\tau_x}/R_x$ and $\sigma^2_{\tau_y}/R_y$, respectively. Thus, the correlation between x and y is

$$\begin{aligned}\rho_{xy} &= \frac{Cov(\tau_x,\tau_y)}{\sigma_x\sigma_y} = \frac{Cov(\tau_x,\tau_y)}{\sigma_{\tau_x}\sigma_{\tau_y}}\sqrt{R_xR_y} \\ &= \rho_{\tau_x\tau_y}\sqrt{R_xR_y}\end{aligned} \tag{2.19}$$

Thus, the correlation between x and y is the correlation between their true scores multiplied by the squared root of the product of their reliabilities. Since $\sqrt{R_xR_y} \le 1$, the correlation is attenuated toward 0 and the reliability of either x or y is reduced.

Unreliability also has important consequences for regression analysis. As an example, for the regression of y on x, measurement errors in the independent variable x will attenuate (i.e., bias toward 0) the estimated slope parameter and may bias the intercept estimator in any direction. To see this, recall that for any two variables x and y, the expected values of the slope and intercept coefficients for regression of y on x are $\beta_{y \cdot x} = \rho_{xy}\sigma_x/\sigma_y$ and $\beta_{0,y \cdot x} = E(x) - \beta_{y \cdot x}E(y)$, respectively. Thus, the slope coefficient, $\beta_{y \cdot x}$, has the form (Biemer and Trewin 1997)

$$\begin{aligned}\beta_{y \cdot x} &= \rho_{xy}\frac{\sigma_x}{\sigma_y} = \rho_{\tau_x\tau_y}\sqrt{R_xR_y}\frac{\sigma_{\tau_x}}{\sigma_{\tau_y}}\sqrt{\frac{R_x}{R_y}} \\ &= \rho_{\tau_x\tau_y}\frac{\sigma_{\tau_x}}{\sigma_{\tau_y}}R_x \\ &= \beta_{\tau_y \cdot \tau_x}R_x\end{aligned} \tag{2.20}$$

which, like ρ_{xy}, is also attenuated toward 0. It can also be shown (Biemer and Trewin 1997) that the intercept has the form

$$\begin{aligned}\beta_{0,y \cdot x} &= \tau_y - \beta_{xy}\tau_x \\ &= \tau_y - \beta_{\tau_y \cdot \tau_x}R_x\tau_x \\ &= \beta_{0,\tau_y \cdot \tau_x} + (1-R_x)\beta_{\tau_y \cdot \tau_x}\tau_x\end{aligned} \tag{2.21}$$

suggesting that the intercept is biased by the term $(1-R_x)\beta_{\tau_y \cdot \tau_x}\tau_x$, which may be either positive or negative.

When there are measurement errors in the dependent variable y, neither the intercept nor the slope coefficients are biased under the measurement error model. This is because the model can be rewritten so that the measurement error term is on the right-hand side of the regression equation where it can be estimated as part of the model residual error term. However, since this increases the model error, the consequence is a reduction in model fit. In fact, it can be shown that the model coefficient of determination cannot exceed the reliability of the dependent variable (i.e., R_y) (Fuller 1987).

Estimating R requires separately estimating the SV and the SRV components. As we shall see, this requires at least two realizations (or replicate measurements) of the characteristic $\mathcal{Y}$. More than that, the measurements must satisfy rather strong assumptions so that the estimation of R is unbiased. Suppose that there are at least two measurements of $\mathcal{Y}$ (i.e., $m \geq 2$) on all n units in the sample. Drawing again on the analogy of our measurement error model to a two-stage sampling process, we can apply the theoretical results for two-stage sampling in Appendix A directly for the estimation of SV and SRV. This is discussed in the next section.

2.2 VARIANCE ESTIMATION IN THE PRESENCE OF MEASUREMENT ERROR

In this section, the formulas for two-stage sampling in Appendix A will be applied to obtain estimators of the variance of the sample mean as well as the reliability ratio. As can be seen from (A.6) and (A.7) in Appendix A, an unbiased estimator of the measurement error variance, σ_e^2 (or SRV) is

$$s_2^2 = \frac{\sum_{i=1}^{n}\sum_{j=1}^{m}(y_{ij} - \bar{y}_i)^2}{n(m-1)} \tag{2.22}$$

provided there are at least two measurements ($m \geq 2$) for each sample unit.[7] When only one measurement is available, which is usually the case in most surveys, estimation of measurement variance is not possible because then s_2^2 is undefined. Nevertheless, if the population size N is quite large or the sampling fraction f is quite small, then the usual textbook estimator of the variance [i.e., $v(\bar{y})$ in (2.13)], will be essentially unbiased for $Var(\bar{y})$ under the present measurement error model (Biemer and Trewin, 1997).

To understand this, note that for $m = 1$, s_1^2 in (A.6) is identical to s^2 in (2.14). Further, setting $m = 1$ in (A.7), we can write

$$E(s^2) = S_1^2 + S_2^2 \tag{2.23}$$

[7]As we shall see in Section 2.3.4, it is usually sufficient to have repeated measurements on a subsample of units in the sample rather than the entire sample.

Now the bias in $v(\bar{y})$ can be written as

$$\begin{aligned} Bias[v(\bar{y})] &= E\left[\frac{(1-f)}{n}s^2\right] - \left[(1-f)\frac{S_1^2}{n} + \frac{S_2^2}{n}\right] \\ &= \frac{(1-f)}{n}(S_1^2 + S_2^2) - \left[(1-f)\frac{S_1^2}{n} + \frac{S_2^2}{n}\right] \\ &= \left[\frac{(1-f)}{n} - \frac{1}{n}\right]S_2^2 = -\frac{1}{N}S_2^2 \doteq 0 \text{ for large } N \end{aligned} \quad (2.24)$$

where the symbol "≐" means "approximately equal to". When the sampling fraction, f, is small (say, less than 0.1), or equivalently, when N is very large, S_2^2/N will be essentially negligible and, therefore, the bias will also be negligible. These results also extend to more complex survey designs as shown in Wolter (2007). In general, when f is negligibly small, a separate estimate of σ_e^2 is not required for efficient estimation of the variance of the mean because the usual estimator is still unbiased under the measurement error model. However, as shown below, for purposes of measurement evaluation and survey improvement, it is still often desirable to have an estimate of the measurement variance.

Let us now consider the estimation of I (or R) from $m > 1$ repeated measurements. An estimator of I will be constructed from separate estimators for the numerator of I, that is, σ_e^2 (or SRV) and denominator, $\sigma_y^2 = \sigma_\tau^2 + \sigma_e^2$ (or SV + SRV).

Let y_{ij} denote the jth measurement on the ith unit, $i = 1,\ldots, n$ and $j = 1,\ldots, m$, and suppose that $m > 1$. Let $\bar{y}_i = \Sigma_j\, y_{ij}/m$ denote the mean of the repeated measurements for the ith unit, and let $\bar{\bar{y}} = \Sigma_i \Sigma_j\, y_{ij}/nm$ denote the sample mean of the $\bar{y}_i$. Further, let

$$s_1^2 = \frac{\Sigma_{i=1}^n(\bar{y}_i - \bar{\bar{y}})^2}{n-1} \quad (2.25)$$

Now assume that $\{y_{i1},\ldots, y_{im}\}$ constitutes a simple random sample from $h_i(\mathcal{Y})$ for $i = 1,\ldots, n$ that is equivalent to assuming the m measurements are parallel (i.e., i.i.d. random variables). Then from (A.7), we obtain

$$E(s_1^2) = \sigma_\tau^2 + \frac{\sigma_e^2}{m} \quad \text{and} \quad E(s_2^2) = \sigma_e^2 \quad (2.26)$$

Thus, a consistent estimator of R is

$$\hat{R} = 1 - \frac{s_2^2}{\hat{\sigma}_y^2} \tag{2.27}$$

where

$$\hat{\sigma}_y^2 = s_1^2 + \left(\frac{m-1}{m}\right) s_2^2 \tag{2.28}$$

The denominator of $\hat{R}$ is an estimator of the total variance given by $\hat{\sigma}_y^2$. The numerator is the an estimator of the true score variance, which can be obtained by subtracting the estimator of the measurement variance s_2^2 from $\hat{\sigma}_y^2$. This estimator can be negative, a consequence that will be discussed subsequently.

As previously noted, when $m = 1$, s_2^2 cannot be computed, but s_1^2 still provides an estimator of σ_y^2. When $m > 1$, s_2^2 is the variance of the repeated measurements on a sample unit averaged over the n sample units. Then s_1^2 will include the extra variance created by the multiple measurements.

Of particular interest in this book are categorical variables. In that regard, the next section looks at the special case of binary variables and obtains a somewhat different form of the reliability ratio.

2.2.1 Binary Response Variables

In this section, we consider special forms of the parameters and estimators for the case of binary variables and subsequently polytomous discrete variables. The translation of the previous formulas to these new forms is straightforward after introducing the following notation.

Suppose that the characteristic $\mathcal{Y}$ is coded $y = 1$ (for a positive response) and $y = 0$ (for a negative response).[8] Further, let $h_i(1) = \Pr(y_i = 1) = P_i$ and $h_i(0) = \Pr(y_i = 0) = Q_i = 1 - P_i$. Note that $P_i = E(y_i | i) = \tau_i$ in the general notation of the previous section. Thus, the true score for unit i is P_i, the probability that the ith individual responds positively where $0 \le P_i \le 1$. It also follows that $\sigma_{ei}^2 = P_i Q_i$ is the variance of the response distribution $h_i(\mathcal{Y})$. For example, if an individual always responds positively or negatively (i.e., with probability 1), the variance of his or her response distribution will be 0. On the other hand, if an individual has a 50–50 chance of responding positively or negatively, the variance of the response distribution is at its maximum.

[8]Later in the book, the digit "2" will denote a negative binary response. In this chapter we use the digit "0" because it simplifies the formulas to follow considerably.

The population mean τ is now the population proportion denoted by $P = E_i(P_i) = \Sigma_i P_i/N$. Now, substituting these symbols for their counterparts in (2.16), we obtain

$$S_1^2 = \frac{N}{N-1}\sigma_\tau^2 = \sum_{i=1}^{N}\frac{(P_i - P)^2}{N-1} \quad \text{and} \quad S_2^2 = \sigma_e^2 = \frac{\Sigma_{i=1}^N P_i Q_i}{N} \tag{2.29}$$

As before, we refer to S_1^2 as the between-unit or true score variance and S_2^2 as the within-unit or measurement error variance.

To define the sample quantities, we again consider a simple random sample of n units from N population units, and let p denote the sample proportion that has the same form as $\bar{y}$ except under binary coding for y. Thus, it follows that $Var(p)$ is still given by (2.15) on substituting the new forms for S_1^2 and S_2^2 in (2.29). This leads to the following form:

$$Var(p) = \left(\frac{N-n}{N-1}\right)\frac{\sigma_\tau^2}{n} + \frac{\sigma_e^2}{n} \tag{2.30}$$

Now consider the estimation of $Var(p)$. First, note that for the cases where $m = 1$ (i.e., one measurement only), we can apply (2.14) to show that $s^2 = npq/(n-1)$, where $q = 1 - p$. It follows from (2.23) and (2.29) that

$$E\left(\frac{n}{n-1}pq\right) = \frac{N}{N-1}\sigma_\tau^2 + \sigma_e^2 \tag{2.31}$$

In other words, $npq/(n-1)$ is approximately unbiased for the total variance of an observation. In practice, the bias in pq as an estimator of the total variance is usually ignored since it is trivial for most applications. The bias is essentially 0 in large populations. Further, the usual estimator of $Var(p)$, given by

$$v(p) = (1-f)\frac{pq}{n-1} \tag{2.32}$$

(Cochran 1977, p. 52), is unbiased for

$$\left(1 - \frac{n-1}{N-1}\right)\left(\frac{\sigma_\tau^2}{n} + \frac{N-1}{N}\frac{\sigma_e^2}{n}\right) \tag{2.33}$$

Compare this expression with (2.30). If N is quite large relative to n, then both (2.30) and (2.33) reduce to $(\sigma_\tau^2 + \sigma_e^2)/n$, the variance of the sample proportion in large populations. In general, the bias in (2.32) is still given by (2.24), except now the expression for S_2^2 in (2.29) can be substituted. This yields a bias of $-\sigma_e^2/N$, which is negligible for most applications. Thus, we say that the usual estimator of the variance $v(p)$ is essentially unbiased under the classical measurement error model.

For estimating σ_τ^2 and σ_e^2, assume that $m > 1$ parallel measurements of $\mathcal{Y}$ are available. Let p_i denote the proportion of 1s among the m measurements for unit i, and let $q_i = 1 - p_i$ denote the proportion of 0s. Thus, $\bar{p} = n^{-1} \sum_i p_i$ is the proportion of 1s among all nm observations. The estimators s_1^2 and s_2^2 can now be rewritten as

$$s_1^2 = \frac{\sum_{i=1}^{n}(p_i - \bar{p})^2}{n-1} \quad \text{and} \quad s_2^2 = \frac{m}{n(m-1)} \sum_{i=1}^{n} p_i q_i \tag{2.34}$$

The estimator of R in (2.27) can still be used after we substitute these new forms for the estimates. We are particularly interested in the case where there are only two measurements for each sample member. This case is examined in some detail next.

2.2.2 Special Case: Two Measurements

The case of two measurements ($m = 2$) on each unit is quite prevalent in survey work because 2 is the minimum number required for estimating SRV and reliability, and the cost and burden of obtaining more than two measurements is often considerable, particularly if remeasurements are obtained via a reinterview survey. Reinterview surveys conducted especially for the purpose of estimating reliability must be designed so that the remeasurements are parallel to the original (first) measurements. This type of study design is often referred to as a *test–retest reinterview* design. The data from reinterview studies can be summarized by a 2 × 2 *interview–reinterview* (or *crossover*) table such as the one shown in Table 2.3. In this table, a represents the number of sample members classified as 1 by both the interview and reinterview; b, the number classified 1 by the interview and 0 by the reinterview; c, the number classified as 0 by the interview and then 1 by the reinterview; and d, the number classified as 0 by both interviews. Let $p_{y_1y_2}$ denote the proportion of the sample classified in cell (y_1,y_2), wherey_1 and y_2 are 0–1 variables denoting the interview and reinterview classifications, respectively. Note that $p_{11} + p_{01} + p_{10} + p_{00} = 1$. Denote the marginal proportions by $p_{y_1+} = p_{y_11} + p_{y_10}$ and $p_{+y_2} = p_{1y_2} + p_{0y_2}$ for $y_1,y_2 = 0,1$, where the "+" denotes summation over the subscript it replaces.

The expression for s_2^2 in (2.34) can be simplified by noting that p_iq_i takes the value 0 when y_{1i} and y_{2i} agree (which happens for $a + d$ cases) and takes

Table 2.3 Crossover Table for Two Dichotomous Measurements of $\mathcal{Y}$

	$y_2 = 1$	$y_2 = 0$	
$y_1 = 1$	$p_{11} = \frac{a}{n}$	$p_{10} = \frac{b}{n}$	$p_{1+} = p_{11} + p_{10}$
$y_1 = 0$	$p_{01} = \frac{c}{n}$	$p_{00} = \frac{d}{n}$	$p_{0+} = p_{01} + p_{00}$
—	$p_{+1} = p_{11} + p_{01}$	$p_{+0} = p_{10} + p_{00}$	—

the value $\frac{1}{4}$ when y_{1i} and y_{2i} disagree (which happens for $b + c$ cases). Thus, we can write

$$\begin{aligned} s_2^2 &= \frac{2}{n(2-1)}\sum_{i=1}^{n} p_i q_i = \frac{2}{n}\left((a+d)\times 0+(b+c)\times\frac{1}{4}\right) \\ &= \frac{b+c}{2n} = \frac{g}{2} \end{aligned} \quad (2.35)$$

where $g = n^{-1}(b + c)$ is the *disagreement rate* or *gross difference rate*, as Hansen et al. (1964) referred to it. Another quantity of interest also defined by Hansen et al. is the *net difference rate* (NDR) defined as

$$\text{NDR} = p_{1+} - p_{+1} = \frac{b-c}{n} \quad (2.36)$$

where $p_{1+} = p_{11} + p_{10}$ is the sample proportion based on y_1 classifications and $p_{+1} = p_{11} + p_{01}$ is the sample proportion based on y_2 classifications. The net difference rate is the difference between two estimates of the population proportion P: one from the interview and the other from the reinterview. As we will see, the NDR can be quite useful for testing the assumption that y_1 and y_2 are parallel measurements.

Recall that when $m = 2$, there is a choice of estimators of σ_y^2 that leads to different estimators of R. In the case of dichotomous variables, a few more estimators are possible. For example, applying (2.31), it follows that both $p_{1+}p_{0+}$ or $p_{+1}p_{+0}$ are essentially unbiased estimators of the total variance SV + SRV for the classical measurement error model. Based upon empirical research conducted by the US Census Bureau (1985), the estimator $(p_{1+}p_{+0} + p_{+1}p_{0+})/2$ is preferred for the denominator of $\hat{I}$. This leads to the following estimator of R:

$$\hat{R} = 1 - \hat{I} = 1 - \frac{g}{p_{1+}p_{+0} + p_{+1}p_{0+}} \quad (2.37)$$

According to the US Census Bureau (1985), this denominator provides much more stability in the estimator than other alternatives that could be considered such as $p_{1+}p_{0+} + p_{+1}p_{+0}$, $2p_{1+}p_{0+}$, or $2p_{+1}p_{+0}$.

Unlike R, which, by definition, must be between 0 and 1, both $\hat{R}$ and $\hat{I}$ can range from −1 to 1. Negative estimates of reliability can occur in rare situations when R is near 0 or when the assumption of parallel measurements is severely violated. In the latter case, $\hat{R}$ is not a valid estimator of R and alternative approaches, discussed in Chapter 3, should be used instead. In either case, it is customary to replace negative values of $\hat{R}$ or $\hat{I}$ with 0.

Hansen et al. (1964) refer to $\hat{I}$ as the *index of inconsistency*. The US Census Bureau (1985) suggests the following rule of thumb for interpreting the magnitude of $\hat{I}$, namely, that $0 \le \hat{I} \le 0.2$ is good, $0.2 < \hat{I} \le 0.5$ is moderate, and $\hat{I} > 0.5$ is poor.[9] This rule is offered merely as a guideline because whether a particular level of reliability is too low depends on the use of the data. Nevertheless, most analysts would agree that an index of inconsistency above 50% or, equivalently, a reliability of less than 0.5, indicates a poorly measured variable.

A second estimator of R can be obtained by noting that

$$\rho_{y_1y_2} = \frac{Cov(y_1, y_2)}{\sqrt{Var(y_1)Var(y_2)}} = R \tag{2.38}$$

under the parallel measurements assumption. To see this, note that $E(y_{i1} | i) = E(y_{i2} | i)$ so that

$$\begin{aligned} Cov(y_1, y_2) &= E_i\,Cov(y_{i1}, y_{i2} \mid i) + Cov_i[E(y_{i1} \mid i), E(y_{i2} \mid i)] \\ &= Var_i[E(y_{i1} \mid i)] \\ &= \text{SV} \end{aligned} \tag{2.39}$$

and $Var(y_1) = Var(y_2) = \text{SV} + \text{SRV}$. Thus, $\rho_{y_1y_2}$ is equal to (2.7). An estimator that is appropriate for either continuous or dichotomous categorical data (provided $s_1 \ne 0$ and $s_2 \ne 0$) is the sample correlation coefficient given by

$$\hat{\rho}_{y_1y_2} = \frac{(1/n)\Sigma_{i=1}^{n}(y_{i1} - \bar{y}_1)(y_{i2} - \bar{y}_2)}{s_1 s_2} \tag{2.40}$$

It can be shown that, in the case of dichotomous measurements, this simplifies to

[9]A similar rule of thumb was developed by Cicchetti (1994) for κ. $\kappa \le 0.20$ is poor agreement, $0.20 < \kappa \le 0.40$ is fair agreement, $0.40 < \kappa \le 0.60$ is moderate agreement, $0.60 < \kappa \le 0.80$ is good agreement, and $0.80 < \kappa \le 1.00$ is very good agreement.

$$\hat{\rho}_{y_1y_2} = \frac{p_{11} - p_{1+}p_{+1}}{\sqrt{p_{1+}p_{0+}p_{+1}p_{+0}}} \tag{2.41}$$

which is an estimator distinctly different from (2.37). The estimator in (2.37) is often more precise because it uses the total number of agreements in the estimation of SV, not just the agreements among positives (i.e., p_{11}). Thus, more information about the response process is incorporated in (2.37).

Another estimator of the reliability ratio is Cohen's kappa (Cohen 1960) defined as

$$\kappa = \frac{p_0 - p_e}{1 - p_e} \tag{2.42}$$

where p_0 is the agreement rate between the interview and reinterview classifications (i.e., $1 - g$) and p_e is an estimate of the probability of agreement by chance alone given by $p_e = p_{1+}p_{+1} + p_{0+}p_{+0}$. Interestingly, kappa was not originally developed as an estimator of R, but coincidentally, it is. In fact, kappa was originally developed as a *chance-adjusted agreement rate*, that is, an agreement rate that removes some number of interview–reinterview response pairs that are expected to agree purely at random or by chance under the parallel assumption. The result is an agreement rate based on response pairs that agree in a nonrandom, substantive way.

To understand this further, suppose that the two measurements, y_1 and y_2, are statistically independent. Let $P_{y_1y_2}$ denote $E(p_{y_1y_2})$, the expected proportion in cell (y_1,y_2). Then, for a given simple random sample of size n, the expected proportion of cases for which $y_1 = y_2 = 1$ is $P_{1+}P_{+1}$. Likewise, the proportion of cases for which $y_1 = y_2 = 0$ is $P_{0+}P_{+0}$. Thus, assuming that y_1 and y_2 are independent, the proportion of cases expected to agree by chance is $P_e = P_{1+}P_{+1} + P_{0+}P_{+0}$.

Next, write the sample agreement rate as

$$\begin{aligned} P_0 &= \Pr(y_{i1} = y_{i2} \mid i = 1,\ldots,n) \\ &= \Pr(y_{i1} = y_{i2} \mid C)\Pr(C) + \Pr(y_{i1} = y_{i2} \mid {\sim}C)\Pr({\sim}C) \\ &= P_e + \kappa(1 - P_e) \end{aligned} \tag{2.43}$$

where C denotes the set of units in the sample that agree by chance, $\Pr(C)$ is an estimate of the proportion of the sample in C (i.e., P_e), and $\Pr(y_{i1} = y_{i2} \mid C)$ is the proportion of the sample in C that agree (which is 1 because all unit in C agree). Cohen's kappa is an estimator of $\kappa = \Pr(y_{i1} = y_{i2} \mid {\sim}C)$, where $\Pr(y_{i1} = y_{i2} \mid {\sim}C)$ is the probability of agreement for units in the complement of C, that is, for the units that do not agree by chance. Let $\Pr({\sim}C) = 1 - \Pr(C)$ denote the probability a unit in the sample is in the complement of C and note that $\Pr({\sim}C) = 1 - P_e$. Solving for κ in (2.43) yields

$$\kappa = \frac{P_0 - P_e}{1 - P_e} \tag{2.44}$$

from whence (2.42) follows after replacing P_0 and P_e by their respective unbiased estimators, p_0 and p_e, respectively.

Note that, in deriving Cohen's kappa, it was not assumed that the two measurements are parallel or that the errors in the measurements are independent. As a measure of chance agreement, these assumptions are not required. However, for interpreting kappa as an estimator of R, the i.i.d. assumptions are still required.

Like $\hat{R}$, the values of κ can range from −1 and 1. When $\kappa = 0$, there is no agreement other than chance agreement. When $\kappa = 1$, there is perfect agreement for the entire sample. When $\kappa < 0$, the rate of agreement is *less* than what would be expected by chance (Cohen 1960). Hess et al. (1999) provide the interesting result that $\hat{R}$ is identical to κ, which is remarkable, given that the two indices were developed from very different theoretical considerations with very different purposes in mind. To see that $\hat{R} = \kappa$, note that the denominator of κ is the 1 minus the chance agreement rate that can be computed as "chance disagreement rate" or the disagreement rate under the assumption that y_1 and y_2 are independent. This can be computed as the probability of classification in the off-diagonal cells of Table 2.3 assuming y_1 and y_2 are independent: $p_{1+}p_{+0} + p_{+1}p_{0+}$. Thus, $1 - p_e = p_{1+}p_{+0} + p_{+1}p_{0+}$, and we have shown that the denominators of $\hat{R}$ and κ are identical. The numerator of κ is

$$\begin{aligned} p_0 - p_e &= (1 - p_e) - (1 - p_0) \\ &= (1 - p_e) - g \end{aligned} \tag{2.45}$$

It follows that

$$\kappa = \frac{(1 - p_e) - g}{1 - p_e} = 1 - \frac{g}{1 - p_e} = 1 - \frac{g}{p_{1+}p_{+0} + p_{+1}p_{0+}} = \hat{R} \tag{2.46}$$

To summarize, κ and $\hat{R}$ both have two very different interpretations in the literature—κ is the chance adjusted agreement rate and $\hat{R}$ is an estimator of the reliability ratio. However, because they are identical, both can be used to estimate the reliability ratio as well as the chance-adjusted agreement rate. In Section 6.2, yet a third interpretation of $\hat{R}$ is considered under the so-called *agreement model* of Guggenmoos-Holzmann and Vonk (1998).

The variance of κ, $\hat{R}$ or $\hat{I}$ can be estimated by the same formula which is provided by Fleiss (1981, p. 221) for κ. In this formula, the alternate notation for $1 - p_{1+}$ and $1 - p_{+1}$ shown in Table 2.2, namely, p_{0+} and p_{+0}, respectively, is used:

$$v(\kappa) = \frac{1}{n}\left\{\frac{p_0(1-p_0)}{(1-p_e)^2} + \frac{2(1-p_0)[2p_0p_e - \sum_{k\in\{0,1\}} p_{kk}(p_{k+}+p_{+k})]}{(1-p_e)^3}\right\}$$
$$+ \frac{(1-p_0)^2[\sum_{k\in\{0,1\}}\sum_{l\in\{0,1\}} p_{kl}(p_{l+}+p_{+k})^2 - 4p_e^2]}{(1-p_e)^4} \tag{2.47}$$

Feder (2007) derived formulas for the estimators of κ and the $Var(\kappa)$ in complex survey designs. In particular, he shows how to appropriately weight the observations to account for unequal probability sampling.

Illustration. To illustrate the use of these formulas, we compute $\hat{I}$, κ, and $\hat{\rho}_{y_1y_2}$ for the data in Table 2.1. To compute $\hat{I}$, we first compute $g = (203 + 11)/954 = 0.2243$. Next, $p_{1+} = (231 + 203)/954 = 0.4549$, $p_{0+} = 1 - p_{1+} = 0.5451$, $p_{+1} = (231 + 11)/954 = 0.2537$, and $p_{+0} = 1 - p_{+1} = 0.7463$. Thus, $p_{1+}p_{+0} + p_{+1}p_{0+} = 0.4778$ and $\hat{I} = 0.2243/0.4778 = 0.4694$. By (2.46), we compute κ as $\hat{R} = 1 - \hat{I} = 0.5306$. Finally, we compute $\hat{\rho}_{y_1y_2}$ using (2.41) as

$$\hat{\rho}_{y_1y_2} = \frac{0.2421-(0.4549)(0.2537)}{\sqrt{(0.4549)(0.5451)(0.2537)(0.7463)}} = 0.5847$$

which is larger than the alternative estimator, κ.

The net difference rate is NDR = 0.4549 − 0.2537 = 0.2012. The variance of the NDR in simple random samples can be estimated using the formula

$$Var(\text{NDR}) \doteq \frac{1}{n}\left(g + \frac{1}{n}\right) \tag{2.48}$$

(US Census Bureau 1985). This yields a standard error estimate of 0.015. Thus, NDR is highly statistically significant from 0 and quite large practically. This means that that the assumption of parallel measurements is not tenable for Q1 (Did you ever receive a grade of "D" or "F" for a class?) and Q2 (Did you ever receive an unsatisfactory or failing grade?) because the two questions produce different true scores. Consequently, $\hat{R}$ cannot be interpreted as an estimator of a common reliability ratio for these two questions. It can, however, be interpreted as the chance agreement rate (i.e., an estimator of κ) since, as noted above, measurements need not be parallel for that interpretation to be valid. Before leaving this topic, it should be noted that while the NDR test can determine that Q1 and Q2 have different $h(\mathcal{Y})$ distributions, it says nothing about whether classification errors for the two questions are independent.

2.2.3 Extension to Polytomous Response Variables

In this section, we show how the ideas developed in the last section for dichotomous data can be easily extended to polytomous data treating the categories as nominal. Suppose that y_i is a response variable with L categories labeled 1, 2,..., L. As an example, the characteristic $\mathcal{Y}$ may be the race of an individual and y_i has three categories corresponding to white, black, and all other races.

To illustrate the basic idea, assume that two measurements are obtained on each unit in the sample (e.g., interview and reinterview measurements). For the ith unit, let y_{ij}, $j = 1,2$ denote the interview and reinterview responses, respectively. The cross-classification of y_{i1} and y_{i2} is displayed in Table 2.4, where $p_{ll'}$ denote the cell proportions and p_{l+} and $p_{+l'}$ denote the row and column marginals, respectively, for $l,l' = 1,\ldots, L$. As before, the "+" notation denotes summation over the indices that it replaces.

One simple method for handling polytomous data is to convert the variable to L dichotomous variables—one associated with each category l of the original variable. Then these L dichotomous variables can be analyzed separately using the techniques of the last section for dichotomous variables. To do this, define the new binary variables, $y_{ij}^{(l)}$, for $l = 1,\ldots, L$, where $y_{ij}^{(l)} = 1$ if $y_{ij} = l$ and $y_{ij}^{(l)} = 0$ if $y_{ij} \neq l$. Thus, $y_{ij}^{(l)}$ is just a binary indicator variable for membership in category l of the parent, polytomous variable. The techniques of Section 2.2.2 can now be applied to the L 2 × 2 tables formed by cross-classifying $y_{i1}^{(l)}$ and $y_{i2}^{(l)}$ to estimate the reliability ratio for each category separately. To compute the index of inconsistency corresponding to a particular category q, let g_q denote the gross difference rate corresponding to the 2×2 $y_1^{(q)} \times y_2^{(q)}$ table. It can be shown that

$$g_q = \sum_{l \neq q} (p_{lq} + p_{ql}) \tag{2.49}$$

$$\hat{I}_q = \frac{g_q}{p_{q+}p_{+q}} \tag{2.50}$$

Table 2.4 ***L*-Fold Interview-Reinterview Cross-Classification Table**

	Reinterview (y_2)				
Interview (y_1)	1	2	...	L	—
1	p_{11}	p_{12}	...	p_{1L}	p_{1+}
2	p_{21}	p_{22}	...	p_{2L}	p_{2+}
⋮	⋮	⋮	...	⋮	⋮
L	p_{L1}	p_{L2}	...	p_{LL}	p_{L+}
—	p_{+1}	p_{+2}	...	p_{+L}	1

Further, an estimator of the reliability ratio for $y_1^{(q)}$ is $\hat{R}_q = 1 - \hat{I}_q$.

For the case of two measurements, the so-called *L-fold index of inconsistency* (US Census Bureau 1985) can also computed. The *L*-fold index of inconsistency is given by

$$\hat{I}_L = \frac{g_+}{p_{eL}} \tag{2.51}$$

where $g_+ = \sum_l g_l = 1 - \sum_l p_{ll} = 1 - p_{0L}$, where p_{0L} is the agreement rate for the $L \times L$ cross-classification table (i.e., the sum of the diagonal cells in Table 2.3) and the denominator is computed as $p_{eL} = p_{1+}p_{+1} + p_{2+}p_{+2} + \cdots + p_{L+}p_{+L}$. It can be shown that I_L is a weighted average the indexes, $\hat{I}_q$, defined in (2.50):

$$\hat{I}_L = \sum_{l=1}^{L} w_l \hat{I}_l, \quad \text{where} \quad w_l = \frac{p_{l+}p_{+l}}{p_{eL}} \tag{2.52}$$

Similarly, $\hat{R}_L = 1 - \hat{I}_L$ is the *L-fold reliability index* that may be interpreted as a weighted average reliability for the original *L*-category variable *y*. In (2.52), the weights w_l are proportional to the total variance of the variable $y_1^{(l)}$. This seems reasonable because the larger the variance of a particular category, the greater is its potential impact on the total reliability of the parent, polytomous *y* variable.

The kappa statistic can also be generalized for *L* categories. Extending the notation in (2.42), we obtain the *L*-fold kappa statistic as

$$\kappa_L = \frac{p_{0L} - p_{eL}}{1 - p_{eL}} \tag{2.53}$$

where p_{0L} and p_{eL} are as defined above. Note from (2.53) and the relationship between κ_L and $\hat{I}_L$ that

$$\hat{I}_L = 1 - \kappa_L = 1 - \frac{p_{0L} - p_{eL}}{1 - p_{eL}} = \frac{1 - p_{0L}}{1 - p_{eL}} \tag{2.54}$$

which provides yet another interpretation for $\hat{I}_L$; specifically, it is the ratio of the observed disagreement rate $(1 - p_{0L})$ and the chance disagreement rate $(1 - p_{eL})$. As in the dichotomous case, this interpretation of κ_L is still valid when the measurements are not parallel. It can be shown that $\hat{I}_L$, $\hat{R}_L$, and κ_L simplify to $\hat{I}$, $\hat{R}$, and κ, respectively, when $L = 2$.

2.3 REPEATED MEASUREMENTS

2.3.1 Designs for Parallel Measurements

As noted in Section 2.2, the reliability of a measurement process cannot be estimated unless at least two measurements are available for each sample unit. Moreover, to be valid as a measure of reliability, the assumption of parallel measurements must hold.[10] This means that the repeated measurements for unit i constitute a simple random sample of measurements from the same response distribution $h_i(\mathcal{Y})$. Stated another way, two conditions must hold for y_1 and y_2:

1. Conditional independence: $E(y_{i1}y_{i2}|i) = E(y_{i1}|i)E(y_{i2}|i)$ for all i
2. Identically distributed responses: $E(y_{i1}|i) = E(y_{i2}|i)$ and $Var(y_{i1}|i) = Var(y_{i2}|i)$ for all i

Assumption 1 essentially means that the measurement errors for the first and second responses are independent. Beginning in Chapter 3, this assumption is referred to as *local independence* and several alternative interpretations are described. Assumption 2 means that the true scores and measurement error variances for the first and second measurements are identical.

A common method for collecting two parallel measurements is the test–retest study. For this design, respondents to the original survey are recontacted and again asked some of the same questions that were asked in the original survey. The primary objective of the reinterview is to collect a second, parallel measurement for all or a subset of characteristics from the main survey. Thus, the reinterview seeks to maintain the same essential survey conditions that existed in the main survey. This usually means that reinterviewers should have essentially the same levels of interviewing proficiency as main survey interviewers and that the original survey respondents and the reinterview survey respondents should be the same. The interview and reinterview questions should use the same wording, except when changes are necessary for the reinterview to reference the same timepoints and time intervals that were referenced in the original interview. In addition, the time interval between the original interview and reinterview should be short enough to minimize real changes in the characteristics but also long enough to avoid the situation where respondents might simply recall their interview responses and simply repeat them in the reinterview, thus resulting in correlated measurement errors.

As Hansen et al. (1964) note, the parallel measurements assumption is seldom satisfied in test–retest reinterview surveys because it is impossible to

[10] In some texts, the definition of *parallel measures* does not explicitly state conditional independence, although it is tacitly assumed. In our definition, conditional independence is explicitly stated since later we consider relaxing this assumption.

identically reproduce the same essential survey conditions of the original survey in the reinterview. For one thing, respondents may have been *conditioned* by the first interview in some way by the very fact that they were interviewed. The reinterview responses may be affected by this conditioning, thus changing the essential survey conditions. As an example, the respondent may have obtained additional information on the survey topics since the interview, and this new knowledge could influence his or her responses in the reinterview. Or he/she may not have thought much about the topic of the question prior to the intial interview but may have considered it more by the time of the reinterview, resulting in a change in responses.

An even more likely occurence for some characteristics is that respondent's true status on the topic may have changed in the intervening period between the interview and reinterview. This, of course, violates assumption 2 above (i.e., a stable response distribution). This problem can be addressed to some extent by either modifying the reinterview questions to refer to the same time period referenced in the interview or by reconciling any discrepancies between the interview and reinterview responses to determine whether a change in characteristics occurred. If so, the question could then be restated to reference the original time period.

An obvious violation of conditional independence is interview–reinterview error correlations. Correlated errors can be induced if respondents tend to simply recall their interview responses and repeat them in the reinterview rather than providing completely independent responses. As noted previously, the reinterview should be timed to allow sufficient time for the respondent to forget their interview responses while minimizing real changes in the characteristics. A time interval between 5 to 10 days seems to work well, all things considered [see, e.g., Bailar (1968)].

The problems associated with conducting test–retest reinterviews that satisfy the basic assumptions seldom can be overcome, leading to biased estimates of reliability. Often what is estimated is not R but some biased version of R that we sometimes refer to as *test–retest reliability*. As we shall see, test–retest reliability is really a vague concept because what it actually measures depends solely on which assumptions are violated, to what extent and how.

A second possibility for collecting repeated measurements is to embed multiple questions about $\mathcal{Y}$ in the same questionnaire. An application of this approach for evaluating questions regarding poor or failing grades in college was given in the introduction to this chapter. In that example, similar but not identical questions were employed, which is typical of the *embedded remeasurements* approach. Besides the obvious cost advantages of eliminating a second interview, an additional advantage of this approach is that it eliminates the time interval between repeated measurements. This means that the risk of the underlying characteristic changing between the first and second measurements is eliminated. An important disadvantage, however, is the increased risk of correlated errors. Obtaining repeated measurements in the same interview makes it is easier for respondents to force their responses to be consistent if they wish. To ameliorate this tendency, the questions are worded somewhat differently so that respondents are not aware that they are being repeatedly asked about the same thing. In doing so, however, another problem arises—different questions are likely to produce different error distributions for the remeasurements, violating the assumption 2 above. In addition, the characteristic measured by different questions may also be somewhat different. This can be viewed as a violation of the parallel measurements assumption since, in that case, the true scores for the response distributions differ.

Given these limitations of test–retest reinterview and embedded remeasurements for producing parallel measurements, one might conclude that the unbiased estimation of the reliability ratio is hopeless. While this may be true to some extent, estimates of R may still be quite useful as long as serious departures from the assumptions described above can be avoided. Although the assumptions will never hold exactly, the biasing effects of small departures from parallel, independent measurements may also be quite small. The next section considers some options for relaxing these assumptions.

2.3.2 Nonparallel Measurements

The parallel measurements response model assumed that the measurements $y_{ij}, j = 1,\ldots, m$ on the unit i constitute a simple random sample from $h_i(\mathcal{Y})$, that is, that $y_{ij}, j = 1,\ldots, m$ are i.i.d. random variables. As discussed in the previous section, it can be very difficult to produce measurements that closely satisfy these assumptions. In this section, we consider relaxing the i.i.d. assumptions and consider the interpretation of estimates of the reliability ratio under these conditions.

Returning to the model in (2.1), suppose that, instead of assuming each remeasurement to be sampled from the same $h(\mathcal{Y})$, we allow the response distribution to change across measurements. To that end, we generalize (2.1) as

$$y_{ij} = \tau_{ij} + e_{ij}, \quad \text{where} \quad E(e_{ij} \mid i) = 0, \quad Var(e_{ij} \mid i) = \sigma^2_{eij} \tag{2.55}$$

where, as before, $Cov(\tau_{ij}, e_{ij} \mid i) = 0$. Parallel measurements assumed

$$\tau_{ij} = \tau_i \tag{2.56}$$

$$\sigma^2_{eij} = \sigma^2_{ei} \tag{2.57}$$

for $j = 1,\ldots, m$. Measurements for which (2.56) holds but (2.57) does not hold are called *τ-equivalent*. For example, in a test–retest reinterview, exactly the same question may be used for both interviews, and so it is reasonable to assume that the true scores for the questions are equal. However, having been *conditioned* by the first interview, responses to the reinterview may have smaller or larger errors. Another way τ-equivalent measures could be produced is the use of more proficient interviewers (e.g., supervisory interviewers) for the reinterview that can result in smaller measurement error variance. Forsman and Schreiner (1991) discuss this type of reinterview in the context of interviewer performance evaluation.

Finally, if neither (2.56) nor (2.57) holds, but it can still be assumed that the correlation between τ_{ij} and $\tau_{ij'}$ is unity for all $j \neq j'$, the measurements are said to be *congeneric*. Such remeasurements occur when different scales are used to measure $\mathcal{Y}$. For example, suppose that the first measurement used a 10-point scale while the second measurement used a 100-point scale. In that case, it may be assumed that $\tau_{i2} = 10\tau_{i1}$ for all i and thus, the correlation between τ_{i1} and τ_{i2} is 1.

In the most general case, suppose that $y_{ij}, j = 1,\ldots, m$ are sampled from different response distributions denoted by $h_{ij}(\mathcal{Y}), j = 1,\ldots, m$, respectively. Thus, rather than two stages of sampling, the situation resembles a three-stage sampling process. As before, the first stage produces a simple random sample of n units from the population of N units. At the second stage, a sample of m *measurement distributions* $[h_{ij}(\mathcal{Y})]$ are selected from a hypothetically infinite number of possible measurement distributions. Each distribution might correspond to a distinct method of questioning, mode of interview, data collection method, and so on for obtaining measurements on $\mathcal{Y}$. Then at the third stage, only one realization, y_{ij}, is obtained from each measurement distribution to produce the m measurements on $\mathcal{Y}$ corresponding to $h_{ij}(\mathcal{Y}), j = 1,\ldots, m$, respectively. This situation is analogous to a three-stage sampling design and, as such, mean and variance estimation formulas are readily provided by sampling theory. For example, Cochran (1977, p. 285) provides formulas for the estimation of the population mean and its standard error in three-stage sampling that are directly applicable to this situation.

Of immediate interest to our study is estimation of the reliability ratio. Now, however, the situation is quite complex because there are m reliability ratios denoted by $R_j, j = 1,\ldots, m$ corresponding to the m measurement distributions. As shown in the previous section, with only one observation from each response distribution, the unbiased estimation of true score variance (or SRV) is not possible, and thus no consistent estimator of R_j exists for any j without invoking additional assumptions on the associations among the $h_{ij}(\mathcal{Y})$. For example, the analysis could proceed as though the measurements were parallel and would produce an estimate of R as defined for parallel measurements. However, this estimate will be biased for anyR_j of interest. Under somewhat restrictive assumptions, the estimate of R so obtained might be interpreted as the average reliability over the m measurements: $\bar{R} = \Sigma_j R_j/m$.

To understand how this works, consider a test–retest reinterview design with measurements y_1 and y_2 with corresponding reliability ratios R_1 and R_2, respectively, and where the goal of the reinterview study is to estimate R_1, the reliability of the characteristic measured by the main survey. In addition to a different response distribution being associated with the reinterview, suppose that measurement errors are correlated between the two trials: $\rho_{12} = E_i Cov(y_{i1}, y_{i2} | i) \neq 0$. It can be shown that the expected value of the gross difference rate g given by

$$E(g) = \text{SRV}_1 + \text{SRV}_2 - 2\rho_{12}\sqrt{\text{SRV}_1\text{SRV}_2} + D_{12}^2 \tag{2.58}$$

(US Census Bureau 1985), where SRV_1 and SRV_2 denote simple response variances for the interview and reinterview measurements, respectively, and $D_{12} = E(y_{i1} - y_{i2}) = \tau_1 - \tau_2$ denotes the unconditional expected difference between the interview and reinterview responses. Under the parallel assumption, $\rho_{12} = D_{12} = 0$, $\text{SRV}_1 = \text{SRV}_2$, and thus g is an unbiased estimator of 2SRV_1. The failure of any of these three conditions to hold will result in $E(g) \neq 2\text{SRV}_1$ and, consequently, $\hat{R}$ will be a biased estimator of R_1. Thus, the test–retest reliability concept that we referred to above, connotes any or all of the biases in (2.58).

If the two measurements are τ-equivalent, then $\rho_{12} = D_{12} = 0$ but $\text{SRV}_1 \neq \text{SRV}_2$. With two measurements, only one part of these assumptions (viz., $D_{12} = 0$) can be tested. Note that $D_{12} = E(\text{NDR})$ where NDR was defined in (2.36). The simple random sampling estimator of the variance of the NDR was given in (2.48). Hence, $\tau = \text{NDR}/\text{s.e}(\text{NDR})$ is approximately a τ-distributed statistic with $n - 1$ degrees of freedom. If the results from this test support the assumption that $D_{12} = 0$, then τ-equivalence is at least plausible. Because $\text{SRV}_1 \neq \text{SRV}_2$, it follows from (2.58) that g is an estimator of $(\text{SRV}_1 + \text{SRV}_2)/2$, the average simple response variance for the two trials. Thus, if $\text{SRV}_1 < \text{SRV}_2$, then $\hat{R}$ will likely overestimate R_1. Thus, greater reliability of the second measurement process will tend to make the reliability for the original interview look better than it really is. Similarly, if $\text{SRV}_1 > \text{SRV}_2$, $\hat{R}$ will likely underestimate R_1 (US Census Bureau 1985). Likewise, then, smaller reliability of the second measurement process will tend to make the reliability for the original interview look worse than it really is.

Extending this argument to m τ-equivalent measurements, the estimator of R estimates the average of the m reliabilities associated with the m measurement distributions. This may not be useful unless the objective is to obtain an average estimate of reliability for the entire set of measurements rather than one particular measurement.

If the measurements are congeneric, then $E(g) = \text{SRV}_1 + \text{SRV}_2 + D_{12}^2$. In this case, it is not possible to predict the direction of the bias of $\hat{R}$ for estimating the reliability of the original interview, R_1, because the bias depends on the direction of the difference $(\text{SRV}_2 + D_{12}^2) - \text{SRV}_1$. For example, if $\tau_2 = \beta\tau_1$ for some constant β, then the squared difference of the two true scores is $D_{12}^2 = \tau^2(1-\beta)^2$. Note that the bias in $\hat{R}$ may be still small provided $\beta \doteq 1$ and $\text{SRV}_2 \doteq \text{SRV}_1$.

Now suppose that SRV is the same for both trials and $D_{12} = 0$ but $\rho_{12} > 0$; that is, the errors in the two measurements are positively correlated. It follows from (2.58) that $E(g) = 2\text{SRV}(1 - \rho_{12})$; thus, SRV will be underestimated and

R_1 will be overestimated. In other words, positive correlations between the errors in the two measurements will make the measurements appear to be more consistent than they are, thus negatively biasing the index of inconsistency. Stated another way, positively correlated errors will result in erroneously attributing greater reliability to the measurements.

To summarize, some violations of the parallel assumptions will result in negative bias while others will result in positive bias in the estimates of R. In general, the bias in $\hat{R}$ is unpredictable. Examples of this unpredictability are provided in the following illustration.

2.3.3 Example: Reliability of Marijuana Use Questions

To illustrate the estimation methodology, we use data from a large, national survey on drug use. In this survey, data on past-year marijuana use were collected in the same interview using three different methods. Responses were recoded to produce three dichotomous measures of past-year marijuana use denoted by y_1, y_2, and y_3, where 1 denotes use and 0 denotes no use. The frequencies for all possible response patterns are shown in Table 2.5. Although the sample was drawn by a complex multistage unequal probability design, simple random sampling will be assumed for the purposes of the illustration. Cell counts have been weighted and rescaled (see Section 5.3.4 for more detail on the approach) to the overall sample size, $n = 18{,}271$.

The parallel measures assumption can be tested by comparing the three proportions, p_{1++}, p_{+1+}, and p_{++1}. (Note that we have extended the "+" notation for two measurements in Section 2.2.2 to three measurements.) If a test of equality is rejected, the assumption of parallel measures is not supported by the data; however, failure to reject does not suggest that y_1, y_2, and y_3 are parallel. We compute the marginal proportion for y_1 as p_{1++} as $(1181 + 96 + 17 + 15)/18{,}269 = 0.072$. In the same way, the marginal proportion for y_2 is $p_{+1+} = 0.084$ and for y_3, it is $p_{++1} = 0.084$. Although the proportions are all close, p_{1++} differs significantly from the other two proportions, and thus the test is rejected,

Table 2.5 Past-Year Marijuana Use for Three Measures

y_1	y_2	y_3	Count
1	1	1	1,181
1	1	0	96
1	0	1	17
1	0	0	15
0	1	1	113
0	1	0	150
0	0	1	229
0	0	0	16,470

indicating that y_1 is not parallel compared to y_2 or y_3. However, that y_2 and y_3 are parallel cannot be rejected on the basis of this test. Thus, we expect our estimates of reliability on the basis of all three measurements to be somewhat biased. Nevertheless, we proceed with the estimation of R using all three measurements.

The estimation of R using (2.27) will be demonstrated with s_2^2 computed as in (2.34). Note that p_i is either 0 (which occurs when $y_1 = y_2 = y_3 = 0$), $\frac{1}{3}$ (when only one of y_1, y_2, or y_3 is 1), $\frac{2}{3}$ (when only one of y_1, y_2, or y_3 is 0), or 1 (when $y_1 = y_2 = y_3 = 1$). Thus, $p_i q_i$ can only take on the values 0 or $\frac{2}{9}$, the latter occurring with frequency 620 (= 96 + 17 + 15 + 113 + 150 + 229). Thus

$$s_2^2 = \frac{3}{18{,}271(3-1)}\left(\frac{2}{9}\times 620\right) = 0.01131$$

To compute $\hat{\sigma}_y^2$ from (2.28), we first compute s_1^2 using (2.34). We compute $\bar{p}$ as

$$\bar{p} = \frac{1\times 1181 + \frac{2}{3}\times(96+17+113) + \frac{1}{3}\times(15+150+229)}{18{,}271} = 0.08$$

and thus

$$s_1^2 = \frac{1181\times(1-0.08)^2 + 226\times(\frac{2}{3}-0.08)^2 + 394\times(\frac{1}{3}-0.08)^2 + 16{,}470\times(0-0.08)^2}{18{,}271-1}$$
$$= 0.06612$$

The estimate of σ_y^2 from (2.28) is therefore $0.06612 + \frac{2}{3}\times 0.01131 = 0.07366$. Combining these results, we find that the estimator of R is

$$\hat{R} = 1 - \frac{0.01131}{0.07364} = 0.8465$$

and the index of inconsistency $\hat{I} = 0.15$.

Using the US Census Bureau's rule of thumb for interpreting the magnitude of $\hat{I}$ (i.e., $0 \le \hat{I} \le 0.2$ is good, $0.2 < \hat{I} \le 0.5$ is moderate, and $\hat{I} > 0.5$ is poor), this result suggests that the average reliability of the three measures of past-year marijuana use is good.

We can use the first two columns of Table 2.5 to demonstrate the calculations for the case where there are only two measurements. Begin by collapsing Table 2.4 over the y_3 measurement to form a 2×2 crossover table having $p_{11} = 0.0699$, $p_{10} = 0.0018$, $p_{01} = 0.0144$, and $p_{00} = 0.9140$. Thus, $g = 0.01615$ and $p_{1+} = 0.072$ and $p_{+1} = 0.084$. The NDR is $0.072 - 0.084 = -0.012$. To test for parallel measurements, we compute the standard error as $\sqrt{Var(\text{NDR})}$, where $Var(\text{NDR})$ is as given in (2.48). This yields $\sqrt{(0.01615 + 1/18{,}271)/18{,}271}$ $\doteq 0.001$, and NDR is highly significant. However, the size of the NDR is small relative to p_{1+} and p_{+1}, so the departure from parallel measures does not seem important for the purposes of reliability estimation.

The denominator of R is $0.072(1 - 0.084) + 0.084(1 - 0.072) = 0.1439$. It follows from (2.37) that $\hat{R} = 1 - (0.0161)/(0.144) = 0.89$, which again indicates good reliability. To calculate κ from (2.42), we first compute $p_0 = p_{11} + p_{00} =$ 0.984 and then $p_e = p_{1+}p_{+1} + p_{0+}p_{+0} = 0.856$. Hence, $\kappa = (0.984 - 0.856)/(1 -$ $0.856) = 0.89$, which, as expected, is the same number produced by the calculation of $\hat{R}$.

As previously noted, both estimates of reliability are biased to some extent because of the failure of the parallel assumptions to hold. The result in (2.58) suggests that SRV may be biased upward because $D_{12}^2 > 0$ and, consequently, estimates of R should be biased downward. This may not be true if $\rho_{12} \neq 0$, but no information on this correlation is available. Because the assumption that $D_{23}^2 = 0$ is plausible is based on the test of equality of p_{+1+} and p_{++1}, we can also compute $\hat{R}$ using y_2 and y_3 and compare this estimate with our previous results. This yields an estimate of $\hat{R} = 0.81$, which is lower than the value computed for y_1 and y_2. One possible explanation is that $\rho_{12}\sqrt{\text{SRV}_1\text{SRV}_2} > D_{12}^2$ which, as can be seen from (2.58), results in a positive bias in $\hat{I}$ computed from y_1 and y_2. Unfortunately, the data in Table 2.5 are not sufficient for resolving this puzzling result.

2.3.4 Designs Based on a Subsample

Reinterviews will usually cost less than the original interviews because they often require less effort to contact and interview the respondent. However, most survey budgets cannot bear the costs of reinterviewing the entire sample, especially considering that the reinterview data are solely for evaluation purposes. Often it is sufficient to reinterview a subsample of between 10% and 50% of the survey respondents to achieve acceptable precision for the estimates of reliability. Let $n' < n$ denote the size of a simple random subsample of the original sample. Because this subsample constitutes a simple random sample from the entire population, the theory developed in the previous sections still holds after replacing n everywhere by n'. However, it is possible to obtain better precision for the estimates of R, I, and κ by computing the denominator of these ratios over the full sample.

An alternative estimator of σ_y^2 can be constructed by using only the main survey (i.e., assuming $m = 1$) and noting from (2.23) that an estimator of the total variance σ_y^2 is $\hat{\sigma}_y^2 = s^2$, where s^2 as defined in (2.14) is based on the n main survey observations ignoring the repeated measurements. Let $q_{1+} = 1 - p_{1+}$ and $q_{+1} = 1 - p_{+1}$. If $p_{1+}q_{1+}$ or $p_{+1}q_{+1}$ calculated from the interview–reinterview table is based on a small sample size, a more efficient estimator can be produced by substituting $p_{1+}q_{1+} + p_{+1}q_{+1}$ by pq, where $p = \Sigma_{i=1}^{n} y_{i1}/n$ is based on all n units in the sample and $q = 1 - p$. Using this denominator, the estimator of the reliability ratio becomes

$$\hat{R}^{I} = 1 - \frac{g}{2pq} \tag{2.59}$$

A somewhat related situation arises when the reinterview subsample is not random. For example, in the CPS described in Section 1.1.1, new sample households and households that are rotated back into the sample after being dormant for 8 months, are not reinterviewed. Also, in many face-to-face surveys, only telephone households are reinterviewed primarily because of the high costs of personal visits to nontelephone households. Another common problem is reinterview nonresponse (as much as 40%), which adds to the main survey nonresponse rate. Because of the nature of the subsampling mechanism, the estimates of reliability based on these reinterview samples may be biased if the proportion of excluded sample members is high and if they have very different error distributions. Despite these risks, reliability estimates may still be computed and quite useful for data quality evaluations. Nevertheless, it is important to realize the risks of bias and to account for them in interpretation of the interview–reinterview analyses.

2.4 RELIABILITY OF MULTIITEM SCALES

Related to the concept of embedded repeated measurements for survey error evaluation is the use of multiitem scale score measures in surveys. Because these are used extensively in some areas of survey research, a discussion of these types of measurements will be included for the sake of completeness and to contrast them with the single-item measurements discussed above. However, the topic is not essential to other chapters, nor will it be considered further beyond the present discussion.

2.4.1 Scale Score Measures

Scale score measurements (SSMs) are typically a standardized set of items embedded in a questionnaire that are assumed to be either parallel or τ-equivalent. The items are summed together and this *sum score*, rather than the individual items, is then the focus of an analysis. SSMs are very common in psychological and social science research. As an example, the *Child Behavior Checklist* (CBCL) is a common SSM for measuring behavior problems in children (Achenbach 1991a, 1991b). It consists of 118 items on behavior problems, each scored on a 3-point scale: 1 = not true, 2 = sometimes true, and 3 = often true of the child. The CBCL *total behavior problem score* is an empirical measure of child behavior computed as a sum of the responses to the 118 items. Other SSMs might measure happiness, quality of life, depression, neighborhood safety, and so on.

The usefulness of an SSM in data analysis depends in large part on its reliability. SSMs with good reliability are relatively free from random measurement error which enhances their utility as an analysis variable [see, e.g., Biemer et al. (2009)]. Thus, assessing scale score reliability is typically an integral and critical step in the use of SSMs in data analysis. In this section, the reliability results from Section 2.2 will be extended to obtain the reliability of an SSM.

Suppose that m measurements, denoted by y_{ij}, $j = 1,\ldots, m$, that constitute a multiitem scale are obtained for each unit i in the sample. For the moment, assume the items to be parallel. Later they will be generalized to τ-equivalent items. Thus, assume that the m measurements on an individual constitute a simple random sample from a single-response distribution $h(\mathcal{Y})$. Define the *scale score* for the ith sample unit as the sum of the responses to the m items: $t_i = \sum_{j=1}^{m} y_{ij}$. The corresponding true score for the ith unit is given by $E(t_i|i) = T_i = \sum_j \tau_i = m\tau_i$. Now, an expression for the reliability ratio for t_i can be obtained by applying (2.7) and noting that

$$\text{SV} = Var_i\, E(t_i \mid i) = Var_i\,(m\tau_i) = m^2\sigma_\tau^2 \tag{2.60}$$

where σ_τ^2 is the true score variance, which is the same for each measurement j. Likewise,

$$\text{SRV} = E_i\, Var(t_i \mid i) = E_i\, Var\left(\sum_{j=1}^{m} y_{ij} \mid i\right) = E_i\,(m\sigma_{ei}^2) = m\sigma_e^2 \tag{2.61}$$

where σ_e^2 is the measurement error variance, which is also the same for each measurement j.

Thus, the scale score reliability of the score t_i, denoted by R_m, may be written as

$$R_m = \frac{\text{SV}}{\text{SV}+\text{SRV}} = \frac{m^2\sigma_\tau^2}{m^2\sigma_\tau^2 + m\sigma_e^2} = \frac{\sigma_\tau^2}{\sigma_\tau^2 + (\sigma_e^2/m)} \tag{2.62}$$

This equation indicates that reliability increases as the number of items in the scale m increases so long as σ_τ^2 and σ_e^2 remain fixed for the different values of m. This is a well-known result in the theory of multiitem scales [see, e.g., Bollen (1989)]. Thus, by (2.62), longer scales for the same SSM are more reliable under the model assumptions. A key assumption here is the conditional independence assumption made for parallel measures. Longer scales having intercorrelated measurement errors may not be more reliable [see, e.g., Bollen 1989, p. 217).

To estimate R_m, note from (2.26) that s_1^2 is an unbiased estimator of the denominator of R_m and s_2^2 still unbiasedly estimates σ_e^2. Therefore, a consistent estimator of the scale score reliability ratio is

$$\hat{R}_m = 1 - \frac{s_2^2}{ms_1^2} \tag{2.63}$$

As described in the next section, this estimator of scale score reliability is still consistent when items in an SSM are only τ-equivalent.

2.4.2 Cronbach's Alpha

The estimator in (2.63) differs considerably from the traditional measure of reliability for SSMs, namely, *Cronbach's alpha* (Cronbach 1951), which is given by

$$\hat{\alpha} = \frac{m}{m-1}\left(1 - \frac{\sum_{j=1}^{m} s_{\langle j\rangle}^2}{m^2 s_1^2}\right) \tag{2.64}$$

where

$$s_{\langle j\rangle}^2 = \frac{\sum_{i=1}^{n}(y_{ij} - \bar{y}_j)^2}{(n-1)} \tag{2.65}$$

Note that, in (2.65), the variance is computed over all units in the sample separately for each item.

As will be shown, Cronbach's alpha is a consistent estimator of R_m when the measurements are τ-equivalent measurements. The main difference between $\hat{\alpha}$ and $\hat{R}_m$ is that the latter is based on the per unit variance of responses among *items* averaged over all units in the sample while alpha is based on the per item variance among *persons* averaged over all items. Therefore, $\hat{\alpha}$ seems a more natural estimator of reliability when the measurements are not parallel but can be assumed to be τ-equivalent.

Suppose that m conditionally independent measurements y_{ij}, $j = 1,\dots, m$ are τ-equivalent. The expression for SV in (2.60) does not change. However, SRV is now given by

$$\begin{aligned} \text{SRV} &= E_i\, Var(t_i \mid i) = E_i\, Var\left(\sum_{j=1}^{m} y_{ij} \mid i\right) \\ &= E_i\left(\sum_{j=1}^{m} \sigma_{ej}^2\right) = m\bar{\sigma}_e^2, \text{ say} \end{aligned} \tag{2.66}$$

where $\bar{\sigma}_e^2 = \Sigma_j \sigma_{ej}^2 / m$ is the per item measurement variance averaged over the m items. Thus, R_m is still given by (2.62) but now σ_e^2 is replaced by $\bar{\sigma}_e^2$. That alpha is still a consistent estimator of R_m under τ equivalence can be verified on noting that, applying (2.23)

$$E\left(\sum_{j=1}^{m} s_{\langle j \rangle}^2\right) = m(\sigma_\tau^2 + \bar{\sigma}_e^2) \tag{2.67}$$

and from (2.26)

$$E(s_1^2) = \sigma_\tau^2 + \frac{\bar{\sigma}_e^2}{m} \tag{2.68}$$

Using a similar approach, one can show that s_2^2 is unbiased for $\bar{\sigma}_e^2$ and, thus, $\hat{R}_m$ is also consistent for R_m.

A form equivalent to Cronbach's alpha for dichotomous variables—the Kuder–Richardson Formula 20 (KD20)—is often used to describe the reliability of multiitem measures [see, e.g., Nunnally and Bernstein (1994)]:

$$\text{KD} = \frac{m}{m-1}\left(1 - \frac{n \Sigma_j p_j q_j}{m^2 \Sigma_i (p_i - \bar{p})^2}\right) \tag{2.69}$$

Illustration. To illustrate how to apply the formulas for $\hat{R}_m$ and $\hat{\alpha}$, we shall use the Stouffer–Toby data in Table 4.1 (in Chapter 4), after replacing the responses coded as "2" by "0." Although the four measures in the table are not parallel or even τ-equivalent, they will be treated as such for the purposes of this exercise. First, consider the computation of $\hat{R}_m$, which can be rewritten as

$$\hat{R}_m = 1 - \left(\frac{n-1}{n(m-1)}\right) \frac{\Sigma_i p_i q_i}{\Sigma_i (p_i - \bar{p})^2} \tag{2.70}$$

To compute $\Sigma_i p_i q_i$, note that $p_i q_i$ takes the values 0.1875, 0.25, and 0 with frequencies 62, 91, and 63, respectively. Therefore, $\Sigma_i p_i q_i = 32.8125$. To compute $\Sigma_i (p_i - \bar{p})^2$, we first compute $\bar{p}$. Note that p_i takes the values 1, 0.75, 0.5, 0.25, and 0 with frequencies 42, 36, 63, 55, and 20, respectively. Therefore, $\bar{p} = (1.0)(42) + (0.75)(36) + (0.5)(63) + (0.25)(55)$ divided by $n = 216$. This yields $\bar{p} = 0.5289$. Thus, $\Sigma_i (p_i - \bar{p})^2 = (1.0 - 0.1322)^2(42) + (0.75 - 0.1322)^2(36) + \cdots + (0 - 0.1322)^2(20) = 21.01$. Finally, the ratio in parentheses in (2.70) is $(216 - 1)/(216 \times 3) = 0.3318$. Thus, $\hat{R}_m = 1 - (0.3318) \times 32.8125/21.01 = 0.48$.

To compute Cronbach's alpha for these same data using (2.69), begin by computing p_1, p_2, p_2 and p_4. This yields 0.7917, 0.5, 0.5139, and 0.3102, respectively, and $\Sigma_j p_j q_j = 0.8787$. Thus, $n\Sigma_j p_j q_j = 189.8009$. In the computation of $\hat{R}_m$ above, we found $\Sigma_i (p_i - \bar{p})^2$ was 55.0 and therefore $m^2 \Sigma_i (p_i - \bar{p})^2 = 880.0$. Collecting these terms yields $\hat{\alpha} = \frac{4}{3}(189.8009/336.1065) = 0.58$, which can be compared with $\hat{R}_m = 0.48$ above. Because these data were not intended to be used as a scale score, neither estimator of SSM reliability is appropriate. However, if these data were to be used as an SSM, one would say that scale score reliability is poor using $\hat{R}_m$ while $\hat{\alpha}$ suggests moderate reliability.

2.5 TRUE VALUES, BIAS, AND VALIDITY

Thus far, the true values associated with the observations have not been discussed because our focus has been on variance and reliability estimation, which does not require true values to even exist. As mentioned in Chapter 1, optimal survey design is aimed at minimizing the total MSE of an estimator that includes both variance and bias components. Estimator bias is the difference between the expected value of the estimator and the true population parameter. In a finite population, the true parameter is defined as the population mean of the true values for the characteristic. Thus, a discussion of estimator bias requires the assumption that true values exist for the characteristics for each population unit.

For most survey characteristics, the existence of a true value underlying the measurements is unequivocal. Personal characteristics such as age, sex, education, marital status, and income are unambiguously defined, verifiable, and, under certain conditions, observable without error. So are behavior characteristics such as voting, crime victimizations, trips to the market, and employment status. On the other hand, the concept of a true attitude, opinion, or psychological or emotional state is more problematic. For example, suppose that a survey question seeks to elicit an individual's stance on capital punishment, that is, whether for it or against it. Because opinions can change and be influenced by external information, including the survey process itself, an individual's true stance cannot be measured or even conceptualized. However, it still makes perfect sense to speak of an individual's true *score* for this characteristic. It can be operationally defined as one's average opinion over many repeated independent trials after fixing the essential survey conditions at each trial.

For some characteristics, whether a true value makes sense is the subject of some debate. For example, an individual's race may fall into this category. This will be discussed further in Section 6.2.3. For the present, the discussion in the next section assumes a true value for the characteristic $\mathcal{Y}$ for each unit in the population.

2.5.1 A True Value Model

Let μ_i denote the true value of the characteristic $\mathcal{Y}$ for the ith population unit for $i = 1,\ldots, N$. Now the target parameter for the survey is the population of the true values given by

$$\mu = \frac{\sum_{i=1}^{N} \mu_i}{N} \tag{2.71}$$

rather than the population mean of the true scores τ. In general, survey estimators of (2.71) will estimate τ. Thus, the difference $\tau - \mu$ is of particular interest in survey evaluations. Denote the population variance by σ_μ^2. Recall that the random measurement error for the ith unit was defined in (2.1) as $e_i = y_i - \tau_i$. Now define the *total measurement error* for the ith unit as $\varepsilon_i = y_i - \mu_i$, which can be rewritten as

$$\begin{aligned} \varepsilon_i &= y_i - \tau_i + \tau_i - \mu_i \\ &= e_i + b_i \end{aligned} \tag{2.72}$$

where $b_i = \tau_i - \mu_i$ is the *measurement bias* of the ith unit. This leads to the following model for an observation on the ith sample unit:

$$\begin{aligned} y_i &= \mu_i + \varepsilon_i \\ &= \mu_i + b_i + e_i \end{aligned} \tag{2.73}$$

On noting that $\tau_i = \mu_i + b_i$, we find that (2.73) is an extended version of the model in (2.1). Let $E(b_i) = B$, $Var(b_i) = \sigma_b^2$, and $Cov(\mu_i, b_i) = \sigma_{\mu b}$. Since, as before, $E(e_i | i) = 0$, $Var(e_i | i) = \sigma_{ei}^2$, it follows that $Cov(\mu_i, e_i) = Cov(b_i, e_i) = 0$. Thus

$$\begin{aligned} Var(y_i) &= Var(\mu_i + b_i + e_i) \\ &= \sigma_\mu^2 + \sigma_b^2 + \sigma_{\mu b} + \sigma_e^2 \end{aligned} \tag{2.74}$$

The expected value of the usual estimator of the population mean assuming simple random sampling is

$$E(\bar{y}) = \mu + B \tag{2.75}$$

where B is the bias in the estimator. Thus, the MSE of $\bar{y}$, denoted by $MSE(\bar{y})$, can be written as

$$\begin{aligned} MSE(\bar{y}) &= E(\bar{y} - \mu)^2 \\ &= B^2 + Var(\bar{y}) \end{aligned} \tag{2.76}$$

The bias is an attribute of an estimator of parameters such as the mean, total, proportion, and regression coefficient. Statisticians are particularly interested in the bias of estimators. By contrast, psychometricians are more concerned with the quality of individual responses as well as alternative measures of $\mathcal{Y}$. Rather than bias, they are concerned with the theoretical and empirical validity of a measure [see, e.g., Lord and Novick (1968)]. *Theoretical validity*, denoted by $\rho_{\tau\mu}$, is the correlation between the true score τ_i and the true value μ_i. *Empirical validity*, denoted by $\rho_{y\mu}$ is the correlation between an observation y_i and the true value μ_i. Because $\sigma_{y\mu} = \sigma_{\tau\mu}$, two concepts are related by the formula

$$\rho_{y\mu} = \frac{\sigma_{y\mu}}{\sigma_y \sigma_\mu} = \frac{\sigma_{\tau\mu}}{\sigma_y \sigma_\mu} = \frac{\sigma_{\tau\mu}}{\sigma_\tau \sigma_\mu} \frac{\sigma_\tau}{\sigma_y} = \rho_{\tau\mu} \sqrt{R} \tag{2.77}$$

Thus, empirical validity is the product of theoretical validity and the square root of reliability. The closer validity is to 1, the more valid is the measure. Note that reliability can be perfect, that is, 1, and yet empirical validity may still be less than 1. This result leads to the conclusion that reliability is essential for valid measures but reliable measures are not necessarily valid.

Since $\tau_i = \mu_i + b_i$, the reliability ratio can be rewritten as

$$R = \frac{\sigma_\tau^2}{\sigma_\tau^2 + \sigma_e^2} = \frac{\sigma_\mu^2 + \sigma_b^2 + \sigma_{\mu b}}{\sigma_\mu^2 + \sigma_b^2 + \sigma_{\mu b} + \sigma_e^2} \tag{2.78}$$

This expression shows the effect of true value variance σ_μ^2 on reliability. In general, the larger the variation in true values, the greater the reliability is, all other components being held constant. In some situations, (2.78) can be simplified by assuming that the covariance between true values and true score bias $\sigma_{\mu b}$ is 0. For many characteristics that are measured on a continuous scale, the magnitude of the bias may be unrelated to the size of the true value. For example, within a fairly wide range of income values, the errors made in income reports may have very little correlation with actual income values. However, in many cases this assumption is unlikely to hold.

For example, for interview surveys, the measurement of sensitive items such as drug use frequency, alcohol consumption, and sexual activity are subject to social desirability biases that tend to repress reporting of these events [see, e.g., Tourangeau et al. (2000)]. The underreporting of events results in negative values for b_i for persons with positive true frequencies and 0 values of b_i for persons with 0 true frequencies. The same can be said for questions about church attendance, voting, and acceptance of immigrants; such questions tend to elicit exaggerated responses under certain conditions, resulting in positive values for b_i. Questions about the number of events (such as trips to the medical clinic, cigarettes smoked in the last week) may be subject to respondents forgetting, which can attenuate the number of recorded events, and this tendency may be greater for larger true values (Schwarz and Sudman 1994).

In other situations, the unit-level biases may be vary considerably yet have essentially 0 expectation (i.e., $B = 0$). As an example, questions requiring the respondent to estimate the number of times that an event occurred may be subject to estimation error that can be positive or negative, but average approximately 0. For these characteristics, the b_i are indistinguishable from random measurement error, e_i. The reliability ratio may be written as

$$R = \frac{\sigma_\mu^2}{\sigma_\mu^2 + \sigma_{e*}^2} \tag{2.79}$$

where σ_{e*}^2 is the variance of $e_i^* = e_i + b_i$, the measurement error in an observation of the true value, μ_i. Now reliability may be interpreted as the proportion of the total variance of a single observation that is *true value* variance.

For dichotomous variables, the circumstances giving rise to (2.79) cannot exist because the true value is either 1 or 0. This necessarily induces a correlation between the true value and the random error. To illustrate, consider the model in (2.73) and suppose that $\mu_i = 1$; then ε_i can take on only the value 0 or −1. Likewise, if $\mu_i = 0$, then ε_i can assume only the value 0 or 1. As shown in the next chapter, this induces a negative correlation between μ_i and ε_i. Consequently, the interpretation of R as a reflection of true value variance can be extremely complex. The next chapter considers some alternative quality indicators for categorical data whose interpretations are much simpler.

Let y_{i1} and y_{i2} denote two measurements on the ith sample unit. Let us assume that $y_{i2} \doteq \mu_i$, that is, that y_{i2}, $i = 1,\ldots,n$ are gold standard measurements. Then an estimate of the bias in the $\bar{y}_1$ is

$$\hat{B} = \bar{y}_1 - \bar{y}_2 \tag{2.80}$$

Likewise, the sample Pearson product moment correlation between y_{i1} and y_{2i} is an estimator of the empirical validity $\rho_{y\mu}$.

To estimate the MSE of $\bar{y}_1$, the following formula may be used:

$$\widehat{\text{MSE}}(\bar{y}_1) = \hat{B}^2 - v(\bar{y}_2) + 2[v(\bar{y}_1)v(\bar{y}_2)]^{1/2} \tag{2.81}$$

(see Biemer 2010).

2.5.2 Obtaining True Values

Estimation of bias or validity using traditional modeling methods relies on knowledge of the true values. This is problematic because the collection of measurements that are essentially infallible can be quite challenging, especially in moderate to large samples. Observations that are obtained through some type of error-free process are often called *gold standard* or *criterion measurements*. They will produce essentially unbiased estimates of measurement bias as long as the assumption $y_i \doteq \mu_i$ holds. Otherwise, the estimates of measurement bias will themselves be subject to biases and misleading inferences.

One approach for obtaining gold standard measurements is the *gold standard reinterview* survey (Biemer and Lyberg 2003; Forsman and Schreiner 1991). This involves using special procdures for reinterviewing a sample of respondents to the main survey to obtain highly accurate responses that are assumed to be error-free. One form of this approach is the *reconciled reinterview*. The reconciled reinterview proceeds much like a test–retest reinterview except that any differences between the interview and reinterview responses are reconciled with the respondent on completion of the reinterview; specifically, the reinterviewer attempts to determine the causes for each discrepancy and, in the process, tries to obtain the most accurate response possible for the survey item (Forsman and Schreiner 1991). Another type of gold standard reinterview uses an in-depth, probing approach rather than reconciliation to obtain much more accurate measurements than the main survey measurements.

Administrative records from population registers, income tax reports, police records, and so on can also provide gold standard measurements. As an example, Körmendi (1988) used income reported to the tax authorities to check the accuracy of income reports in a Danish survey. Marquis and Moore (1990) obtained administrative data from eight income programs including social security income, food stamps, and unemployment insurance income to evaluate program participation and income reports in the US Survey of Income and Program Participation. There are also numerous examples where blood, urine, hair, or other biological specimen collections were used to check the accuracy of self-reported drug use [see Harrison (1997) for an overview of this methodology].

The literature also provides a number of examples where so-called gold standard measurements have led to poor estimates of measurement bias. For example, a number of articles show that reconciled reinterview data can be as erroneous as the original measurements that they were intended to evaluate [see, e.g., Biemer and Forsman (1992), Sinclair and Gastwirth (1996), and Biemer et al. (2001)]. Administrative records data can often be quite inaccurate and difficult to use (Jay et al. 1994, Marquis 1978) as a result of differences in time reference periods and operational definitions, as well as errors in the records themselves. Even biological measures such as hair analysis and urinalysis used in studies of drug use contain substantial false-positive and false-negative errors for detecting some types of drug use [see, e.g., Visher and McFadden (1991)].

The next chapter discusses some alternative methods based on latent class models for estimating measurement bias and validity. These estimates do not suffer from the same limitations as do estimates based on gold standard measurements.

2.5.3 Example: Poor- or Failing-Grade Data

We conclude this chapter with further analysis of the data that we used to introduce it. Consider the data in Tables 2.1 and 2.2. In Table 2.1, data from a student questionnaire were cross-classified by two questions about whether the student had ever received a poor grade. Table 2.2 cross-classified the response from each of the two questions according to whether a poor grade was found on the student's transcript. Treating the transcript data as the truth, the NDR for the cross-classification of the question response and the transcript classification provides an estimate of the measurement bias for the question. From Table 2.2, the NDR values for Q1 and Q2 are, respectively

$$\text{NDR}_1 = \frac{11-150}{954} = -0.15 \quad \text{and} \quad \text{NDR}_2 = \frac{7-338}{954} = -0.35$$

With standard errors between 1 and 2 percentage points, these differences are highly significant and the biases are quite substantial. The biases are negative, which implies that both Q1 and Q2 tend to underestimate the proportion of students with poor or failing grades. Question 1 results in an estimated bias of –15%, while question 2 is more than twice this value: –35%. We can decompose the bias into components for true positives and true negatives in order to better understand the nature of the biases.

Let $\phi = \Pr(y_i = 1 \mid \mu_i = 0)$ and $\theta = \Pr(y_i = 0 \mid \mu_i = 1)$ denote the *false-positive* and *false-negative* probabilities, respectively, for a measure y . Since we know the true value μ_i, we can estimate these probabilities for Q1 and Q2 from the data in Table 2.2. Consider a 2×2 cross-classification of dichotomous variables $y_i = l$ and $\mu_i = l'$ with cell proportions $p_{ll'}$, for l and l' taking the values 0 and 1. We can easily show that

$$\hat{\phi} = \frac{p_{10}}{p_{10} + p_{00}} \quad \text{and} \quad \hat{\theta} = \frac{p_{01}}{p_{01} + p_{11}}$$

are unbiased estimators of ϕ and θ, respectively. Applying these formulas for Q1, we obtain an estimated false-positive probability of $\hat{\phi}_1 = 11/(11+370) = 0.029$ and an estimated false-negative probability of $\hat{\theta}_1 = 150/(150+423) = 0.26$. These estimates suggest than while about 3% of students having no poor or failing grade respond that they do, 26% who do have a poor or failing grade respond that they do not. Perhaps students have trouble admitting or have forgottten that they have ever received a failing grade. The situation is worse for Q2, where the false-positive rate is $\hat{\phi}_2 = 0.018$ and the false-negative rate is $\hat{\theta}_2 = 0.59$; in other words, almost 60% of students who have a poor or failing grade deny that they do. In addition to the issues with Q1, perhaps students also misunderstand Q2. The term "unsatisfactory" may be misleading because even a "C" grade may be unsatisfactory if one was expecting an "A" or "B." However, that would seem to lead to more false-positive responses. Perhaps a number of students who have a "D" grade on their transcripts view that grade as satisfactory, especially if a failing grade was expected.

One can only speculate as to why students with bad grades are misclassified at such high rates and why the situation is worse for Question 2 than for Question 1. To obtain more information, cognitive interviews (Lessler et al. 1992; Forsyth and Lessler 1991) could be conducted with students to examine the social undesirability issues associated with the topic and the potential comprehension errors that may cloud the meaning of the questions. If the problem is the former, more private methods of interviewing or greater assurances of confidentiality may be the remedy. If the problem is misinterpretation of the questions, revisions of the question that better convey the meaning as intended by the researcher could be tested. As this example illustrates, after demonstrating that a question elicits inaccurate responses and that the magnitude of the problem is unacceptable, questions as to the causes of the errors still remain. Still, attention can now be directed toward a data quality issue that heretofore remained obscured in the data. Now the work of improving quality can be more precisely directed and, hence, more effective.

The next chapter develops the ideas introduced in this example in much greater detail. It shows that knowledge of the true values are not really needed to evaluate false-positive and false-negative probabilities when a good model of the response process is available. The response probability models introduced in the next chapter are the essential building blocks of the latent class modeling framework.

CHAPTER 3

Response Probability Models for Two Measurements

3.1 RESPONSE PROBABILITY MODEL

The previous chapter introduced the concept of the true score τ_i for the ith sample unit. For dichotomous variables, τ_i is defined as the probability of a positive response and is conceptually similar to the following weighted-coin experiment. Suppose that the ith respondent carries a weighted coin with probabilities τ_i of a "head" and $1 - \tau_i$ of a "tail." For responding to a dichotomous survey item, the respondent flips the coin to determine a response. If the result is "head," the respondent responds "yes" ($y_i = 1$) and if "tail," the respondent responds "no" ($y_i = 0$). So, for example, if $\tau_i = 0.8$, then the respondent has an 80% chance of responding "yes" to the question; thus, over many repeated, independent trials under the same essential survey conditions, about 80% of the respondent's responses will be "yes" and 20% will be "no." The respondent has many such coins, one for each dichotomous question on the questionnaire and each with different probabilities of a "head."

This experiment, although unrealistic, embodies the essential assumptions of classical test theory. Note that in this discussion, there is no mention of the respondent's true characteristic since the existence of a true value is not required in classical test theory, nor is it needed to estimate the reliability of a characteristic. However, the concept of measurement bias requires a true value. The next section introduces Bross' model, which can be viewed as an extension of classical test theory to incorporate a possibly unknown true value of the characteristic. This simple model will be used to introduce a number of new concepts pertaining to the bias in a response or an estimator.

Latent Class Analysis of Survey Error By Paul P. Biemer

3.1.1 Bross' Model

Let μ_i denote the (unknown) true value of the characteristic measured by y_i. Both μ_i and y_i are assumed to be dichotomous but later will be extended to polytomous measures. As an example, suppose that the respondent is asked "Have you ever visited Paris?" Suppose that the respondent once took a connecting flght through Paris but stayed in the Charles de Gaulle airport for only 2 hours. Otherwise, he/she has never been to Paris. Considering the researcher's intent for the question, the respondent should answer "no." However, the respondent is confused by the term "visited" and will answer "yes" with probability 0.20. In other words, over many hypothetical independent repetitions of the survey question, the respondent's answer will be "yes" 20% of the time and "no" 80% of the time. Thus, $y_i = 1$ ("yes") with probability $\tau_i = 0.2$ even though $\mu_i = 0$ ("no") by the researcher's definition. If *theoretical validity* is high, that is, if $Corr(\mu_i,\tau_i)$ is near 1, then one would expect τ_i to be close to 0 whenever $\mu_i = 0$ and close to 1 whenever $\mu_i = 1$. In other words, *validity* implies that true negative respondents should have a small probability of responding positively to the question, while *true-positive* respondents should have a small probability of responding negatively.

We can express these concepts mathematically, using notation established by Bross (1954). Let ϕ_i denote the probability of a false-positive error [i.e., $\phi_i = \Pr(y_i = 1 \mid \mu_i = 0)$] and θ_i denote the probability of a false-negative error [i.e., $\theta_i = \Pr(y_i = 0 \mid \mu_i = 1)$]. We shall refer to θ_i and ϕ_i as *error probabilities*.[1] The probability of a correct positive observation is $1 - \theta_i$ and of a correct negative observation is $1 - \phi_i$. Now we can write

$$\begin{aligned}\tau_i &= P(y_i = 1 \mid i) \\ &= \begin{cases}(1-\theta_i) & \text{if} \quad \mu_i = 1 \\ \phi_i & \text{if} \quad \mu_i = 0\end{cases} \\ &= \mu_i(1-\theta_i) + (1-\mu_i)\phi_i\end{aligned} \tag{3.1}$$

This result shows that the true score is a linear combination of the false-negative and false-positive probabilities. If unit i is a true positive, the true score is $(1 - \theta_i)$. Likewise, if unit i is a true negative, the true score is ϕ_i. This notation is summarized in Table 3.1.

For both true positives and true negatives, μ_i and τ_i will tend to agree if the error probabilities are small. This means that $Corr(\mu_i,\tau_i)$, or theoretical validity, will be near its maximum value of 1 when both θ_i and ϕ_i are quite small. As either θ_i and ϕ_i increases, theoretical validity will be reduced.

It will be shown that, for binary data, the error probabilities, θ_i and ϕ_i are the essential components not only for validity but also for the concepts of

[1] In the literature on latent class analysis, these probabilities are often referred to as *response probabilities*. We prefer to refer to them as *error probabilities* to emphasize their interpretation for survey error evaluation.

Table 3.1 Bross' Notation for Response Probabilities

	True Positive ($\mu_i = 1$)	True Negative ($\mu_i = 0$)
Observed positive ($y_i = 1$)	No error ($\varepsilon_i = 0$) $1 - \theta_i = \Pr(y_i = 1 \mid \mu_i = 1)$	False positive ($\varepsilon_i = 1$) $\phi_i = \Pr(y_i = 1 \mid \mu_i = 0)$
Observed negative ($y_i = 0$)	False negative ($\varepsilon_i = -1$) $\theta_i = \Pr(y_i = 0 \mid \mu_i = 1)$	No error ($\varepsilon_i = 0$) $1 - \phi_i = \Pr(y_i = 0 \mid \mu_i = 0)$

reliability, simple response variance, measurement bias, and misclassification. The error probability approach can easily be extended to polytomous data as well. For polytomous data, error probabilities must be defined for misclassifying individuals from their true category to any other possible category of the dependent variable. This extension is considered in Section 3.5 and presented in more detail in Chapter 4. As noted in Section 1.3, the focus of LCA for survey evaluation is the estimation of these error probabilities. Thus, Bross' model can be viewed as the foundation of LCA for survey error evaluation.

To see how classification probabilities relate to measurement error, consider a simple model that equates y_i to the corresponding true value, μ_i plus an error term

$$y_i = \mu_i + \varepsilon_i \tag{3.2}$$

where ε_i is a random measurement error component, which takes the value 1 if there is false-positive error (i.e., $y_i = 1$ when $\mu_i = 0$), the value -1 if there is a false-negative error (i.e., $y_i = 0$ when $\mu_i = 1$), and the value 0 if no error (i.e., $y_i = \mu_i$).

Bross' framework for dichotomous variables provides more detail about an individual's response distribution than does the classical test theory model. It distinguishes between two types of units in the population: true positives and true negatives. A unit's true score is the probability of responding positively, which can be expressed as a function the unit's true value and the conditional probability of responding positively given this true value. Using this model, formulas for estimator bias, variance, and reliability given in Chapter 2 can be restated by substituting the expression in (3.1) for τ_i and simplifying the results. This will be demonstrated in the following text.

Consider the effect of misclassification for statistical inference for the population proportion π, defined as

$$\pi = \frac{1}{N}\sum_{i=1}^{N}\mu_i \tag{3.3}$$

Define the average false-negative and false-positive probabilities, respectively, as

$$\theta = \frac{1}{N_1}\sum_{i=1}^{N_1}\theta_i \quad \text{and} \quad \phi = \frac{1}{N_0}\sum_{i=1}^{N_0}\phi_i \tag{3.4}$$

where the first sum extends over individuals in the population who are true positives (i.e., $\mu_i = 1$) and the second sum is over individuals who are true negatives (i.e., $\mu_i = 0$), $N_1 = \Sigma_i \mu_i$ is the number of positives and $N_0 = N - N_1$ is the number of negatives in the population. Recall from the last chapter [equation (2.1)] that the classical test theory model for y_i is

$$y_i = \tau_i + e_i \tag{3.5}$$

But now τ_i can be rewritten as

$$\tau_i = \Pr(y_i = 1) = \mu_i(1-\theta_i) + (1-\mu_i)\phi_i \tag{3.6}$$

The population mean of the true scores τ_i can written in terms of the μ_i as

$$\begin{aligned}
E(\tau_i) &= \frac{1}{N}\sum_{i=1}^{N}\tau_i \\
&= \frac{1}{N}\sum_{i=1}^{N}[\mu_i(1-\theta_i) + (1-\mu_i)\phi_i] \\
&= \frac{1}{N}\left[\sum_{i=1}^{N}[\mu_i(1-\theta_i)] + \sum_{i=1}^{N}(1-\mu_i)\phi_i\right] \\
&= \frac{1}{N}\left[\sum_{i=1}^{N_1}(1-\theta_i) + \sum_{i=1}^{N_0}\phi_i\right] \\
&= \frac{N_1}{N}(1-\theta) + \frac{N_0}{N}(1-\phi) \\
&= \pi(1-\theta) + (1-\pi)\phi
\end{aligned} \tag{3.7}$$

Thus, the population mean of true values will differ from the prevalence probabability π when the average false-negative probability θ or the average false-positive probability ϕ is positive.

It follows from (3.5) and (3.7) that expected value of the sample proportion p, which is the estimator of π in for simple random sampling, is

$$\begin{aligned}
E(p) &= E\left(\frac{1}{n}\sum_{i=1}^{n}y_i\right) \\
&= \frac{1}{n}\sum_{i=1}^{n}E(\tau_i + e_i) \\
&= \frac{1}{n}\sum_{i=1}^{n}E(\tau_i) \\
&= \pi(1-\theta) + (1-\pi)\phi
\end{aligned} \tag{3.8}$$

An expression for the bias in p as an estimator of π can now be written as

$$\begin{aligned} B(p) &= E(p) - \pi \\ &= -\pi\theta + (1-\pi)\phi \end{aligned} \tag{3.9}$$

Using this result, we see that there are two ways inference about the population proportion can be unbiased. The bias is 0 if either (1) both θ and ϕ are 0 (i.e., no misclassification) or (2) $\pi\theta = (1 - \pi)\phi$ (i.e., the expected number of false negatives and the expected number of false positives are equal). Condition (1) is unlikely to hold in most situations, although both θ and ϕ may be quite small for some characteristics. Condition (2) is also unlikely to hold in most practical situations, particularly when π is quite small or large. Note that if false negatives outnumber false positives, the bias will be negative. If the opposite is true, the bias will be positive.

To illustrate the effect of the value of π on the bias $B(p)$, suppose that $\theta = \phi = \lambda$, say. Then (3.9) can be written as $B(p) = \lambda(1 - 2\pi)$. Now it is obvious that the bias is 0 when $\pi = 0.5$. When π is quite small, even a very small value of λ can produce a large bias relative to the size of the population proportion. As an example, suppose that $\lambda = 0.01$, a very small error rate, and define the *relative bias* (RB) in p by RB $= B(p)/\pi$. When $\pi = 0.01$, the relative bias is 0.98% or 98%. In other words, on average, p is almost twice the value of π even if the error probability is miniscule (i.e., only one-tenth of 1%). For larger proportions, the relative bias due to this level of error becomes inconsequential. For proportions that are greater than $\pi = 0.5$, $(1 - 2\pi)$ is negative, and so is the bias. For small λ, the relative bias is also small. But for larger values of λ, the effect of misclassification on the estimates of π, particularly when π is small, can be considerable.

Next consider the variance of the estimator, p. Writing $Var(p)$ in Bross' framework is fairly straightforward by applying (2.18). Here we show how to derive new expressions for σ_τ^2 and σ_e^2 using (3.1). First, write $\sigma_\tau^2 = E(\tau_i^2) - \tau^2$ and expand the term $E(\tau_i^2)$ as follows

$$\begin{aligned} E(\tau_i^2) &= \frac{1}{N}\sum_{i=1}^{N}[\mu_i(1-\theta_i) + (1-\mu_i)\phi_i]^2 \\ &= \frac{1}{N}\left[\sum_{i=1}^{N_1}(1-\theta_i)^2 + \sum_{i=1}^{N_0}\phi_i^2\right] \\ &= \frac{N_1}{N}\frac{\sum_{i=1}^{N_1}(1-\theta_i)^2}{N_1} + \frac{N_0}{N}\frac{\sum_{i=1}^{N_0}\phi_i^2}{N_0} \\ &= \frac{N_1}{N}\left(Var(1-\theta_i) + (1-\theta)^2\right) + \frac{N_0}{N}\left(Var(\phi_i) + \phi^2\right) \\ &= \pi\left[\sigma_\theta^2 + (1-\theta)^2\right] + (1-\pi)(\sigma_\phi^2 + \phi^2) \end{aligned} \tag{3.10}$$

where $\sigma_\theta^2 = Var(\theta_i \mid \mu_i = 1) = N_1^{-1}\sum_{\mu_i=1}(\theta_i - \theta)^2$ and $\sigma_\phi^2 = Var(\phi_i \mid \mu_i = 0) = N_1^{-1}\sum_{\mu_i=0}(\phi_i - \phi)^2$. Thus

$$\begin{aligned}\sigma_\tau^2 &= E(\tau_i^2) - \tau^2 \\ &= \pi\left[\sigma_\theta^2 + (1-\theta)^2\right] + (1-\pi)(\sigma_\phi^2 + \phi^2) - [\pi(1-\theta) + (1-\pi)\phi]^2\end{aligned} \tag{3.11}$$

Simplifying this expression, we obtain

$$\sigma_\tau^2 = \pi(1-\pi)(1-\theta-\phi)^2 + \gamma_{\theta\phi} \tag{3.12}$$

where $\gamma_{\theta\phi} = \pi\sigma_\theta^2 + (1-\pi)\sigma_\phi^2$. The term $\gamma_{\theta\phi}$ reflects the variation of the classification error rates in the population. When $\gamma_{\theta\phi} = 0$, all true positives have the same error probability, namely, θ, and all true negatives have the same error probability, namely, ϕ. As $\gamma_{\theta\phi}$ becomes large, the variation in the error probabilities in the population increases. As an example, the misclassification of y may be related to an individual's demographic or other personal characteristics. Thus, in heterogeneous populations, $\gamma_{\theta\phi}$ will not be 0. When $\gamma_{\theta\phi} = 0$, we say that the population is *homogeneous* with respect to the error probabilities. That the population is homogeneous is an important assumption of LCA that will be explored further in the next chapter.

Turning to σ_e^2 in (2.18), note that

$$\begin{aligned}\sigma_e^2 &= E_i\, Var(y_i \mid i) \\ &= E_i\, Var(\mu_i + \varepsilon_i \mid i) \\ &= E_i\, Var(\varepsilon_i \mid i)\end{aligned} \tag{3.13}$$

because, after conditioning upon the unit i, the true value for unit i is invariant and thus $Var(\mu_i \mid i) = 0$. Further, it is easy to show $E(\varepsilon_i \mid i) = -\mu_i\theta_i + (1-\mu_i)\phi_i$ and $E(\varepsilon_i^2 \mid i) = \mu_i\theta_i + (1-\mu_i)\phi_i$. Using these results, it follows that

$$\begin{aligned}Var(\varepsilon_i \mid i) &= E(\varepsilon_i^2 \mid i) - [E(\varepsilon_i \mid i)]^2 \\ &= \mu_i\theta_i + (1-\mu_i)\phi_i - [-\mu_i\theta_i + (1-\mu_i)\phi_i]^2 \\ &= \mu_i\theta_i + (1-\mu_i)\phi_i - \mu_i\theta_i^2 - (1-\mu_i)\phi_i^2 \\ &= \mu_i(\theta_i - \theta_i^2) + (1-\mu_i)(\phi_i - \phi_i^2) \\ &= \mu_i\theta_i(1-\theta_i) + (1-\mu_i)\phi_i(1-\phi_i)\end{aligned} \tag{3.14}$$

In other words, for true positives, $Var(\varepsilon_i \mid i) = \theta_i(1-\theta_i)$ and for true negatives $Var(\varepsilon_i \mid i) = \phi_i(1-\phi_i)$. Hence, we can write the error variance as

$$\begin{aligned}\sigma_e^2 &= \frac{1}{N}\sum_{i=1}^{N}[\mu_i\theta_i(1-\theta_i)+(1-\mu_i)\phi_i(1-\phi_i)] \\ &= \frac{1}{N}\sum_{\mu_i=1}(\theta_i-\theta_i^2)+\frac{1}{N}\sum_{\mu_i=0}(\phi_i-\phi_i^2) \\ &= \frac{N_1}{N}[\theta-(\sigma_\theta^2+\theta^2)]+\frac{N_0}{N}[\phi-(\sigma_\phi^2+\phi^2)] \\ &= \pi\theta(1-\theta)+(1-\pi)\phi(1-\phi)-\gamma_{\theta\phi}\end{aligned} \tag{3.15}$$

Substituting these new expressions for σ_τ^2 and σ_e^2 in (2.18) produces an expression for *Var*(p) in terms of the parameters π, θ, ϕ, and $\gamma_{\theta\phi}$. Therefore

$$\begin{aligned}Var(p) &= \left(1-\frac{n-1}{N-1}\right)\frac{\sigma_\tau^2}{n}+\frac{\sigma_e^2}{n} \\ &= \left(1-\frac{n-1}{N-1}\right)\frac{\pi(1-\pi)(1-\theta-\phi)^2+\gamma_{\theta\phi}}{n} \\ &\quad +\frac{\pi\theta(1-\theta)+(1-\pi)\phi(1-\phi)-\gamma_{\theta\phi}}{n}\end{aligned} \tag{3.16}$$

which, using (2.18), can be rewritten approximately as

$$Var(p) \doteq \frac{1}{R}\frac{\pi(1-\pi)(1-\theta-\phi)^2+\gamma_{\theta\phi}}{n} \tag{3.17}$$

In this expression, the reliability ratio R has the form

$$\begin{aligned}R &= \frac{\sigma_\tau^2}{\sigma_\tau^2+\sigma_e^2} \\ &= \frac{\pi(1-\pi)(1-\theta-\phi)^2+\gamma_{\theta\phi}}{\pi(1-\pi)(1-\theta-\phi)^2+\pi\theta(1-\theta)+(1-\pi)\phi(1-\phi)}\end{aligned} \tag{3.18}$$

The expression for the variance of *Var*(p) shows the complex relationship between the components of error π, θ, ϕ, and $\gamma_{\theta\phi}$ and the sampling variance for the estimator of π. The next section discusses the implications of this form of the variance with particular emphasis on the reliability ratio.

3.1.2 Implications for Survey Quality Investigations

The form of *Var*(p) in (3.17) suggests that, in the Bross framework, the effects of measurement errors on the precision of the estimator p is quite complex.

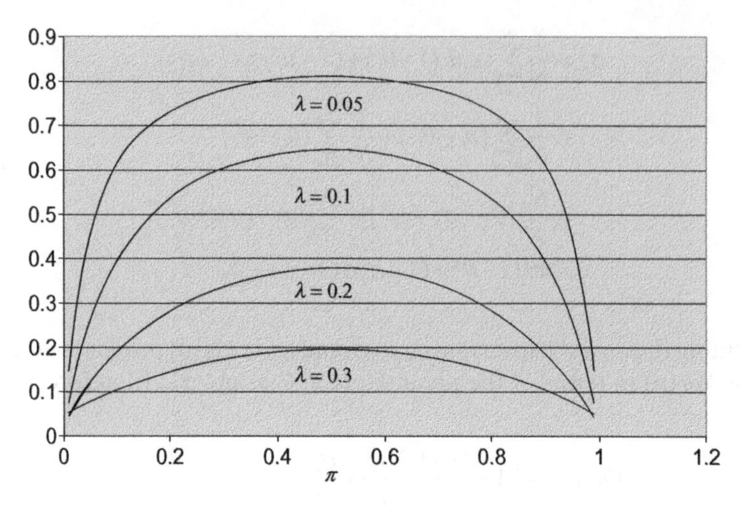

Figure 3.1 The reliability ratio as a function of π and $\theta = \phi = \lambda$.

It is not obvious how an increase in classification error will affect sampling variance. In fact, the variance can increase or decrease as classification error increases, which is quite different from the way measurement error affects the sampling variance for continuous measurements. The interpretation of R is also much more complex for dichotomous variables. For dichotomous variables, R is a nonlinear function of the parameters π, θ, ϕ, and $\gamma_{\theta\phi}$. As such, it is no longer clear how reliability can be interpreted as an indicator of data quality.

To understand why, consider the situations depicted in Figure 3.1, where R is plotted as a function of the true population proportion π for various values of the misclassification probabilities for the case where $\theta = \phi = \lambda$. The curvilinear shape of relationship suggests that reliability varies with prevalence even when the misclassification probabilities are held constant. This fact makes it very difficult to interpret reliability across subgroups of the population that have very different prevalence rates.

As an example, suppose that we know that the reliabilities for two population subgroups (say, males and females) are 0.4 and 0.6, respectively. A typical misuse of R in this situation is to say that males have larger error (i.e., higher rates of misclassification) than do females. However, as is clear from Figure 3.1, another explanation is also plausible. It is quite possible that the two groups have exactly the same (or very similar) error rates but that the prevalence of the characteristic differs in the two groups. For example, for the curve corresponding to $\lambda = 0.1$, a reliability of 0.4 occurs around $\pi = 0.1$ and a reliability of 0.6 occurs around $\pi = 0.3$. Thus, the differing reliability estimates may have little to do with levels of classification error and instead reflect the different prevalence rates.

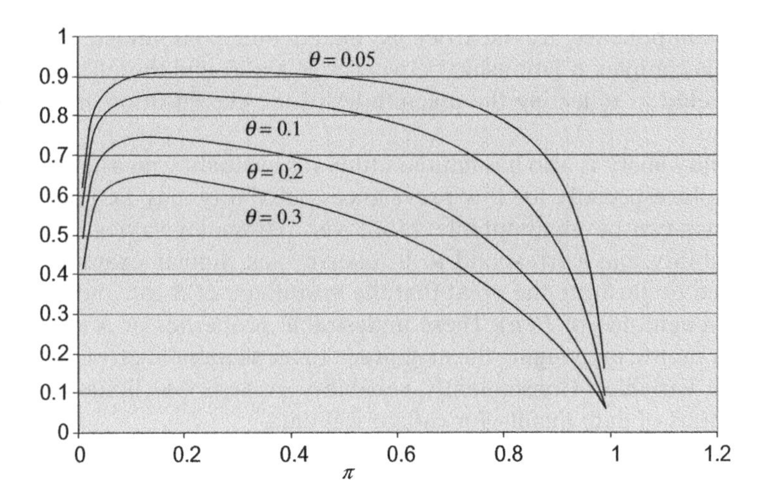

Figure 3.2 The reliability ratio as a function of π and θ for $\phi = 0.005$.

The same point can be made for Figure 3.2, which allows the false-negative and false-positive probabilities to differ. In this figure, the false-positive probability ϕ is fixed at 0.005 while both π and θ are allowed to vary. This situation is not unusual since it could pertain to the measurement of any highly sensitive and stigmatized characteristic such as illegal drug use. For such characteristics, the false-negative probability (i.e., denying the illegal behavior) may be quite high while the false-positive probability (falsely claiming the behavior) is practically nil. In addition, stigmatized characteristics tend to have low prevalence—say, in the range of 0.01–0.10. In this range, reliability can vary by 20 percentage points or more for the same values of θ and ϕ as shown in Figure 3.2. Two subgroups with exactly the same rates of misclassification can have very different reliabilities when their prevalence probabilities differ. This is more or less true across the entire range of π; reliability varies considerably although classification error remains constant.

This discussion suggests that R may be a poor metric for gauging measurement quality for categorical variables since lower values of R do not necessarily imply greater misclassification; nor do high values of R imply that misclassification is low. One may ask "Is there a useful interpretation of R for categorical responses?" From Chapter 2 we know that R is the ratio of true score variance over total variance—that is, true score variance plus measurement error variance. This is still true for dichotomous data. However, for dichotomous data, the total variance does not necessarily increase as measurement variance increases. In addition, the true score variance does not remain fixed as measurement error varies since both true score variance and measurement variance depend on the same three parameters, π, θ, and ϕ. Thus, the two variance components are inextricably linked and confounded. Reliability reflects this complex relationship between π, θ, and ϕ, and therefore R cannot be interpreted as reflecting the magnitudes of the classification error parameters alone.

Reliability analysis also has limited utility for evaluating measurement bias. For example, especially for low prevalances, reliability may be good yet the relative bias can be unacceptable. Moreover, parameter values that yield a poor reliability may correspond to 0 relative bias. Similar examples can be constructed to illustrate the point that the magnitude of R can and often does belie the magnitude of $B(p)$. These undesirable properties of R suggest that it is a poor metric for gauging the magnitude of measurement error in categorical survey variables. Consequently, reliability analysis has limited utility in investigations of data quality for categorical data.

On the other hand, the parameters π, θ, and ϕ provide all the information one requires to know the bias, the total variance, the sampling variance, and the simple response variance for the estimator p. For categorical variables, the primary goal of survey measurement error evaluation is to estimate π, θ, and ϕ or comparable parameters whenever possible. The next section discusses methods for obtaining estimates of these parameters.

3.2 ESTIMATING π, θ, AND ϕ

The parameters π, θ, and ϕ are the basic building blocks for constructing estimates of the mean-squared error of p and its components, including bias, relative bias, true score variance (SV), measurement error variance (SRV), and the ratios I and R. Likewise, estimates of π, θ, and ϕ can be combined to obtain estimates of these bias and variance components. In fact, one could say that the remainder of this book is all about how to estimate these three essential parameters or similar parameters in more complex models. Two methods for estimating π, θ, and ϕ will be considered in this section beginning with a very simple method based on gold standard measures.

In Section 2.5, we discussed a method for estimating measurement error parameters based on gold standard measurements. In the present context, gold standard measurements are observations or determinations of μ_i denoted by y_i that have the property that $y_i \doteq \mu_i$, where the symbol " $\doteq$ " denotes approximate equality. In other words, the y_i are virtually error-free. As noted there, such measurements are very difficult to obtain in practice, yet a number of studies have attempted to obtain them. Gold standard measurements may come from reconciled reinterviews surveys (Forsman and Schreiner 1991); in-depth, probing reinterviews; record check studies (Biemer 1988); blood, urine, hair, or other biological specimen collections (Harrison 1997); or any other method that yields measurements that can be regarded as gold-standard variables in an error evaluation. Also cited in Section 2.5.2

were a number of articles showing that gold standard data can be as erroneous as the original measurements that they were intended to evaluate. Errors in criterion variables used in an evaluation imply that gold standard estimates of measurement error parameters (namely, π, θ, and ϕ) will be biased. The extent of the bias depends on the magnitudes of the errors in criterion variables.

A key limitation in the use of gold standard measurements is the unverifiable assumption that they are highly accurate measurements. Often this assumption is justified by arguing that the process used to collect them was ideally designed, tightly controlled, and carefully monitored so that measurement errors were unlikely to occur. A careful description of the measurement process is usually provided to support these claims and to convince skeptics. This might include an analysis showing the errors found by the gold standard process and the plausibility of the results. In the end, however, there is seldom any statistical evidence of the validity of the results. Indeed, there are many examples in the literature where such gold standard measurements were later found to be quite erroneous.

One example is the US Current Population Survey (CPS), a household sample survey conducted monthly by the US Census Bureau to provide estimates of employment, unemployment, and other characteristics of the general US labor force population. Given the importance of the CPS data series to public policy, there have been numerous evaluations of the accuracy of the data. For example, since the early 1950s, the Census Bureau has conducted the CPS Reinterview Program for the purpose of evaluating the quality of the labor force data. The CPS Reinterview Program selects a subsample of about 5% of the CPS respondents and assigns 25% of these to a test–retest (unreconciled) reinterview and the other 75% to a reconciliation-type reinterview. The former reinterview survey attempts to obtain parallel measurements for estimating test–retest reliability, while the latter attempts to obtain the most accurate responses [see Forsman and Schreiner (1991) for a detailed description of these procedures].

For many years, the Census Bureau had used reconciled reinterviews for evaluating the CPS labor force classifications and has published annual reports on classification bias. However, this practice was discontinued when it was discovered that the reconciled reinterview data were subject to considerable classification error (Biemer and Forsman 1992). The estimated error parameters were biased and misleading to the point where they were not at all useful for evaluating measurement bias. Biemer and Forsman speculated that the primary cause of the error in the reconciled reinterview data was the reinterview process itself, which suffered from important design flaws.

A major problem with the design was the use of supervisory field representatives (SFRs) to conduct the reinterviews. The SFRs used the reconciled reinterview program as a means of evaluating the performance of the field representatives (FRs) or interviewers in their employ. The use of the results to estimate measurement bias was considered to be a secondary purpose. In that regard, the SFRs placed a higher priority on determining whether their interviewers were the source or cause of the errors that were identified during the reconciliation process rather than obtaining the "best" answer to a question. Consequently, the final reconciled reinterview response was often in error since the SFRs tended to default to the original item response when a discrepancy arose. Although these problems could have been remedied, a reconciled reinterview process, regardless of how it is designed, is still an interviewing process and, as such, subject to many of the same measurement issues associated with any interview process.

Still, reconciled reinterview should not be disregarded as a quality evaluation methodology. As an example, Fecso and Pafford (1988) used reconciled reinterviews apparently quite successfully to evaluate the bias in agricultural data. The data collection mode was computer-assisted telephone interviewing (CATI)—a new data collection technology being evaluated by the US National Agricultural Statistics Service (NASS). Reinterviews of about 1000 CATI respondents were conducted by personal visit and reconciled. Regarding the reinterview responses as a gold standard, they found that corn and soybean stocks were substantially underestimated by the CATI survey. Using the reasons for the discrepancies collected through the reconciliation process, they discovered that definitional problems were the largest contributors to bias. This work led to a revised questionnaire, improved training of CATI interviewers, and a shift to more face-to-face interviews. Despite success stories such as this one, the gold standard approach for estimating measurement bias still remains quite limited, and alternatives for estimating measurement accuracy are needed. In the next section, we discuss one such alternative method.

3.2.1 Maximum-Likelihood Estimates of π, θ, and ϕ

Maximum-likelihood (ML) estimation of the misclassification parameters can be used whenever two or more measurements of μ_i are available. In can be used as an alternative to or in conjunction with the gold standard approach; for example, as an alternative method for analyzing the gold standard data. ML estimation employs a model for expressing the likelihood of observing the sample in terms of π, θ, and ϕ. It then uses an optimization algorithm to find the values of the parameters that maximize this likelihood function. These optimal values of the parameters are referred to as maximum-likelihood estimates (MLEs).

For example, consider the likelihood of observing interview–reinterview data under the assumption of simple random sampling with replacement sampling and parallel measurements. Table 3.2 summarizes the notation for the 2×2 interview–reinterview table. Let P_{jk} denote the probability of an observation falling into cell (j,k) of the table [i.e., $\Pr(y_1 = j, y_2 = k)$] for $j,k = 0,1$. Assume the following:

Table 3.2 Interview–Reinterview Table

	$y_2 = 1$	$y_2 = 0$
$y_1 = 1$	a	b
$y_2 = 0$	c	d

1. For each measurement, the error probabilities are *homogeneous* (i.e., $\gamma_{\theta\phi} = 0$). This means that every unit in the sample has the same probability of misclassification:

$$\theta_{1i} = \theta_1 \quad \text{and} \quad \phi_{1i} = \phi_1 \quad \text{for} \quad i = 1, \ldots, n \tag{3.19}$$

$$\theta_{2i} = \theta_2 \quad \text{and} \quad \phi_{2i} = \phi_2 \quad \text{for} \quad i = 1, \ldots, n \tag{3.20}$$

2. The measurements are *locally independent*

$$\Pr(y_1 = j, y_2 = k \mid \mu = l) = \Pr(y_1 = j \mid \mu = l)\Pr(y_2 = k \mid \mu = l) \tag{3.21}$$

   for $j,k,l = 0,1$.[2]
3. The measurements are parallel (i.e., $\theta_1 = \theta_2$ and $\phi_1 = \phi_2$).

These three assumptions form the core set of assumptions embodied in LCA, which will be discussed in greater detail in Section 4.2.1. Assumptions 1 and 3 seem fairly straightforward; however, understanding assumption 2 (local independence) may require further discussion. It states that given ith unit's true value is l, classification of the unit by the interview and reinterview is done independently. Another way of stating this is that the classification errors for interview and reinterview are independent since otherwise (3.21) does not hold. This assumption can be violated in other ways, as shown in Section 5.2.

Under these assumptions, the distribution of the cell count vector (a, b, c, d) is multinomial with parameters $n = a + b + c + d$ and cell probabilities P_{jk}, for $j,k = 0,1$. It therefore follows that the likelihood function corresponding to Table 3.2 is

$$\mathcal{L}(\pi,\theta,\phi \mid a,b,c,d) = \frac{n!}{a!b!c!d!} P_{11}^{a} P_{10}^{b} P_{01}^{c} P_{00}^{d} \tag{3.22}$$

[2]In the language of LCA, the event $\{\mu_i = l\}$ is stated as "unit i belongs to *latent class* l." This emphasizes the tacit assumption that μ_i is an unknown (*latent*) variable.

where

$$
\begin{aligned}
P_{11} &= \pi(1-\theta)^2 + (1-\pi)\phi^2 \\
P_{10} &= \pi(1-\theta)\theta + (1-\pi)(1-\phi)\phi \\
P_{01} &= P_{10} \\
P_{00} &= \pi\theta^2 + (1-\pi)(1-\phi)^2
\end{aligned}
\tag{3.23}
$$

To simplify the subsequent formulas, the logarithm of the likelihood function will be used and will be denoted by l. Since the logarithm function is monotone, the maximum of $\mathcal{L}(a,b,c,d \mid \pi,\theta,\phi)$ is the same as the maximum of $\log[\mathcal{L}(a,b,c,d \mid \pi,\theta,\phi)] = \ell(a,b,c,d \mid \pi,\theta,\phi)$. Working with the loglikelihood simplifies many estimation problems that will be considered in this book.

Since the primary purpose of deriving the likelihood is to estimate the error parameters, we need only consider that part of the likelihood that contains the parameters to be estimated which is referred to as the *loglikelihood kernel.* The kernel of the logarithm for the likelihood in (3.22) with constraints in (3.23) is given by

$$
\begin{aligned}
\ell(\pi,\theta,\phi \mid a,b,c,d) = {} & a\log\left[\pi(1-\theta)^2 + (1-\pi)\phi^2\right] \\
& \times (b+c)\log[\pi(1-\theta)\theta + (1-\pi)(1-\phi)\phi] \\
& \times d\log\left[\pi\theta^2 + (1-\pi)(1-\phi)^2\right]
\end{aligned}
\tag{3.24}
$$

The next step would be to find the values of π, θ, and ϕ that maximize this function. Unfortunately, in this situation, a unique maximum does not exist. There are many values of the parameters that will result in the same maximum value of (3.24). To understand why this is true, let us try to apply the usual method for finding the MLEs: differentiating ℓ with respect to π, θ, and ϕ to obtain the three likelihood equations, setting the derivatives to 0, and solving for the three parameters. However, because the sum of the four cell probabilities is 1 and because the two off-diagonal cells (P_{10} and P_{01}, say) have exactly the same probabilities (i.e., $P_{10} = P_{01}$), it follows that $P_{00} = 1 - P_{11} - P_{10} - P_{01} = 1 - P_{11} - 2P_{10}$ so that there are only two unique cells. The other cells can be obtained if P_{11} and P_{10} are known. This means that differentiation of (3.24) will yield only two unique likelihood equations. Two equations are not sufficient for obtaining a solution for three unknown parameters. Thus, no unique maximum exists for this likelihood and the likelihood is said to be *nonidentifiable.* (Identifiability will be discussed in some detail in Section 5.1.) A necessary condition for a model to be identifiable is that the number of parameters to be estimated must not exceed the number of degrees of freedom for the data table. Since Table 3.2 has only 2 degrees of freedom (or two unique likelihood equations) and there are three unknown parameters, the model is nonidentifiable.

The likelihood can be made identifiable by reducing the number of parameters by one by either (1) setting one of the error probabilities to 0 or (2) equating two of the three parameters; for example, setting $\theta = \phi$. By reducing the number of parameters to 2, the model becomes *just identified*; thus, the number of parameters and the number of degrees of freedom (unique likelihood equations) are the same. Another term that is often used to describe this situation is to say that the model is *fully saturated* (in other words, there are zero model degrees of freedom).

To illustrate, let us set $\phi = 0$, which, as discussed above, might be a plausible assumption for some data; in particular, for sensitive questions where the probability of false positive reports is negligible. Then (3.23) becomes

$$\begin{aligned} P_{11} &= \pi(1-\theta)^2 \\ P_{10} &= \pi(1-\theta)\theta \\ P_{01} &= P_{10} \\ P_{00} &= \pi\theta^2 + (1-\pi) \end{aligned} \tag{3.25}$$

Setting these probabilities equal to the observed cell proportions and solving for the parameters, we obtain

$$\begin{aligned} \hat{\theta} &= \frac{b+c}{2a+b+c} \\ \hat{\pi} &= \frac{(2a+b+c)^2}{4an} \end{aligned} \tag{3.26}$$

The values of a, b, c, and n can be substituted into (3.26) to obtain the MLEs for this model.

Another possible constraint is to set $\phi = \theta = \lambda$, say, which is satisfied when false positives and false negatives are equally probable. This constraint seems implausible for many applications. But it could be reasonable in the situation where a poorly written question confuses respondents to the point where they might just guess its meaning and respond accordingly. In that case, persons who are true positives are equally likely to guess correctly as persons who are true negatives. This means that true positives and negatives are misclassified at about the same rate. Although this model has limited utility, let us proceed to illustrate how this constraint would be applied for the ML estimation method.

Table 3.3 Contrived Interview–Reinterview Data

	$y_2 = 1$	$y_2 = 0$
$y_1 = 1$	1000	50
$y_2 = 0$	100	500

The cell probabilities under this assumption are

$$
\begin{aligned}
P_{11} &= \pi(1-\lambda)^2 + (1-\pi)\lambda^2 \\
P_{10} &= \pi(1-\lambda)\lambda + (1-\pi)(1-\lambda)\lambda = (1-\lambda)\lambda \\
P_{01} &= P_{10} \\
P_{00} &= \pi\lambda^2 + (1-\pi)(1-\lambda)^2
\end{aligned}
\tag{3.27}
$$

Differentiation of the likelihood is not required to determine the MLEs for π and λ since a closed-form solution exists. First, an expression for λ can be obtained by solving the quadratic equation for P_{10}. This estimator of λ can then be substituting this into the expression for P_{11}. Solving for π yields an expression for the MLE of π. Alternatively, a numerical method can be used to obtain the MLEs by maximizing the likelihood with respect to λ and π, as will be illustrated next.

A simple method can be used for maximizing the likelihood in (3.24) with respect to λ and π using the cell probabilities in (3.27) and the values of a, b, and c in Table 3.3. The first step is to express the likelihood as a function of λ and π and then to compute the value of this likelihood for all possible values for λ and π using the so-called *grid search* method. The MLEs are the values of λ and π for which the likelihood kernel $\ell(\pi,\lambda)$ is largest.

The likelihood kernel $\ell(1000,50,100,500 \mid \pi,\lambda)$ is given by

$$
\begin{aligned}
\ell(\pi,\lambda) = &\ 1000\ln\left[\pi(1-\lambda)^2 + (1-\pi)\lambda^2\right] \\
&+ (100+50)\ln[(1-\lambda)\lambda] \\
&+ 500\left[\pi\lambda^2 + (1-\pi)(1-\lambda)^2\right]
\end{aligned}
\tag{3.28}
$$

This function is then evaluated for values of λ and π in the range [0,1] in steps of 0.01 or some other suitable step size depending on the numerical precision desired. Table 3.4 shows a portion of the grid used in determining the estimates. The maximum value of $\ell(\pi,\lambda)$ is –1562.9, which occurs at $\hat{\lambda} = 0.05$ and $\hat{\pi} = 0.67$. Greater precision in the estimates can be obtained by reducing the step size in the grid search to something less than 0.01.

Table 3.4 Illustration of the Grid Search Method of Maximum-Likelihood Estimation

	λ									
π	0.01	0.02	0.03	0.04	0.05	0.06	0.07	0.08	0.09	0.1
0.57	−1706.5	−1634.3	−1605.4	−1594.4	−1593.2	−1598.3	−1607.8	−1620.6	−1635.9	−1653.3
0.58	−1700.8	−1628.6	−1599.7	−1588.7	−1587.5	−1592.6	−1602.1	−1614.8	−1630.0	−1647.3
0.59	−1695.8	−1623.6	−1594.7	−1583.6	−1582.4	−1587.4	−1596.9	−1609.5	−1624.7	−1641.9
0.6	−1691.3	−1619.1	−1590.2	−1579.1	−1577.8	−1582.8	−1592.2	−1604.8	−1620.0	−1637.1
0.61	−1687.5	−1615.2	−1586.3	−1575.2	−1573.9	−1578.8	−1588.2	−1600.7	−1615.8	−1632.8
0.62	−1684.2	−1611.9	−1583.0	−1571.8	−1570.5	−1575.4	−1584.7	−1597.2	−1612.2	−1629.1
0.63	−1681.5	−1609.2	−1580.3	−1569.1	−1567.7	−1572.6	−1581.8	−1594.3	−1609.2	−1626.0
0.64	−1679.5	−1607.2	−1578.2	−1567.0	−1565.6	−1570.4	−1579.6	−1591.9	−1606.8	−1623.5
0.65	−1678.0	−1605.7	−1576.7	−1565.5	−1564.1	−1568.8	−1577.9	−1590.2	−1605.0	−1621.6
0.66	−1677.3	−1605.0	−1575.9	−1564.7	−1563.2	−1567.9	−1576.9	−1589.1	−1603.8	−1620.4
0.67	−1677.1	−1604.8	−1575.8	−1564.5	**−1562.9**	−1567.6	−1576.6	−1588.7	−1603.2	−1619.7
0.68	−1677.7	−1605.4	−1576.3	−1565.0	−1563.4	−1568.0	−1576.9	−1588.9	−1603.4	−1619.7
0.69	−1679.0	−1606.6	−1577.5	−1566.1	−1564.5	−1569.1	−1577.9	−1589.8	−1604.2	−1620.4
0.70	−1681.0	−1608.6	−1579.5	−1568.1	−1566.4	−1570.8	−1579.6	−1591.4	−1605.6	−1621.7

Although the data in Table 3.3 are completely contrived, they could have been produced by an interview–reinterview process having an overall error probability of 0.05 and population proportion of 67%. Since the model is fully saturated, these parameter estimates fit the data perfectly. In addition, there is no measure of lack of fit for the model or other means of testing whether the model is statistically acceptable. Since the data are contrived, there is no *correct* model underlying the data. However, we can say that the models above *could have* generated the data; that is, the model and the data are consistent.

For models involving more than just a few parameters, the grid search method is extremely inefficient and complex. Other much more efficient methods include the scoring algorithm (Haberman 1979), Newton–Raphson (Everitt and Hand 1981; Vermunt 1997a), and the expectation–maximization (EM) algorithm (Hartley 1958; Dempster et al. 1977; McLachlan and Krishnan 1997) are available and work much more efficiently. The latter two methods are more popular; the EM algorithm is the most widely used. The Newton–Raphson method is more complex to program and numerically intensive since it involves numerical differentiation of the likelihood to obtain the Hessian matrix (i.e., the matrix of second order partial derivatives with respect to parameter pairs). Its advantage is that it produces the variance–covariance matrix of the parameters as a byproduct of the estimation process since this is simply the inverse of the information matrix (i.e., –1 times the Hessian matrix). The EM algorithm is much simpler to program since it does not require derivatives. But neither does it produce standard errors of the estimates as a byproduct of the maximization process. This method is discussed in greater detail in the next section.

3.2.2 The EM Algorithm for Two Measurements

Most commercially available software used for estimating the parameters of probability models with missing or unobserved data implements the *expectation–maximization* (EM) algorithm. This algorithm is a very general numerical analysis procedure that can be used to estimate all the models that will be considered in this book. For that reason, it is important that we understand a little about how it works, especially for the advanced user of LCA. However, the casual reader may wish to skip this section since it is quite technical and not critical for acquiring a basic understanding of applied LCA.

The EM algorithm is an iterative optimization program that alternates systematically between an expectation (E) step and a maximization (M) step until some specified convergence criterion is reached. The E-step computes a *complete data* table using starting values for the parameters that are supplied by the user or by some other process to fill in the missing information. This complete data table leads to the complete data likelihood which, in the M-step, can now be maximized to produce new values of the parameters. These parameters then replace the original starting values and a new E-step begins. This process is repeated many, perhaps thousands, of times until the estimates

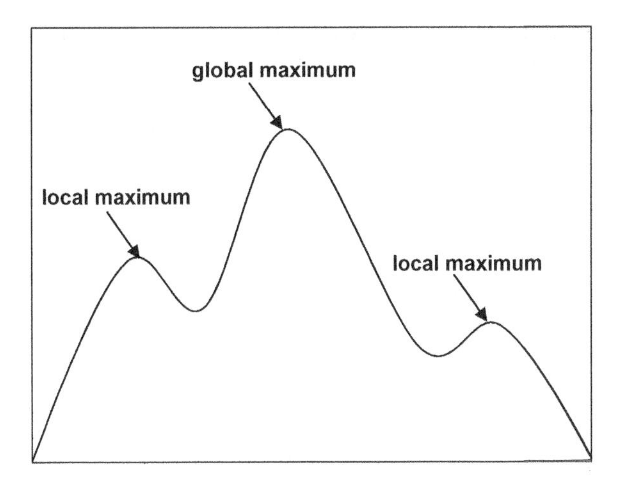

Figure 3.3 Local and global maxima for a likelihood function.

from two successive M-steps differ by no more then the limits specified by the convergence criterion. At that point, the algorithm is then said to have *converged*.

The choice of starting values for the model parameters is usually arbitrary (e.g., selected completely at random from the parameter space). Sometimes, however, implausible starting values can lead to a *local maximum* for the likelihood and other convergence issues. At a local maximum, the set of parameter estimates maximizes the likelihood within a small neighborhood or region surrounding the solution set (like cresting a small hill as one travels up a mountain road). There may be many local maxima and, depending on the starting values supplied to it, the EM algorithm can produce many different estimates of the same parameter. By definition, MLEs are the estimates corresponding to the maximum of all local maxima, called the *global maximum*. The global maximum is the unique set of estimates that maximizes the likelihood across all possible values of the parameters, that is, over the entire parameter space (see Figure 3.3).

Finding the global maximum can sometimes be difficult, requiring numerous (perhaps 100 or more) repetitions of the algorithm using very different starting values for the parameters at each repetition. If the same maximum is obtained for the different starting values, that maximum is quite likely the global maximum and the corresponding solution set can be regarded as the MLE. On the other hand, if more than one maximum is obtained from different starting values, then one or more local maxima exist. The usual approach is to take the solution corresponding to the maximum of the local maxima found by repeated runs of the EM algorithm with different starting values. However, this approach does not always yield the MLE. These issues are discussed in greater detail in Section 5.1.6.

To illustrate the EM algorithm, we shall apply the procedure to the model in (3.28) to find the MLEs for π and λ. In this model, the parameters θ and ϕ are constrained to be equal to λ. Let $\pi^{(1)}$ and $\lambda^{(1)}$ denote initial starting values for π and λ, respectively, selected by some arbitrary (or random) process:

E-Step. Compute the complete data table (i.e., the three-way table cross-classifying y_1, y_2, and the missing variable μ) using the starting values of the parameters. We do this by using the model to obtain an expression for the cell probabilities P_{jkl}, that is, the proportion in cell $y_1 = j$, $y_2 = k$, and $\mu = l$ for j, k, and $l = 0,1$ of the complete data table. These probabilities can be expressed as

$$\begin{aligned} P_{jkl} &= \Pr(y_1 = j, y_2 = k, \mu = l) \\ &= \Pr(y_1 = j, y_2 = k \mid \mu = l)\Pr(\mu = l) \\ &= \Pr(y_1 = j \mid \mu = l)\Pr(y_2 = k \mid \mu = l)\Pr(\mu = l) \\ &= \begin{cases} (\lambda)^2 \pi^l (1-\pi)^{(1-l)} & \text{for } j \neq l \text{ and } k \neq l \\ (\lambda)(1-\lambda)\pi^l (1-\pi)^{(1-l)} & \text{for } j \neq l \text{ or } k \neq l \\ (1-\lambda)^2 \pi^l (1-\pi)^{(1-l)} & \text{for } j = l \text{ and } k = l \end{cases} \end{aligned} \tag{3.29}$$

Now define $P_{l|jk} = (P_{jkl}/P_{jk})$, where $P_{jk} = \Sigma_l P_{jkl}$. Note that P_{jk} are the expected proportions of the observed two-way table. Let $P_{l|jk}^{(1)}$ denote $P_{l|jk}$ after substituting $\pi^{(1)}$ for π and $\lambda^{(1)}$ for λ in (3.29) and compute the expected cell frequencies

$$f_{jkl}^{(1)} = P_{l|jk}^{(1)} n_{jk} \tag{3.30}$$

M-Step. Compute new values of λ and π by applying ML estimation to the complete data table. It can be shown that the complete data MLEs at iteration $t + 1$ are given by

$$\begin{aligned} \lambda^{(t+1)} &= \frac{1}{n}\left(f_{110}^{(t)} + f_{001}^{(t)} + \frac{f_{101}^{(t)} + f_{011}^{(t)} + f_{100}^{(t)} + f_{010}^{(t)}}{2} \right) \\ \pi^{(t+1)} &= \frac{1}{n}\sum_j \sum_k f_{jk1}^{(t)} \end{aligned} \tag{3.31}$$

Armed with new values of λ and π, denoted by $\lambda^{(2)}$ and $\pi^{(2)}$, we repeat the E-step to obtain a new complete data table with cell frequencies $f_{jkl}^{(2)}$ by applying (3.29) and then (3.30) with the new parameter values. We then repeat the M-step to obtain $\lambda^{(3)}$ and $\pi^{(3)}$, another E-step, and so on. This iterative process continues until $|\lambda^{(t+1)} - \lambda^{(t)}| \leq \varepsilon$ and $|\pi^{(t+1)} - \pi^{(t)}| \leq \varepsilon$ for some tolerance level ε and iteration t.

To illustrate, let us choose arbitrary starting values $\pi^{(1)} = 0.1$ and $\lambda^{(1)} = 0.1$. (These values could also be chosen at random.) The E-step uses these values to complete Table 3.2 by adding the cross-classification of μ_i to the $y_1 \times y_2$ cross-classification. A single interation will be performed to show the computations.

E-Step. Compute the expected frequencies in (3.30). This requires that we first compute the joint probabilities P_{jki} using (3.29) and inserting $\pi^{(1)}$ and $\lambda^{(1)}$ as follows:

$$
\begin{aligned}
P_{111} &= (1-\lambda)^2\pi = (1-0.1)^2 0.1 = 0.081 \\
P_{101} &= \lambda(1-\lambda)\pi = 0.1(1-0.1)0.1 = 0.009 \\
P_{011} &= (1-\lambda)\lambda\pi = (1-0.1)(0.1)0.1 = 0.009 \\
P_{001} &= \lambda^2(1-\pi) = (0.1)^2 0.1 = 0.081 = 0.001 \\
P_{110} &= \lambda^2(1-\pi) = (0.1)^2(1-0.1) = 0.009 \\
P_{100} &= (1-\lambda)\lambda(1-\pi) = (1-0.1)(0.1)(1-0.1) = 0.081 \\
P_{010} &= \lambda(1-\lambda)(1-\pi) = 0.1(1-0.1)(1-0.1) = 0.081 \\
P_{000} &= (1-\lambda)^2(1-\pi) = (1-0.1)^2(1-0.1) = 0.081 = 0.729
\end{aligned}
\tag{3.32}
$$

Next, the conditional probabilities $P_{l|jk}$ are computed using the formulas in (3.32). Computations for just half of the table are shown. The second half is computed similarly.

$$
\begin{aligned}
P_{1|11} &= \frac{P_{111}}{P_{11}} = \frac{0.081}{0.081+0.009} = 0.9 \\
P_{1|10} &= \frac{P_{101}}{P_{10}} = \frac{0.009}{0.009+0.081} = 0.1 \\
P_{1|01} &= \frac{P_{011}}{P_{01}} = \frac{0.009}{0.009+0.081} = 0.1 \\
P_{1|00} &= \frac{P_{001}}{P_{00}} = \frac{0.001}{0.001+0.729} = 0.00137
\end{aligned}
\tag{3.33}
$$

Finally, the cell frequencies are computed using (3.30). Again, the computations for only the first half of the table are shown:

Table 3.5 Iteration 1 of the E-Step

	$\mu = 1$		$\mu = 0$	
	$y_2 = 1$	$y_2 = 0$	$y_2 = 1$	$y_2 = 0$
$y_1 = 1$	900.0	5.0	100.0	45.0
$y_1 = 0$	10.0	0.7	90.0	499.3

$$
\begin{aligned}
f_{111}^{(1)} &= P_{111|1}^{(1)} n_{11} = 0.9(1000) = 900 \\
f_{101}^{(1)} &= P_{101|1}^{(1)} n_{10} = 0.1(50) = 5 \\
f_{011}^{(1)} &= P_{011|1}^{(1)} n_{01} = 0.1(100) = 10 \\
f_{001}^{(1)} &= P_{001|1}^{(1)} n_{00} = 0.00137(500) = 0.685
\end{aligned}
\tag{3.34}
$$

The second half is computed similarly. These computations yield the data in Table 3.5, thus completing the first iteration of the E-step.

M-Step. Using the completed data in Table 3.5, we can now compute MLEs for π and λ. From Table 3.5, $f_{111}^{(1)} = 900.0$, $f_{110}^{(1)} = 100.0$, $f_{101}^{(1)} = 5.0$, $f_{100}^{(1)} = 45.0$, $f_{011}^{(1)} = 10.0$, $f_{010}^{(1)} = 90.0$, $f_{001}^{(1)} = 0.7$, and $f_{000}^{(1)} = 499.3$. Now, substituting these values in (3.31) with $n = 1650$, we obtain new values of the parameters:

$$
\begin{aligned}
\lambda^{(2)} &= \tfrac{1}{1650}\left(100 + 0.7 + \frac{5.0 + 10.0 + 90.0 + 45.0}{2}\right) = 0.106 \\
\pi^{(2)} &= \tfrac{1}{1650}(900.0 + 5.0 + 10.0 + 0.7) = 0.555
\end{aligned}
\tag{3.35}
$$

This completes the first iteration of the M-step as well as one E–M cycle of the algorithm.

Repeating the E-step and M-step with these new values of λ and π in (3.35), we obtain $\lambda^{(3)} = 0.058$ and $\pi^{(3)} = 0.654$. The details of these computations will not be shown but is left as an exercise for the reader. After seven cycles, the EM algorithm converges, with the difference between the parameters values for two consecutive E–M cycles no larger than $\varepsilon = 10^{-6}$. At iteration 7, the parameter values are $\hat{\lambda} = \lambda^{(7)} = 0.0477$ and $\hat{\pi} = \pi^{(7)} = 0.668$. We can then say these estimates are the MLEs for the "incomplete" data table shown in Table 3.3 under the model.

A general formula for the EM algorithm for any probability model will be developed in the next chapter. The EM algorithm has been implemented in a number of software packages for maximizing likelihoods such as the one in

(3.22). One that is freely available (currently available from `http://www.uvt.nl/faculteiten/fsw/organisatie/departementen/mto/software2.html`) and easy to use is ℓEM (Vermunt 1997b), which is used extensively for illustrations in this book. Other population commercial software packages are PANMARK, MPlus, and Latent GOLD. These packages and others will be discussed further in Section 4.1.3.

In the next section, we discuss an important model that can be applied when only two measurements $\mathcal{Y}$ are available. This case is important in survey evaluation since situations in which three or more measurements are available are extremely rare. This model, referred to as the *Hui–Walter model*, requires that two groups be found that are believed to have different values of π but the same values of θ and ϕ. Estimation then proceeds under these constraints. This is the so-called Hui–Walter model.

3.3 HUI–WALTER MODEL FOR TWO DICHOTOMOUS MEASUREMENTS

3.3.1 Notation and Assumptions

An important method for estimating π, θ, and ϕ when only two measurements of the variable $\mathcal{Y}$ are available is due to Hui and Walter (1980). This method was developed for evaluating medical diagnostic tests in the absence of any gold standard measurements. The method is quite general in that is does not require that the two measurements, y_1 and y_2, are parallel although local independence is still required. Another key assumption for the method is that the population can be divided into two groups (say, males and females) such that the misclassification probabilities, θ and ϕ, are the same for both groups while π differs between groups. With these constraints, the model is just identified and ML estimation can be applied to estimate the model parameters.

Before considering this model further, a new system notation is introduced that will be used for the remainder of the book. This notation, which was developed by Leo Goodman, is more flexible with regard to the number of variables in the model and their number of categories. Models of much greater complexity than have thus far been considered can be handled quite simply with this notation.

Let X_i denote the true value (previously denoted by μ_i) for the ith unit. As a simplification, the subscript i will be dropped since we are not particularly concerned about the values associated with any particular unit. Instead, an arbitrary true positive in the population is denoted by $X=1$ and an arbitrary true negative, by $X=2$. Note that now we use the digit "2" to denote a negative where previously the digit "0" was used. Let A and B denote the first and second measurements, respectively (previously denoted by y_1 and y_2), which also can take the values 1 for a positive classification and 2 for a negative classification.

Next, we introduce a third variable, referred to as a *grouping variable*, denoted by G, to identify an individual's membership in two or more population subgroups. For example, G could denote the respondent's gender with $G=1$ for males and $G=2$ for females. Grouping variables will be discussed in considerable detail in the next chapter. However, for now, we use them strictly as a device for achieving an identifiable model with only two measurements.

The parameters of the model are probabilities associated with either a latent class or a *manifest* (i.e., observed) variable and will be denoted by the symbol π with a superscript to denote the variable and a subscript to denote the value of the variable. Thus, π_1^X denote $\Pr(X=1)$ or the population prevalence probability (previously denoted by π with no subscripts or superscripts). Likewise, π_1^G and π_2^G denote the proportion of the population belonging to subgroups 1 and 2, respectively. These two probabilities can also be written as π_g^G, $g=1{,}2$. For the classifier variables, the marginal probabilities are denoted by π_a^A and π_b^B and the observed cell probabilities by π_{ab}^{AB}, where $a,b=1{,}2$. Extending this notation, we can write the conditional probability $\Pr(X=1 \mid G=g)$ as $\pi_{1|g}^{X|G}$, which is the prevalence probability for group $G=g$ and $\pi_{2|g}^{X|G}=1-\pi_{1|g}^{X|G}$ is the negative value prevalence probability. The conditional probabilities are $\pi_{a|x}^{A|X}$ and $\pi_{b|x}^{B|X}$ for the interview and reinterview, respectively, and are referred to as *response probabilities* in the LCA literature. In our applications, they are called *error probabilities* since $\pi_{1|2}^{A|X}$ and $\pi_{2|1}^{A|X}$ are the same as ϕ (false-positive probability) and θ (false-negative probability) in the old notation. Now, however, we have corresponding error probabilities for B denoted by $\pi_{1|2}^{B|X}$ and $\pi_{2|1}^{B|X}$. More generally, we can write

$$\pi_{a|xg}^{A|XG}=\Pr(A=a \mid X=x,G=g) \quad \text{and} \quad \pi_{b|xg}^{B|XG}=\Pr(B=b \mid X=x,G=g) \tag{3.36}$$

to denote the conditional classification probabilities for group g where $g=1{,}2$.

So far, we have considered only two class models with dichotomous manifest variables. However, the notation is easily generalized to any number of classes for X and categories for G, A and B. This flexibility will be exploited in Section 3.5 and following sections, where polytomous measures are considered in some detail.

Consider the three-way cross-classification of the group G and the classifiers A and B shown in Table 3.6 where n_{gab} denotes the number of observations classified in cell $G=g$, $A=a$, and $B=b$, for $g,a,b=1{,}2$. Under the assumption of simple random sampling with replacement (or without replacement from a very large population), the vector of cell counts denoted by $\mathbf{n}=[n_{111}\ldots n_{222}]'$ is distributed as a multinomial distribution with associated probabilities $\boldsymbol{\pi}=[\pi_{111}^{GAB},\pi_{112}^{GAB},\ldots,\pi_{222}^{GAB}]'$ where π_{gab}^{GAB} is the joint probability that a sample member is classified into the cell (g,a,b). The loglikelihood kernel corresponding to Table 3.4 is

$$\ell(\boldsymbol{\pi} \mid \mathbf{n})=\sum_{g=1}^{2}\sum_{a=1}^{2}\sum_{b=1}^{2} n_{gab}\log(\pi_{gab}^{GAB}) \tag{3.37}$$

Table 3.6 Data Display for Hui–Walter Method

	$G = 1$			$G = 2$	
	$B = 1$	$B = 2$		$B = 1$	$B = 2$
$A = 1$	n_{111}	n_{112}	$A = 1$	n_{211}	n_{212}
$A = 2$	n_{121}	n_{122}	$A = 2$	n_{221}	n_{222}

Table 3.6 has eight cells providing a total of 7 degrees of freedom for parameter estimation because the sum of the eight cell counts must be n. At most, only seven parameters can be estimated yet the unrestricted model contains 11 parameters. These parameters are $\pi_1^G, \pi_{1|1}^{X|G}, \pi_{1|2}^{X|G}, \pi_{1|21}^{A|XG}, \pi_{1|22}^{A|XG}, \pi_{2|11}^{A|XG}, \pi_{2|12}^{A|XG}, \pi_{1|21}^{B|XG}, \pi_{1|22}^{B|XG}, \pi_{2|11}^{B|XG}$, and $\pi_{2|12}^{B|XG}$. Hui and Walter determined a meaningful and plausible way to reduce the number of parameters to seven such that a restricted model can be estimated. They assumed the following:

A1. The misclassification probabilities for both measures are the same in both groups:

$$\pi_{a|x1}^{A|XG} = \pi_{a|x2}^{A|XG} = \pi_{a|x}^{A|X}, \text{ say} \tag{3.38}$$

$$\pi_{b|x1}^{B|XG} = \pi_{b|x2}^{B|XG} = \pi_{b|x}^{B|X}, \text{ say} \tag{3.39}$$

for $a \neq x$ and $b \neq x$.

A2. The prevalence probability differs by group:

$$\pi_{1|1}^{X|G} \neq \pi_{1|2}^{X|G} \tag{3.40}$$

A3. A and B are *locally independent* within each group G:

$$\pi_{ab|xg}^{AB|XG} = \pi_{a|xg}^{A|XG} \pi_{b|xg}^{B|XG} \tag{3.41}$$

Assumption A3 is often restated as *uncoditional* rather than *conditional* (i.e., within group g) local independence. The unconditional local independence assumption can be stated as follows:

A3′. A and B are locally independent; meaning that: $\pi_{ab|x}^{AB|X} = \pi_{a|x}^{A|X} \pi_{b|x}^{B|X}$.

In general, assumption A3′ is unnecessarily strict since it implies A3 but the converse is not true. It can be shown that, for the Hui–Walter model, A3 and A3′ are equivalent because of A1. Note that it is not acceptable to define G

by a random assignment of units to two groups. Randomly generated groups would satisfy A1 but not A2. With these assumptions, there are only seven parameters (viz., π_1^G, $\pi_{1|1}^{X|G}$, $\pi_{1|2}^{X|G}$, $\pi_{1|2}^{A|X}$, $\pi_{2|1}^{A|X}$, $\pi_{1|2}^{B|X}$, $\pi_{2|1}^{B|X}$), and Hui and Walter proved the model is identifiable.[3] Since the model is fully saturated, there are no degrees of freedom for testing model fit; however, standard errors of the model estimates can still be obtained.

Whether the Hui–Walter assumptions are satisfied for any given application is contingent on the choice of grouping variable. This can be a problem in applying the method because there is no statistically rigorous way to test whether a particular choice of G satisfies the assumptions. Instead, G is based largely on subjective criteria, by considering whether the assumptions are plausible for a particular choice of grouping variable. In some cases, a plausible choice for G may not be available precluding use of the method.

Assumption 1 (A1) can be checked to some extent by comparing estimates of reliability for each group under the assumption that A and B are parallel measurements using the methods in Chapter 2. If Assumption 1 holds, the reliabilities for each group should be approximately equal provided the prevalence probabilities are not too different. However, as noted in Section 3.1.2, two groups may have very nearly the same reliabilities and still have very different misclassification probabilities if their prevalence probabilities are quite different, so this comparison is at best a rough guideline.

If classification error is small for A, then $\pi_a^A \doteq \pi_x^X$ for $a = x$. The same argument holds for B. Therefore, Assumption 2 (A2) can be checked to some extent by testing whether $\pi_{1|1}^{A|G} = \pi_{1|2}^{A|G}$ and/or $\pi_{1|1}^{B|G} = \pi_{1|2}^{B|G}$. Although rejecting these hypotheses is no assurance that $\pi_{x|1}^{X|G} \neq \pi_{x|2}^{X|G}$, failing to reject could suggest a poor choice of grouping variable.

To obtain the maximum-likelihood estimates of the model parameters, we begin by rewriting (3.37) in terms of the seven parameters. Note that the probability of classifying a sample unit into cell (g,a,b) can be written as

$$\begin{aligned}\Pr(G=g, A=a, B=b) &= \Pr(G=g)\Pr(A=a, B=b \mid G=g) \\ &= \pi_g^G \pi_{ab|g}^{AB|G}\end{aligned} \tag{3.42}$$

Further, we can write

$$\pi_{ab|g}^{AB|G} = \pi_{1|g}^{X|G}\pi_{ab|1g}^{AB|XG} + \pi_{2|g}^{X|G}\pi_{ab|2g}^{AB|XG} \tag{3.43}$$

Now, invoking local independence, we write $\pi_{ab|xg}^{AB|XG}$ for $x = 1,2$ and $g = 1,2$ as

[3]As will be discussed in Chapter 4, limiting the number of parameters to be no more than the number of degrees of freedom is a necessary but not sufficient condition for estimability of this model. See also Goodman (1974a) for a discussion of this topic.

$$\pi_{ab|xg}^{AB|XG} = \pi_{x|g}^{X|G} \pi_{a|xg}^{A|XG} \pi_{b|xg}^{B|XG} \tag{3.44}$$

Combining (3.43) and (3.44) into (3.42), we obtain

$$\pi_{gab}^{GAB} = \pi_g^G \left(\pi_{1|g}^{X|G} \pi_{a|1g}^{A|XG} \pi_{b|1g}^{B|XG} + \pi_{2|g}^{X|G} \pi_{a|2g}^{A|XG} \pi_{b|2g}^{B|XG} \right) \tag{3.45}$$

Substituting this expression for π_{gab}^{GAB} in the loglikelihood in (3.37), we can obtain an expression for the loglikelihood kernel in terms of the seven model parameters:

$$\ell(\boldsymbol{\pi} \mid \mathbf{n}) = \sum_{g=1}^{2} \sum_{a=1}^{2} \sum_{b=1}^{2} n_{gab} \left[\log \pi_g^G + \log \left(\pi_{1|g}^{X|G} \pi_{a|1g}^{A|XG} \pi_{b|1g}^{B|XG} + \pi_{2|g}^{X|G} \pi_{a|2g}^{A|XG} \pi_{b|2g}^{B|XG} \right) \right] \tag{3.46}$$

The MLEs can be obtained numerically using the EM algorithm or other methods discussed in the last section. Closed-form expressions of the estimates were derived by Hui and Walter (1980) as follows.

Let $p_{ab|g} = n_{gab}/n_g$, where $n_g = \sum_{a,b} n_{gab}$; that is, $p_{ab|g}$ denotes the proportion of the group g sample in cell (a,b). Further, let $p_{a+|g} = p_{a1|g} + p_{a2|g}$ and $p_{+b|g} = p_{1b|g} + p_{2b|g}$. Now, define the following quantities:

$$\begin{aligned}
\mathcal{C}_{1|2}^{A|X} &= p_{1+|1} p_{+1|2} - p_{+1|1} p_{1+|2} + p_{11|2} - p_{11|1} \quad \text{and} \\
\mathcal{C}_{1|2}^{B|X} &= p_{1+|2} p_{+1|1} - p_{+1|2} p_{1+|1} + p_{11|2} - p_{11|1} \\
\mathcal{C}_{2|1}^{A|X} &= p_{+2|1} p_{2+|2} - p_{2+|1} p_{+2|2} + p_{22|1} - p_{22|2} \quad \text{and} \\
\mathcal{C}_{2|1}^{B|X} &= p_{+2|2} p_{2+|1} - p_{2+|2} p_{+2|1} + p_{22|1} - p_{22|2} \\
\mathcal{D} &= \pm \Big[\left(p_{1+|1} p_{+1|2} - p_{1+|2} p_{+1|1} + p_{11|1} - p_{11|2} \right)^2 \\
&\qquad - 4 \left(p_{1+|1} - p_{1+|2} \right) \left(p_{11|1} p_{+1|2} - p_{11|2} p_{+1|1} \right) \Big]^{1/2} \\
\mathcal{E}_A &= p_{+1|2} - p_{+1|1} \quad \text{and} \quad \mathcal{E}_B = p_{1+|2} - p_{1+|1}
\end{aligned} \tag{3.47}$$

Hui–Walter provide the following MLEs for the model parameters:

$$\begin{aligned}
\hat{\pi}_{a|x}^{A|X} &= \frac{\left(\mathcal{C}_{a|x}^{A|X} + \mathcal{D} \right)}{2\mathcal{E}_A}, \quad a \neq x \\
\hat{\pi}_{b|x}^{B|X} &= \frac{\left(\mathcal{C}_{b|x}^{B|X} + \mathcal{D} \right)}{2\mathcal{E}_B}, \quad b \neq x \\
\hat{\pi}_{1|g}^{X|G} &= \frac{1}{2} + \frac{\left[p_{1+|g} \left(p_{+1|1} - p_{+1|2} \right) + p_{+1|g} \left(p_{1+|1} - p_{1+|2} \right) + p_{11|2} - p_{11|1} \right]}{2\mathcal{D}}
\end{aligned} \tag{3.48}$$

As Hui and Walter note, if either $\mathcal{E}_A$, $\mathcal{E}_B$, or $\mathcal{D}$ is 0, numerical methods must be used to obtain the estimates to avoid division by 0 in the closed-form formulas. Further, from the form of $\mathcal{D}$ in (3.47), two distinct sets of solutions exist depending on whether the positive or negative value of $\mathcal{D}$ is selected. In practical situations, both should be computed and one solution set will provide plausible estimates, and the other will not. Knowledge of what are plausible estimates is required to select the appropriate solution. This will be illustrated in the example in the next section.

Hui and Walter also provide formulas for estimating the standard errors of the estimates in (3.48). However, these will not be discussed here.

3.3.2 Example: Labor Force Misclassifications

Sinclair and Gastwirth (1996) and Biemer and Bushery (2001) applied the Hui–Walter method to data from the US Current Population Survey (CPS) Reinterview Program (the test–retest reinterview component) in order to evaluate the errors in classifying individuals as either "in the labor force" or "not in the labor force." In both analyses, respondent gender was the grouping variable. Table 3.7 provides a typical data table from the CPS 1995–1996 test–retest reinterview data. Since the reinterviews were conducted by field supervisors rather than interviewers, the assumption of parallel measurements does not seem plausible. In the next chapter, we show how the assumption of parallel measurements can be tested within the Hui–Walter framework. It will be shown that, for the data in Table 3.7, the assumption does not hold.

In Table 3.7, NLF denotes persons that are "not in the labor force" according to the CPS classification process. These are persons who are neither employed nor unemployed (i.e., neither looking for work nor laid off). LF means "in the labor force" and includes all persons classified as either employed or unemployed. The total sample size is 14,619, and the data are unweighted for this illustration.

Table 3.7 Interview–Reinterview Data for the 1995–1996 CPS

	Reinterview (B)	
Interview (A)	NLF	LF
	Males ($G = 1$)	
NLF	1841	176
LF	121	4722
	Females ($G = 2$)	
NLF	3160	212
LF	171	4216

To apply the Hui–Walter method to these data, let A and B denote the interview and reinterview classifications, respectively, and let X denote the latent (true) classifications. Thus, $A = 1$ denotes NLF and $A = 2$ denotes LF for the interview and $B = 1$ and $B = 2$ are defined analogously for the reinterview. The true, unobserved labor force status, X, is also coded analogously while $G = 1$ and $G = 2$ denote males and females, respectively. Corresponding to Assumptions 1 and 2 above, we assume that

1. Labor force participation rates differ by gender.
2. Males and females have the same probabilities of being misclassified as either NLF or LF by both the CPS and the reinterview survey.

Assumption 1 seems plausible for this application on the basis of subjective criteria. In addition, the gross difference rates are similar for males and females (4.3% vs. 4.9%), which provides some support for the assumption. It is widely accepted that males and females have different labor force participation rates, so Assumption 2 is satisfied and is consistent with the observed data on the basis of the rough test of this assumption described above. For example, using the interview classifications in Table 3.4, the NLF rate for males is 29% compared to about 43% for females. As noted previously, that the observed prevalence rates differ is no proof that Assumption 2 holds but it is nonetheless reassuring.

The ℓEM software (Vermunt 1997b) was used to fit the Hui–Walter model to the data in Table 3.7 via the EM algorithm. The ℓEM input statements are shown in Figure 3.4.[4] In the next chapter, we will provide the form of the E- and M-steps for this optimization. Of particular interest in this analysis are the conditional classification probabilities for both interview and reinterview; in particular, the error probabilities associated with measures A and B. From the results shown in Table 3.8 we see that the false-negative probability for A is estimated to be 4.0% [standard error (s.e.) = 1%]. This means that, of the persons "truly" NLF, the CPS main survey misclassifies about 4% as in the labor force (LF). This result is similar for the reinterview classification process, which misclassifies NLF persons at the rate of 3.5% (s.e. = 1.2%). For persons who are truly in the labor force (LF), the estimated error probabilities tend to be somewhat smaller. For the CPS interview the probability of misclassification as NLF is 2.3% (s.e. = 0.7%) but less than 1% for the reinterview (s.e. = 0.6%). The standard errors are quite small for this analysis because of the relatively large sample sizes.

This analysis suggests that the error probabilities for both CPS and the reinterview are not large, indicating that the surveys do fairly well in classify-

[4]Some familiarity with the ℓEM software is assumed here. The reader is encouraged to download the ℓEM instruction manual (Vermunt 1997b), which is available at `http://spitswww.uvt.nl/web/fsw/mto/lem/manual.pdf`, and become familiar with the simple commands used in this example.

```
lat 1
man 3
dim 2 2 2 2
lab X G A B
mod G
    X|G    *By including G, HW Assumption 1 is imposed
    A|X    *By omitting G, HW Assumption 2 is imposed for A
    B|X    *By omitting G, HW Assumption 2 is imposed for B
dat [1841 176 121 4722   * Data for Males from Table 3.7
     3160 212 171 4216]  * Data for Females from Table 3.7
**************************************************************
* Comment: As will be clearer later, there are a number of ways to specify this
* model in the LEM software. The above model statement is equivalent to both:
* mod XG  A|X   B|X
* or
* mod (GX AX BX}.
* The primary difference is the way conditional probabilities are reported in
* the LEM output.
* `Our preference for mod G X|G A|X B|X is because the conditional
* probabilities X|G are reported instead of the joint probabilities XG.
*******************************************************************************
```

Figure 3.4 ℓEM input statements for fitting the basic Hui–Walter model to the data in Table 3.7.

Table 3.8 Hui–Walter Estimates of Misclassification Probabilities with Standard Errors Shown in Parentheses

Classification	Estimated Misclassification Probabilities (in percent)	
Interview	$\pi_{1\|2}^{A\|X}$ = Pr (false positive)	2.3 (0.7)
	$\pi_{2\|1}^{A\|X}$ = Pr (false negative)	4.0 (1.0)
Reinterview	$\pi_{1\|2}^{B\|X}$ = Pr (false positive)	0.9 (0.6)
	$\pi_{2\|1}^{B\|X}$ = Pr (false negative)	3.5 (1.2)

ing persons as NLF or LF. False negatives (i.e., classifying persons who are truly NLF as LF) are more prevalent than false positves (i.e, classifying persons who are truly LF and NLF). For example, the CPS false-negative rate is about 4%, while the false-positive rate is about 2%. Reinterview classifications are slightly more accurate than the interview classifications, particularly for the false positives. The CPS interview false positive rate is 2.3 compared to 0.9 for the reinterview. The difference of 1.4 percentage points is not significant, however.

The Hui–Walter estimate of the true prevalence rate for males is $\hat{\pi}_{1|1}^{X|G} = 0.29$ and for females, $\hat{\pi}_{1|2}^{X|G} = 0.44$. This can be compared with the observed CPS prevalence rates for these two groups, computed directly from Table 3.4 (namely, $\hat{\pi}_{1|1}^{A|G} = 0.29$ and $\hat{\pi}_{1|2}^{A|G} = 0.43$). The bias in the CPS for males is estimated by $\hat{\pi}_{1|1}^{A|G} - \hat{\pi}_{1|1}^{X|G} = 0.0$ and for females by $\hat{\pi}_{1|2}^{A|G} - \hat{\pi}_{1|2}^{X|G} = 0.01$, which are obviously not statistically significant. Thus, we can say there is no evidence from this analysis of bias in the proportions of NLF for males and females. However, there is evidence of misclassification of persons who are not in the labor force as in the labor force; the misclassification rate is around 4% for both interview and reinterview.

It is worth reiterating that the validity of this analysis depends critically on the assumption that the Hui–Walter model assumptions hold. To the extent that the model assumptions are violated, the model estimates will be biased. The effects of model assumption violations on the estimates will be discussed in Section 3.4.3.

Also later in this chapter, an even more informative analysis of the CPS labor force classification process will be conducted using these data by expanding the number of labor force categories from two to three. First we present a useful variation on the Hui–Walter model to estimate mode bias in an application to the US National Health Interview Survey.

3.3.3 Example: Mode of Data Collection Bias

This illustration presents a useful modification of the Hui–Walter model for estimating mode effects in a split-ballot experiment when a subsample of respondents from each mode has been reinterviewed. This application, taken from Biemer (2001), compares the quality of data obtained by face-to-face interviewing with that obtained by telephone interviewing for the 1994 National Health Interview Survey (NHIS) (National Health Interview Survey 2008). Classification error was estimated separately for the two interview modes in order to determine which mode of interview provides the most accurate responses. The study was not originally designed with LCA in mind, but later it was discovered that various latent class models could be fit to the data by exploiting a Hui–Walter-type grouping variable in the analysis. The two modes under study were centralized computer-assisted telephone interviewing (CATI) and the usual NHIS production face-to-face interviewing. The study evaluated two sources of error: measurement error and nonresponse error. In this example, we consider some of the models that were used to evaluate the former error.

The study was conducted in two states: Texas (TX) and California (CA). Face-to-face interviews were conducted as part of the usual NHIS production survey in those states. Simultaneously, random-digit dialing (RDD) samples were selected in each state and interviewed using a supplement to the usual NHIS questionnaire called the Healthy People 2000 Supplement. Within 2 weeks of their first or original interview, both NHIS and RDD respondents were reinterviewed by telephone. All telephone interviews—the RDD interviews and the RDD and NHIS rreinterviews—were conducted in the same centralized telephone facility. In the RDD sample, the original interviewer was not allowed to conduct the reinterview. Table 3.9 shows the design of the study.

The NHIS response rate for the combined TX–CA sample was 81% for the main survey and 69% for the reinterview, while the RDD survey response

Table 3.9 Design of the NHIS RDD Experiment

Study Component	Description	Mode
NHIS in TX and CA only	NHIS supplement collected as part of the usual national NHIS sample in each state	Face-to-face
NHIS reinterview	Reinterview survey administered to approximately a half-sample of the NHIS respondents in each state	Telephone
RDD survey	Same NHIS information collected for an independently selected RDD sample in Texas and California	Telephone
RDD reinterview	Reinterview administered to an approximately 25% subsample of RDD respondents in each state	Telephone

rates were 60% for the main survey and 74% for the reinterview. Only cases that responded to both the interview and the reinterviews were retained in the analysis. Responses to 20 health characteristics were evaluated. Data for one characteristic (smoking in the home) are presented in Table 3.9 to illustrate the modeling approach. Full results are available in Biemer (2001).

When working with any latent variable model, model identifiability is a central concern. As such, a primary consideration in this analysis is the number of degrees of freedom available for estimation. Table 3.9 contains eight cells, which equates to only 6 degrees of freedom because there are two design constraints: (1) the NHIS sample must sum up to the NHIS sample size and, likewise (2) the four RDD cells must sum to the RDD sample size. Consequently, no more than six parameters can be estimated for the data in this table.

A key requirement of the Hui–Walter method is to identify a dichotomous grouping variable satisfying the Hui–Walter model constraints. The grouping variable that was used in this application is a dichotomous indicator for two survey samples (i.e., the NHIS and RDD). Let $G = 1$ denote the NHIS sample and $G = 2$ denote the RDD sample. As in previous examples, A denotes the interview classification and B the reinterview classification. The true prevalence rate for target population members in group g is $\pi_{1|g}^{X|G}$, and $\pi_{2|1g}^{A|XG}$ and $\pi_{1|2g}^{A|XG}$ are the false-negative and false-positive probabilities, respectively, for group g for the interview (A). Likewise, $\pi_{2|1g}^{B|XG}$ and $\pi_{1|2g}^{B|XG}$ are defined analogously for the reinterview (B). The joint probability $\pi_{ab|g}^{AB|G}$ is the probability a unit is group g is assigned to category a by the interview and category b by the reinterview. Thus, $n\pi_{ab|1}^{AB|G}$ are the expected cell counts for the NHIS interview–reinterview table and $n\pi_{ab|2}^{AB|G}$ are the corresponding RDD expected cell counts.

The Hui–Walter model assumes that the prevalence of the latent variable X differs by levels of G. Since both the NHIS and the RDD surveys are based on random samples of the same population, it would seem that the two prevalence probabilities (i.e., $\pi_{1|2}^{X|G}$ and $\pi_{1|1}^{X|G}$) do not differ. If so, the assumption of different prevalence probabilities will not hold and the Hui–Walter mode will not perform well. Fortunately, $\pi_{x|2}^{X|G}$ and $\pi_{x|1}^{X|G}$ may still differ in this application because the population responding to the NHIS may be quite different than the population responding to the RDD survey. There are two reasons for this. First, the response rates for the two surveys are quite different—about a 20 percentage point difference. This suggests that nonresponse bias may be larger for the RDD survey than for the NHIS. Futhermore, it is well known from the survey methods literature that persons who respond to face-to-face surveys are quite different from those who respond to RDD surveys for many characteristics [see, e.g., Groves (1989), Groves and Couper (1998), and Biemer and Lyberg (2003)]. RDD surveys tend to underrpresent persons living in small households, high-crime areas, persons with low income and low education levels, males, and several minority races. On the other hand, face-to-face interviews better represent these groups. Therefore, to the extent that the charactertics for this analysis are biased differentially because of unit nonresponse, this Hui–Walter assumption will be satisfied. If there is insufficient separation between the prevalences of the two populations, the model will be *weakly identifiable* (discussed in Section 5.1.2) and the estimates will be unstable. Thus, checking the identifiability of the models using the techniques described in the Chapters 4 and 5 is important in this analysis.

The Hui–Walter model also assumes that that the error probabilities are the same for the two groups defined by the variable G. In this example, the two groups are the NHIS sample in TX and CA and the RDD sample in these states. Since the NHIS and RDD modes differ, the assumption of equal error probabilities may not hold since face-to-face interviewing is likely to have different error probabilities than telephone interviewing. Indeed, this possibility is precisely what the study aims to investigate. However, it seems plausible that all interviews conducted by the same mode should have very similar error rates for any given survey item. In particular, all telephone interviews or reinterviews should have the same general error probabilities for a given item. Indeed, as we shall see, the validity of the results of this analysis hinges on the plausibility of this assumption. The primary model for the analysis assumes that error probabilities are the same for all interviews conducted by telephone regardless of whether it is the original interview or the reinterview.

In that regard, the model shall constrain the misclassification probabilities for the NHIS reinterview (measurement B and $g = 1$) to be equal to the corresponding misclassification probabilities for the RDD interview (measurement A and $g = 2$). These probabilities are also constrained to be equal to the corresponding probabilities for the RDD reinterview (measurement B and $g = 2$). Thus, we assume that misclassification probabilities for all interviews conducted by telephone form an *equivalence set of probabilities*. Let $\pi_{t|x}^{T|X}$

denote the error probability for $T = A$ or B, where T denotes telephone mode. The equality constraints can be written concisely as, say

$$\pi_{a|x2}^{A|XG} = \pi_{b|x2}^{B|XG} = \pi_{b|x1}^{B|XG} = \pi_{t|x}^{T|X} \tag{3.49}$$

where $\pi_{a|x2}^{A|XG}$ corresponds to the RDD interview, $\pi_{b|x2}^{B|XG}$ the RDD reinterview, and $\pi_{b|x1}^{B|XG}$ the NHIS reinterview. The misclassification probability for the NHIS interview is denoted by $\pi_{a|x1}^{A|XG}$.

In order to determine the joint likelihood of the data in Table 3.9, we need an expression for the expected cell counts for each cell of the GAB table. For the NHIS sample ($g = 1$), the probability of being classified in cell (a,b) of the GAB table is

$$\begin{aligned}\pi_{ab|1}^{AB|G} &= \pi_{1|1}^{X|G}\pi_{a|11}^{A|XG}\pi_{b|11}^{B|XG} + \pi_{2|1}^{X|G}\pi_{a|21}^{A|XG}\pi_{b|21}^{B|XG} \\ &= \pi_{1|1}^{X|G}\pi_{a|11}^{A|XG}\pi_{t|1}^{T|X} + \pi_{2|1}^{X|G}\pi_{a|21}^{A|XG}\pi_{t|2}^{T|X}\end{aligned} \tag{3.50}$$

This equation states that $\pi_{ab|1}^{AB|G}$ is the sum of two terms—the first corresponding to true positives ($X = 1$) and the second to true negatives ($X = 2$). The probabilities are, in turn, decomposed into classification probabilities for the interview (A) and reinterview (B). The reinterview classification probability is then replaced by the equivalent probability denoted by T. Similarly, for the RDD sample ($g = 2$), the corresponding probability is

$$\begin{aligned}\pi_{ab|2}^{AB|G} &= \pi_{1|2}^{X|G}\pi_{a|12}^{A|XG}\pi_{b|12}^{B|XG} + \pi_{2|2}^{X|G}\pi_{a|22}^{A|XG}\pi_{b|22}^{B|XG} \\ &= \pi_{1|2}^{X|G}\pi_{t|1}^{T|X}\pi_{t|1}^{T|X} + \pi_{2|2}^{X|G}\pi_{t|2}^{T|X}\pi_{t|2}^{T|X} \\ &= \pi_{1|2}^{X|G}(\pi_{t|1}^{T|X})^2 + \pi_{2|2}^{X|G}(\pi_{t|2}^{T|X})^2\end{aligned} \tag{3.51}$$

The likelihood kernel can be written as $\pi_{g=1}^{G}\pi_{ab|1}^{AB|G} + \pi_{g=2}^{G}\pi_{ab|2}^{AB|G}$, which, on combination of (3.50) with (3.51) and simplification, is

$$\pi_{gab}^{GAB} = \pi_1^G(\pi_{1|1}^{X|G}\pi_{a|11}^{A|XG}\pi_{t|1}^{T|X} + \pi_{2|1}^{X|G}\pi_{a|21}^{A|XG}\pi_{t|2}^{T|X}) + \pi_2^G\left[\pi_{1|2}^{X|G}(\pi_{t|1}^{T|X})^2 + \pi_{2|2}^{X|G}(\pi_{t|2}^{T|X})^2\right] \tag{3.52}$$

Note that the prevalence rates ($\pi_{1|1}^{X|G}$ and $\pi_{1|2}^{X|G}$) are assumed to differ because the response rates for the NHIS and the RDD datasets differed substantially; thus, the two achieved samples may represent different responding populations with different prevalence rates.

For the data in Table 3.10, the model in (3.52) is saturated since there are 6 degrees of freedom and six parameters: $\pi_{1|1}^{X|G}$, $\pi_{1|2}^{X|G}$, $\pi_{1|21}^{A|XG}$, $\pi_{2|11}^{A|XG}$, $\pi_{2|1}^{T|X}$, and $\pi_{1|2}^{T|X}$. The parameters represent, respectively, the prevalence probabilities for the NHIS and RDD responding populations, the error (false positive and false negative) probabilities for the face-to-face survey, and the error (false positive

Table 3.10 Typical Data Table for the 1994 NHIS Mode of Interview Evaluation[a]

	Does Anyone Smoke Inside the Home?				
	$G = 1$ (NHIS sample)			$G = 2$ (RDD sample)	
	$B = 1$	$B = 2$		$B = 1$	$B = 2$
$A = 1$	334	70	$A = 1$	282	20
$A = 2$	29	1233	$A = 2$	9	931

[a]NHIS interviews were face-to-face. All other interviews and reinterviews were by telephone.

```
* Group G=1 is the NHIS survey
* Group G=2 is the RDD survey
* A is the interview, B is the reinterview
* H-W Method Applied to NHIS Mode Effects Study
* Assume Group 1-B, Group 2-A and 2-B have same error rates
lat 1
man 3
dim 2 2 2 2
lab X G A B          *    G =1 for NHIS, G =2 for RDD
mod G
    X|G              *    P(X|G) differs by survey due to nonresponse
    A|GX eq2         *    For G=1 A = ff and B=CATI
    B|GX eq2         *    For G=2 A & B are both CATI
sta X|G [.3 .4 .7 .6]
des [0 1 0 2        * X=1 F-F false negative and RDD false negative
     3 0 4 0        * X=2 F-F false positive and RDD false positive
     0 2 0 2        * X=1 Reint: False negative same as RDD false negative
     4 0 4 0]       * X=2 Reint: False positive same as RDD false positive
dat [ 334 70 29 1233 282 20  9  931]
seed 485            * This seed will produce the estimates in Table 3.11
```

Figure 3.5 ℓEM input statements for fitting the model in (3.52).

and false negative) for all interviewing conducted by telephone. The ℓEM software was used to fit the model using the input statements in Figure 3.5. The constraints described above are imposed on the ℓEM algorithm through the `eq2` command and its assocated design matrix specified with the `des` command. The model constrains these error probabilities to be equal: the face-to-face reinterview false positive ($\pi_{1|21}^{B|XG}$), the RDD survey interview false positive ($\pi_{1|22}^{A|XG}$), and the RDD survey reinterview false positive ($\pi_{1|22}^{B|XG}$). Likewise, the face-to-face reinterview false-negative ($\pi_{2|11}^{B|XG}$), the RDD survey interview false negative ($\pi_{2|12}^{A|XG}$), and the RDD survey reinterview false-negative ($\pi_{2|12}^{B|XG}$) probabilities are constrained to be equal.

Maximizing the likelihood in (3.52) yielded the estimates of these parameters expressed as percentages, listed in Table 3.11.

According to the model, the NHIS smoking prevalence is 22.7% compared with 24.8% for the RDD survey. The difference 22.7 – 24.8 = –2.1 can be attributed to both sampling variance and nonresponse bias. No standard errors are produced by the software; however, the standard error can be computed

Table 3.11 Parameter Estimates for the Model in (3.52)[a]

$\pi_{1\|1}^{X\|G}$	$\pi_{1\|2}^{X\|G}$	$\pi_{1\|21}^{A\|XG}$	$\pi_{2\|11}^{A\|XG}$	$\pi_{1\|2}^{T\|X}$	$\pi_{2\|1}^{T\|X}$
22.7	24.8	4.3	7.9	4.3	0.0

[a]Standard errors are not provided by ℓEM for this model.

roughly by the simple random sampling formula, $\sqrt{p(1-p)/n}$, where p is a proportion and n is the sample size upon which p is based. Using this rough calculation of the standard error and the sample sizes in the table, the difference is not statistically significant. The model estimates of the false-negative probabilities are quite different for the two modes. The false-negative rate for the face-to-face interview is almost 8% while that for the telephone model is approximately 0. It should be noted that these estimates differ somewhat from those in Biemer (2001) since the model used here is somewhat simplified.

For the current model, the TX and CA population differences have been ignored. However, it is quite likely that the prevalence probabilities as well as the NHIS face-to-face error probabilities differ for two states. Biemer (2001) considered a number of more general models that expand (3.52) to include a second grouping variable, say, S, denoting the state and allowed all the model parameters to differ by S. He also tested hypotheses of no difference by setting some parameters equal across states. As an example, it may be plausible to consider the case where the error parameters associated with telephone interviewing are the same for Texas (TX) and California (CA) since all telephone interviews in those states were conducted from the same centralized telephone facility using the same staff, interviewers, supervisors, questionnaires, and survey procedures. Thus, letting the subscript s denote parameters specific to the state ($s = 1$ for TX and $s = 2$ for CA), the assumption of state equality for telephone interview misclassification can be expressed as $\pi_{t|x1}^{T|XAS} = \pi_{t|x2}^{T|XAS}$. Biemer tested this assumption as well as a number of alternative model specifications for each characteristic in the study.

Biemer concluded that error probabilities differed by state for three of the 14 characteristics in his analysis. He further found that the misclassification, averaged across all the 14 questions was significantly greater for the NHIS than the RDD survey. He speculated that this may be the result of smaller interviewer effects in the RDD survey due to centralized call monitoring and quality control.

3.4 FURTHER ASPECTS OF THE HUI–WALTER MODEL

3.4.1 Two Polytomous Measurements

The Hui–Walter method can easily be extended to two polytomous measures, A and B, of the polytomous latent variable X. Suppose that A, B, and X

each consists of K categories. Likewise, suppose that the grouping variable G partitions the population into L groups. The true prevalence for each category by group is $\pi_{x|g}^{X|G}$ for $x = 1,\ldots,K$, where $\sum \pi_{x|g}^{X|G} = 1$ for all g. The misclassification probabilities for measurements A and B are $\pi_{a|xg}^{A|XG}$ and $\pi_{b|xg}^{B|XG}$ for $a \neq x$, $b \neq x$ for all g. As before, we assume that measurement errors are independent (i.e., possess local independence) and express the loglikelihood kernel for the GAB table as

$$\ell(\boldsymbol{\pi}|GAB) = \sum_g \sum_a \sum_b n_{gab} \log(\pi_{gab}^{GAB}) \tag{3.53}$$

where

$$\pi_{gab}^{GAB} = \sum_x \pi_g \pi_{x|g}^{X|G} \pi_{a|xg}^{A|XG} \pi_{b|xg}^{B|XG} \tag{3.54}$$

When no constraints are placed on the model, there are $LK^2 - 1$ degree of freedom for estimating $LK(2K - 1) - 1$ model parameters. By assuming $\pi_{a|xg}^{A|XG} = \pi_{a|x}^{A|X}$ and $\pi_{b|xg}^{B|XG} = \pi_{b|x}^{B|X}$ (corresponding to assumption 1 above), the number of parameters is reduced to $LK + 2K(K - 1) - 1$.

For example, when $L = 2$ and $K = 3$, there are 17 degrees of freedom and 29 parameters in the unconstrained model. With constraints the number of parameters is reduced to 17, which is just identified. In fact, the model will always be saturated for $L = 2$ groups. Hui and Walter (1980) provide closed-form solutions for the MLE of the model parameters for this case. For $L \geq 3$ groups, the models are unsaturated and a test of model fit can be conducted. Since no closed-form solution for the parameters exists for $L \geq 3$, numerical methods (e.g., the EM algorithm) must be employed to determine the MLEs.

A final condition for identifiability for all models is that $\pi_{x|g}^{X|G} \neq \pi_{x|g'}^{X|G}$ for some $g' \neq g$ corresponding to assumption 2 above. Now π_{gab}^{GAB} can be rewritten as

$$\pi_{gab}^{GAB} = \sum_x \pi_g \pi_{x|g}^{X|G} \pi_{a|x}^{A|X} \pi_{b|x}^{B|X} \tag{3.55}$$

The following example illustrates the use of the Hui–Walter method for $L = 2$ and $K = 3$.

3.4.2 Example: Misclassification with Three Categories

The data in Table 3.12 will be used to illustrate the Hui–Walter method for $K = 2$, $L = 3$. In this table, the LF (labor force) category in Example 3.3.2 has been split into two categories corresponding to employed (EMP) and

Table 3.12 Interview–Reinterview Data for the 1995–1996 CPS

	Reinterview Response (A)		
Interview Response (B)	EMP	UNE	NLF
	Males		
EMP	4487	30	64
UNE	26	179	57
NLF	132	44	1841
	Females		
EMP	4044	21	106
UNE	19	132	65
NLF	157	55	3160

```
lat 1
man 3
dim 3 2 3 3
lab X G A B
mod G
    X|G
    A|X
    B|X
dat [4487 30  64 26 179 57 132 44 1841
     4044 21 106 19 132 65 157 55 3160]
sta A|X [.9 .05 .05 .05 .9 .05 .05 .05 .9]
sta B|X [.9 .05 .05 .05 .9 .05 .05 .05 .9]
```

Figure 3.6 ℓEM input statements for fitting the Hui–Walter model to the data in Table 3.12.

unemployed (UNE) in order to estimate the misclassification probabilities for all these categories.

As before, A and B denote the interview and reinterview classifications, respectively, and X denotes the true, latent classifications. Corresponding to assumptions 1 and 2 above, we assume that (1) males and females have the same probabilities of being misclassified in the categories EMP, UNE, and NLF by the CPS; and (2) they differ in the labor force classification rates for these three categories. As a rough check on Assumption 1, the gross difference rates were computed from the table and found to be similar for males and females (5.1% vs. 5.4%). To check Assumption 2, the labor force rates for the three categories were compared for males and females. These were 67%, 4%, and 29% for males compared with 54%, 3%, and 43% for females. Thus, both assumptions are at least plausible and consistent with the observed data.

The ℓEM software was used to fit the Hui–Walter model to these data using the input statements in Figure 3.6. Model estimates of the misclassification probabilities for the CPS interview are displayed in Table 3.13. These results

Table 3.13 Hui–Walter Estimates of Classification Probabilities (in percent) for the CPS Interview Classifier[a]

	Observed Status			
True Status	EMP	UNE	NLF	True Prevalence
EMP	98.2 (0.6)	0.0 (0.0)	1.8 (0.6)	59.4
UNE	12.4 (1.8)	78.5 (5.2)	9.1 (5.6)	3.6
NLF	3.1 (0.3)	1.2 (0.5)	95.7 (0.6)	37.1
Observed prevalence	59.9	3.3	36.9	—

[a]Standard errors are shown in parentheses.

suggest that only 78.5% of truly unemployed (UNE) persons are classified correctly by the CPS. However, persons truly EMP or NLF are classified with high accuracy. Note that about 3% of persons who are truly NLF are classified as EMP by the CPS. In addition, the CPS classifies about 12% of truly UNE persons as EMP and about 9% as NLF. Finally, the model estimate of the bias in the CPS unemployment estimate is –0.3. Since the results are unweighted, they may not reflect the bias in the reported CPS estimates.

Several features of the CPS labor force analysis lend validity to these results. Historical data on the reliability of the CPS data suggest that the classifications of respondents as EMP and NLF are highly reliable and the UNE is highly unreliable [see, e.g., Poterba and Summers (1995)]. The estimates in Table 3.13 are quite consistent with these prior findings, as are the magnitudes of the classification probabilities in the table.

The results are also plausible from an operational perspective. It is generally agreed that the process of classifying individuals as UNE is relatively more prone to error than is EMP or NLF. Determining whether an individual is EMP or NLF is not complicated in most cases. The former requires determining whether an individual is working or has an income. The latter requires determining whether the individual is retired, not working by choice, or not able to work. These concepts are fairly unambiguous. However, the concept of unemployment is often more difficult, involving somewhat vague and complex criteria such as whether a nonworking individual is looking for work, not looking for work because of layoff with the expectation of being recalled, or not looking for work with no desire to find work. The process for determining whether a person who is not working was actually looking for work during the reference week is particularly subjective as it involves a series of open-ended questions regarding what the individual actually did to find work.

Regarding the validity of the Hui–Walter method for estimating CPS labor force classification error, Biemer and Bushery (2001) showed that the Hui–Walter estimates agree quite well with estimates obtained using traditional methods. Sinclair and Gastwirth (1996, 1998, 2000) also report results that strongly support the validity of the Hui–Walter estimates for the CPS. As a followup to the Biemer and Bushery (2001) work, Biemer (2004a) presents a detailed analysis of the CPS classification error problem that suggests that confusion regarding the "looking for work" and "layoff" concepts are a major source of error in the CPS unemployment classification system (see Section 7.1.7 for more details about these findings).

3.4.3 Sensitivity of the Hui–Walter Method to Violations in the Underlying Assumptions

As shown in this chapter, evaluating classification error when only two measurements of a latent variable X are available requires rather strong assumptions and most of these assumptions cannot be tested for a given data set. A number of authors [e.g., Vacek (1985), Sinclair (1994), Sinclair and Gastwirth (2000), Biemer and Bushery (2001)] considered the validity of the Hui–Walter method when some of the assumptions are not satisfied. Vacek (1985) considered the bias in the estimates when the assumption of local independence fails (i.e., when A and B are *locally dependent*). He provides expressions for the biases in the estimates in the dichotomous case when the conditional covariances $Cov(A,B|X)$ are not 0. The expressions are quite complex, but several general observations are possible. Positive covariances [i.e., either $Cov(A,B|X=1)$ or $Cov(A,B|X=2)$ or both are larger than 0] will usually result in underestimation of the error probabilities. This is expected since errors in the same direction imply classification agreement, which is consistent with smaller classification error. Further, positive values of $Cov(A,B|X=0)$ will cause π to be overestimated while positive values of $Cov(A,B|X=1)$ will cause π to be underestimated. Thus, when both $Cov(A,B|X=0)$ and $Cov(A,B|X=1)$ are positive, the associated biases in the estimation of π tend to cancel one another somewhat. Generally, the sign and magnitude of the bias depend on the true prevalence and the relative magnitude of the two conditional covariances.

In a study by Sinclair and Gastwirth (2000), the robustness of the Hui–Walter estimates to violations of Assumptions 1 and 2 above was investigated. As for Vacek's study, the simple case of two groups and dichotomous measurements was assumed. Sinclair and Gastwirth formed a large number of $2 \times 2 \times 2$ GAB tables where the prevalence rates and the classification error rates differed between the two groups to varying degrees. To the extent that the classification error rates $\pi_{1|2}^{A|X}$ and $\pi_{2|1}^{A|X}$ or $\pi_{1|2}^{B|X}$ and $\pi_{2|1}^{B|X}$ varied between the two groups, robustness to Assumption 1 was evaluated. Likewise, by creating tables where the difference $\pi_{1|1}^{X|G} - \pi_{1|2}^{X|G}$ was small, the sensitivity of the estimates to assumption 2 was tested.

They found that the Hui–Walter method yields a robust estimator of the prevalence rates when the two groups have quite different prevalence rates (Assumption 2). However, the estimates of false-positive and false-negative probabilities were more affected by modest violations of assumption 1. Nevertheless, the greater the difference $\pi_{1|1}^{X|G} - \pi_{1|2}^{X|G}$, the less affected were the estimates of $\pi_{a|x}^{A|X}$ and $\pi_{b|x}^{B|X}$ to violations of Assumption 1. All of their analysis assumed that A and B were locally independent.

3.4.4 Hui–Walter Estimates of Reliability

Biemer and Bushery (2001) provided some evidence of the validity of the Hui–Walter method for the CPS labor force application presented in Section 3.3.2. Using the expression for the reliability ratio in (3.18) and the Hui–Walter estimates of π_x^X and $\pi_{a|x}^{A|X}$, they computed an alternative estimator of the inconsistency ratio for each CPS labor force category and compared them to the traditional test–retest reinterview estimates. To see how the Hui–Walter reliability estimates can be derived, let $\hat{\pi}_x^X$ and $\hat{\pi}_{a|x}^{A|X}$ denote the Hui–Walter estimates for the corresponding CPS interview parameters. Using the expression of R in (3.18) and ignoring the variance term $\gamma_{\theta\phi}$, it can be shown that the inconsistency ratio has the form

$$\begin{aligned} I &= 1-R \\ &= \frac{\pi\theta(1-\theta)+(1-\pi)\phi(1-\phi)}{P(1-P)} \end{aligned} \tag{3.56}$$

using Bross' notation where $P = E(p)$. For a polytomous variable A with K categories, let I_{Ak} denote the inconsistency ratio for the interview category k, for $k = 1,\ldots, K$ (see Section 2.2.3). Under the assumption of two parallel measurements, it can be shown that a consistent estimator of I_{Ak} is given by

$$\hat{I}_{Ak} = \frac{\sum_x \hat{\pi}_x^X \hat{\pi}_{k|x}^{A|X}\left(1-\hat{\pi}_{k|x}^{A|X}\right)}{\hat{\pi}_k^A\left(1-\hat{\pi}_k^A\right)} \tag{3.57}$$

where

$$\hat{\pi}_k^A = \sum_x \hat{\pi}_x^X \hat{\pi}_{a|x}^{A|X}. \tag{3.58}$$

This estimator can be compared with the traditional estimator of I_{Ak} [i.e., the index of inconsistency given by (2.50) in Chapter 2]. The extent the two methods of estimation agree garners support for the validity of both methods.

The inconsistency ratio was estimated using the traditional index of inconsistency and $\hat{I}_{Ak}$ for the data in Table 3.12. The results are displayed in Table 3.14. Of particular interest in this comparison is the UNE category since it is had the largest classification error in Table 3.13. Standard errors are not available for the Hui–Walter estimates of the index, so formal tests of hypothesis are not possible in these comparisons. However, standard errors

Table 3.14 Comparison of Hui–Walter and Traditional Estimates of Inconsistency Ratio for the 1995–1996 CPS Labor Force Classification[a]

	Labor Force Classification		
Estimation Method	EMP	UNE	NLF
Index of inconsistency	7.9 (0.3)	34.9 (2.0)	10.0 (0.4)
Hui–Walter	10.6	33.1	12.3

[a]The estimates of I in this table differ slightly from those of Biemer and Bushery (2001). The differences are due to using a somewhat different dataset and method of estimation.

for the traditional estimates are provided and can also be used as rough approximations of the standard errors for the Hui–Walter estimates.

Although there are small discrepancies between the two methods, the methods produce very similar results, which suggests that both methods are valid for these data. To the extent that the two estimates disagree, one or more assumptions of either or both methods could be violated by the data. Since both methods require the assumption of local independence, failure of that assumption to hold may not explain the discrepancies. Another possible explanation for the differences between the two methods is the assumption of parallel measurements. Note that this assumption, while required for the index of inconsistency, is not required for the Hui–Walter method. In that regard, the latter method is more robust to violations in that assumption. As noted in Chapter 2, when the interview and reinterview measurements are not parallel, the index of inconsistency reflects the average SRV for both measurements.

3.5 THREE OR MORE POLYTOMOUS MEASUREMENTS

When three or more measurements of the characteristic $\mathcal{Y}$ are available, the assumptions required for identifiability are much less restrictive than in the case of only two measurements. For example, suppose that three measurements, A, B, and C, of the latent variable X are available, each having K categories corresponding to the K categories of X. Assume that the three measurements are locally independent:

$$\pi_{abc|x}^{ABC|X} = \pi_{a|x}^{A|X} \pi_{b|x}^{B|X} \pi_{c|x}^{C|X} \tag{3.59}$$

For the likelihood kernel, we write

$$\pi_{abc}^{ABC} = \pi_x^X \pi_{a|x}^{A|X} \pi_{b|x}^{B|X} \pi_{c|x}^{C|X} \tag{3.60}$$

The ABC cross-classification table contains K^3 cells, and thus $K^3 - 1$ degrees of freedom are available for parameter estimation. The unrestricted model contains $(K - 1)(3K + 1)$ parameters. When $K = 2$, the model is just identified; for $K = 3$, there are 26 degrees of freedom but only 20 parameters leaving 6 additional degrees of freedom for defining additional parameters. No grouping variable need be defined for identifiability for any value of K, and the restrictive assumptions of the Hui–Walter method are not needed. Some of the surplus degrees of freedom available when $K \geq 2$ might be used to test model fit, to relax the local independence assumption for some combinations of variables, or perhaps to explore the assumption of unidimensionality (or *univocality* as it is called in Chapters 4 and 5) of the measurements. These degrees of freedom can be multiplied when grouping variables are added to the model with or without parameter restrictions. The possibilities for model specification seem endless.

This class of model and its variants, referred to as the *latent class* models, will be the focus of the remainder of this book. In the next section, we consider the essential elements of these models and the conditions for their identifiability under more general assumptions and provide some applications that demonstrate their use and potential for survey error analysis.

CHAPTER 4

Latent Class Models for Evaluating Classification Errors

Chapter 3 discussed the essential ideas associated with the evaluation of measurement error in a categorical response variable for two fallible measurements. In this chapter, these concepts are extended for three or more measurements. The concept of a true value associated with the characteristic of interest ($\mathcal{Y}$) is retained in this and the following chapters. As in previous chapters, our goal is to evaluate the error in the imperfect measurements of $\mathcal{Y}$. The true underlying variable is referred to as a *latent variable* (denoted by X), and the variables designed to measure this latent variable are referred to as *indicator variables* (denoted by A, B, C, etc.). Latent class analysis (LCA) provides an ideal framework for modeling the relationships between the latent variable and its indicators. LCA is a general methodology that can be applied in situations other than those that will be discussed in this book. For example, it can be used to explore the question about whether a latent variable can explain the correlations among a set of measured variables. In those applications, the latent variable need not be interpreted as the true underlying variable. This chapter is more narrowly focused. Here we are concerned with evaluating the quality of a set of indicators designed to measure a prespecified, measurable construct. The distinction between our use of LCA and more general applications of this powerful modeling methodology will be discussed further.

4.1 THE STANDARD LATENT CLASS MODEL

The π-probability notation for representing the model parameters introduced in Chapter 3 will be continued in this and the remaining chapters. Suppose that there are three measurements, A, B and C, intended to measure some unknown true value, X. We can suppose that X is measurable under ideal conditions, but in a typical survey setting, X is measured with error. Under the assumption of local independence introduced in Chapter 3, we can write the likelihood kernel for the ABC cross-classification table in terms of X as

$$\pi_{abc}^{ABC} = \sum_x \pi_x^X \pi_{a|x}^{A|X} \pi_{b|x}^{B|X} \pi_{c|x}^{C|X} \tag{4.1}$$

This model is one of the simplest forms of the general latent class (LC) model that is the subject of this chapter. In the LCA framework, X is referred to as a *latent variable* and the categories of X are referred to as *latent classes*. By contrast, observed variables such as A, B, and C are referred to as *manifest* variables. If the manifest variables are designed specifically to measure X, then they are called *indicators* of X. Other manifest variables may appear in the model for describing the variation in the latent variable or the error probabilities. For example, the grouping variable G used in the Hui–Walter model is an example of a manifest variable that is not an indicator variable. Although correlated with X, G is not intended to measure X, whereas indicators are proxies for X.

The conditional probabilities of the indicators given the latent variable are called *response probabilities* in the LCA literature. In this book, these quantities are referred to as *error probabilities* since we are assuming that X is the true value of the characteristic. For example, $\pi_{a|x}^{A|X}$ is an error probability for all $a \neq x$ since it is the probability that X and its indicator A disagree. An *accuracy probability* is a response probability where the category of the indicator variable and the class of the latent variable are the same; for example, $\pi_{a|x}^{A|X}$ for $a = x$ is the accuracy probability for classifying persons in latent class x by the indicator A. Note that the accuracy probability for an indicator is 1 minus the sum of the error probabilities for the indicator for each latent class: $\sum_{a \neq x} \pi_{a|x}^{A|X} = 1 - \pi_{a|x=a}^{A|X}$. In that sense, response probabilities and error probabilities are synonymous.

4.1.1 Latent Variable Models

The latent class models represent a large class of models for the analysis of categorical data when some of the variables are unobserved or latent. The general idea can be traced back to the early work of Lazarsfeld (1950a, 1950b); however, widespread use of the method by practitioners was delayed until critical problems in the estimation of the model parameters could be resolved. This came when Goodman (1974a, 1974b) proposed the use of maximum-likelihood estimation, which addressed most of the problems. However, it was not until Haberman (1979) showed how the LC model could be reparameterized as a loglinear model with latent variables that the field quickly expanded. Although somewhat dated, there are a number of excellent textbooks describing the essential ideas of LCA such as those by Lazarsfeld and Henry (1968), Haberman (1979), McCutcheon (1987), Hagenaars (1990), and Heinen (1996). These books can be credited with spreading knowledge of the LC models and their versatility in the analysis of categorical data. Some of the early software programs that boosted the popularity of LCA are LAT and Newton (Haberman 1979), MLSSA (Clogg 1979), and LCAG (Hagenaars 1988a). Today, there is still a growing literature in this field and new applications of LCA are being continually discovered. Edited volumes by Hagenaars and McCutcheon (2002), Langeheine and Rost (1988), and Rost and Langeheine (1997) demonstrate the wide applicability of these models. The most recent published book by Collins and Lanza (2010) suggests that LCA is continuing to attract converts.

Traditionally, LC models have been employed in typology, which is the study of population subtypes or domains. Typological analysis and applications differ considerably from the survey error evaluation applications that are the focus of this book. Let us consider how they differ. In traditional applications, a researcher theorizes that the population consists of certain subtypes of individuals that are difficult if not impossible to measure and classify directly. To determine which subtype best describes the individual, a series of questions is asked and the responses are analyzed using LCA to classify the individual. In this application, LCA is used as a kind of factor analysis to estimate the probability an individual belongs to each subtype.

As an example, a researcher wishing to test a theory about smoking behavior may measure a number of behavioral characteristics related to cigarette smoking, such as

"Have you ever smoked?"
"Do you now smoke?"
"Have you ever tried to quit smoking?"
"Do you plan to smoke in the future?"

These questions yield four dichotomous variables that jointly define 2^4 or 16 patterns of 1s (for positive responses) and 2s (for negative responses). Associated with each pattern is a type of smoker or a nonsmoker. Attempting to analyze 16 types of smokers requires a large sample, and the level of detail may be too great. Instead, the researcher may attempt to reduce the number of subtypes (or classes) to a few general types that describe the dominant behaviors in the population. An important question in these applications is "How many classes are needed to describe the subtypes of interest?" LCA can help answer that question.

A key distinction between these applications and those considered in this book is that the class (or subtype) membership for an individual is not directly measurable. Rather, traditional LCA combines the information from the manifest variables using a triangulation approach to determine the probability that an individual belongs to each latent class. To determine the number of classes, LC models with varying numbers of categories of the latent variable can be fitted successively and tested for their abilities to capture the variation in the 16 response patterns from the four questions. For typological analysis, determining the number of classes is an important aspect of the model selection process.

To see how this works, suppose that the LCA suggests that there are three main types of smoking behaviors. By examining the relationships between the four indicators and the assumed latent variable, the researcher might even name the three classes, such as nonsmokers, light or experimental smokers, and regular, frequent smokers. This reduction of population types from 16 to 3 can substantially simplify theories regarding the correlates of cigarette smoking or whatever phenomenon is under study. In that sense, LCA is akin to exploratory factor analysis [see, e.g., Thurstone (1947)] and, indeed, it is often referred to as the "categorical data analog to factor analysis."

Contrast this use of LCA with our applications to the evaluation of classification error. One major difference is that the latent variable X in our applications is directly measurable. The definition of X and the number of classes it includes are known a priori. As an example, the questionnaire may contain four dichotomous questions about cigarette smoking such as "Do you currently smoke cigarettes?" All four questions are similar and are designed to classify an individual as either a nonsmoker or a smoker. Here, LCA is focused on estimating the prevalence of smoking as well as the error in each of the four indicators of X. One purpose of this type of analysis is to determine which of the four indicators is best for measuring smoking behavior. Thus, one might say our use of LCA is similar to *confirmatory factor analysis*.

To interpret the LC response probabilities as error probabilities, the assumption that X is a well-defined, measurable construct with a known number of classes is rather crucial. Otherwise, the definition of X is subjective and X cannot be regarded as the true characteristic. In the error evaluation usage of LCA, X is predefined to be 1 if the respondent smokes and 2 if not. The number of classes contained in X is not a research question; rather, it is predefined by the questionnaire designer. The primary reason for introducing multiple indicators of X is either to improve our measurement of X or as a device for estimating the classification error for one or more of the indicators. Although the goals are different, survey error evaluation and typological analysis use LCA as a tool, albeit in a slightly different way. Still, much of the literature on LCA aimed at improving its use as a type of factor analysis is still relevant and applicable for its use in survey error evaluations.

Latent class analysis occupies a relatively small corner of a much larger field referred to in the literature as *latent structure* (or *variable*) *analysis*. In that literature, at least four types of models can be distinguished in terms of the assumptions underlying the latent variables and their indicators (see Figure 4.1). As shown in the figure, LCA is appropriate when the latent variable and all the indicator variables are discrete. In this book, we will consider methods

	LATENT VARIABLE(S)	
Indicator Variables	***Discrete***	***Continuous***
Discrete	Latent class analysis	Latent trait analysis
Continuous	Latent profile analysis	Factor analysis

Figure 4.1 Four types of latent variable analysis. [*Source*: Bartholomew and Knott (1999).]

appropriate for both nominal and ordinal categorical variables, including some special methods that have been developed specifically for ordinal data. By contrast, factor analysis was developed for situations where the latent and the indicator variables are continuous (usually normally distributed). Collins and Lanza (2010) discuss the similarities and differences between LCA and factor analysis in some detail.

Latent trait analysis (Heinen 1996) is appropriate when the categorical manifest variables are assumed to be indicators of one or more continuous latent variables. As an example, this is often the case when the continuous variable of interest is measured on a discrete scale; for example, income, although a continuous variable, is often measured by discrete categories. Likewise, one might assume that any item measured with three or more ordinal categories represents a continuous latent variable. This is the basic assumption of *item response theory* (IRT) [see, e.g., Heinen (1996)], which is a special case of latent trait analysis. Finally, *latent profile analysis* may be appropriate when a discrete number of population subtypes determine the categories of a discrete latent variable whose classes are characterized by manifest variables measured on a continuous scale.

Skrondal and Rabe-Hesketh (2004) show that all four cases shown in Figure 4.1 are simply special cases of a general modeling framework that they refer to as *generalized linear latent and mixed models* (GLLAMM). Their framework unifies and extends latent variable models, including multilevel or generalized linear mixed models, longitudinal or panel models, item response or factor models, latent class or finite mixture models, and structural equation models. They also provide software for estimating all the models using this general framework (Rabe-Hesketh et al. 2004).

4.1.2 An Example from Typology Analysis

Using a famous study by Stouffer and Toby (1951), the use of LCA as a means of studying population subtypes can be illustrated. In this study, respondents were given four situations to consider:

A. An automobile–pedestrian accident to which the respondent is a witness and in which the faulty driver is a friend.

B. A questionable physical exam that, if reported honestly, could forfeit a friend's eligibility for insurance.

C. A Broadway play authored by a friend that the respondent is reviewing and feels is a bad play.

D. A secret meeting that yields information related to the future price of a stock that a friend owns.

For each situation, respondents were asked whether they would act in favor of their friend, although it may be unethical to do so (called *particularistic behavior*) or act ethically recognizing their responsibilities to society at large (called *universalistic* behavior). Four dichotomous manifest variables (A, B, C, and D) were created corresponding to the four situations above. If the respondent chose a "universalistic" response to a particular situation, a "1" is recorded for that situation. Otherwise, if a "particularistic" response is chosen, a "2" is recorded. The data for 216 observations appears in Table 4.1.

In addition to those due to Stouffer and Toby (1951), these data were analyzed by a number of researchers, including Lazarsfeld and Henry (1968), Goodman (1974a, 2002), and McCutcheon (1987). The LCA focuses on the nature of the latent variable, which is defined as a hypothetical construct for grouping respondents into some number of latent classes representing various degrees of particularistic or universalistic tendencies. In this example, the latent variable ("type of behavior exhibited") is not directly observable; nor is the number of classes the latent variable contains prespecified. Rather, the number of classes is determined by comparing alternative models that specify two, three, or four classes. Among these models, the model exhibiting the best

Table 4.1 Stouffer and Toby's Data

Response				
A	*B*	*C*	*D*	Frequency
1	1	1	1	42
1	1	1	2	23
1	1	2	1	6
1	1	2	2	25
1	2	1	1	6
1	2	1	2	24
1	2	2	1	7
1	2	2	2	38
2	1	1	1	1
2	1	1	2	4
2	1	2	1	1
2	1	2	2	6
2	2	1	1	2
2	2	1	2	9
2	2	2	1	2
2	2	2	2	20

fit of the data is selected as the best model. Using this best model, the categories of X can interpreted and named through an examination of the response probabilities, $\pi_{a|x}^{A|X}$, $\pi_{b|x}^{B|X}$, $\pi_{c|x}^{C|X}$, and $\pi_{d|x}^{D|X}$. Here we referred to the conditional probabilities of the manifest variable given X as response probabilities because there is no assumption that X represents the true value for the measurements A, B, C, and D.

Determining the "optimal" number of classes for these data begins by fitting a two-class model. Any of the LCA software packages described in section 4.1.3 can be used to fit this simple model. Doing so will yield the following MLEs of the response probabilities for the four indicators: $\hat{\pi}_{1|1}^{A|X} = 0.99$, $\hat{\pi}_{1|1}^{B|X} = 0.94$, $\hat{\pi}_{1|1}^{C|X} = 0.93$, and $\hat{\pi}_{1|1}^{D|X} = 0.77$ for $X = 1$ and $\hat{\pi}_{2|2}^{A|X} = 0.29$, $\hat{\pi}_{2|2}^{B|X} = 0.67$, $\hat{\pi}_{2|2}^{C|X} = 0.65$, and $\hat{\pi}_{2|2}^{D|X} = 0.87$ for $X = 2$. For all four manifest variables, the probability of "1" is very high for class $X = 1$. Thus, the latent class $X = 1$ could represent persons tending toward universalistic decisions. Likewise, when $X = 2$, the probability of responding "2" is high suggesting that $X = 2$ could represent the particularistic group.

Next, the three class LC model can be considered. This is essentially the same model used above except with "number of latent classes" set to "3" in the software. For this model, all manifest variables have a very high probability of "1" when $X = 1$, which suggests that the first class should be labeled "strictly universalistic." Likewise, the probability of responding "2" given $X = 3$ was very high for all four manifest variables, suggesting that the third class should be labeled "strictly individualistic." In the middle category of X, the probability of a universalistic response varies from 0.83 (for A) to 0.19 (for D) suggesting that $X = 3$ represents a "mixed group."

In this example, LCA is used to test theories about the population and the subpopulations contained within it. This goal is quite different that the goal of survey error evaluation. LCA as it is applied to typology analysis is not concerned with the accuracy of A, B, C, and D as indicators of some true value, X. Further, these applications of LCA do not regard the four indicators as observations on X that are subject to misclassification. Instead, the four manifest variables are used to uncover the very nature of X and to theorize about the composition of the population with regard to X. The latent variable was not prespecified, but rather deduced through analysis. In fact, the mere existence of X in this application is purely hypothetical, and its interpretation is only implied by correlations among the manifest variables. Although the manifest variables are called *indicators* of X in the LCA literature, each variable individually is insufficient for determining X. Rather, multiple indicators are required to *deduce* X. Furthermore, the nature and definition of X are highly dependent on the choice of manifest variables for the LCA. This is very different from our use of LCA in this book.

The remainder of this book is confined to the study of LC models as they pertain to the analysis of survey classification error. In our applications, X is a prespecified, well-defined variable whose distribution in the population is to be measured via one or more indicator variables specifically designed to measure X. As an example, an individual's labor force status is observable and can be directly assessed through survey questions. In addition, we can define labor force status in terms of three categories: employed, unemployed, and not in the labor force. Survey questions are developed for classifying each sample person into one and only one of these three categories. Since our ability to measure X is not perfect, the indicators from a survey will be contaminated by classification error. We will show how to apply LCA to estimate the true distribution of X and the magnitude of the errors in its indicators.

Latent class analysis can, therefore, be regarded as a generalization of the measurement error theory discussed in Chapter 3. In fact, all the models considered in Chapter 3 can all be specified as latent class models of some sort. The LCA methodology encompasses a wide range of applications and models. The more recent literature has expanded the LCA modeling framework by incorporating loglinear models with latent variables (discussed later in the chapter). This development has greatly increased the generality of the methodology. In this chapter, we provide an introduction to the field of latent class analysis (LCA) with more advanced topics to follow in the remaining chapters of the book.

4.1.3 Latent Class Analysis Software

The software for fitting LC models can be divided into two types: freeware and commercial/shareware. The disadvantage of the freeware programs is that they have fewer capabilities than does the commercial software and provide very limited or no user support. However, being monofunctional and widely accessible, they are relatively simple to use and excellent for teaching purposes. The following is a list of the most popular freeware programs ordered more or less as to their popularity.

- ℓEM (`http://www.uvt.nl/faculteiten/fsw/organisatie/departementen/mto/software2.html`). This Windows-based program is perhaps the best of the freeware packages and one that has been used for most of the examples and illustrations in this book. ℓEM will fit virtually fully any loglinear model with or without latent variables and with or without equality restrictions. In addition, it will fit ordinal models, association models, hazard models, latent trait models, and many other types of categorical data models. The program provides a wide range of model fit statistics and diagnostics, provides identifiability checks via information matrix eigenvalues, and computes standard errors. The ℓEM manual is fairly readable with the exception of some of the special, boutique-type model options. Although the software is no longer being maintained, it can still be downloaded for free.
- MLLSA (`http://ourworld.compuserve.com/homepages/jsuebersax/mllsa.zip`). This package was the mainstay of latent class analysts for years before ℓEM came along. Judging by the many early

LCA publications that reference its use, it was quite widely used. However, it lacks many of ℓEM's useful features and versatility. MLLSA uses an outdated card-oriented format, but it is still an excellent program, for LCA. MLLSA is no longer being maintained now that it has been subsumed into the CDAS program, which is available for a small fee.

- WinLTA (`http://methodology.psu.edu/index.php/downloads/winlta`). WinLTA is a freestanding Windows application for conducting LCA as well as latent transition analysis (LTA). (LTA is discussed further in Section 7.1.3.) Although it is no longer maintained, it is still available free of charge. As an option, the software uses data augmentation (DA), a Markov chain Monte Carlo procedure, for obtaining standard errors of parameter estimates. This makes it possible to construct hypothesis tests for standard model parameters as well as combinations of parameters, affording greater flexibility than with traditional methods (Lanza et al. 2005). WinLTA is an early version of the twin packages SAS Proc LCA and SAS Proc LTA.

The shareware and other commercially available packages for LCA are not free, but a few are quite inexpensive and feature-rich. A key advantage of these products is that they are currently supported and maintained with new features being added continually as newer versions are released. The following list, ordered alphabetically, contains the major commercial packages.

- Latent GOLD (`http://www.statisticalinnovations.com/products/latentgold_v4_aboutlc.html`). Latent GOLD is a Windows-based program for LCA that is quite general and laden with features such as Bayesian constraints for avoiding boundary estimates, multiple starting values, automatic identification checking, survey weighting and complex sampling designs, and bootstrap chi-square testing.
- MPlus (`http://www.statmodel.com/`). MPlus is currently the software package of choice among social scientists, primarily because of its generality for many types of latent variable modeling, not only LCA. It can be used for structural equation modeling and multilevel modeling as well as for regression analysis and ANOVA. Like Latent GOLD, it has many features to aid the analyst, including appropriately handling complex survey designs and Monte Carlo estimation and power analysis.
- SAS Proc LCA (`http://methodology.psu.edu/index.php/downloads/proclcalta`). Proc LCA is a SAS procedure for LCA. Although the module is free, it is not a standalone package and requires the SAS software to run. PROC LCA reads and writes SAS datasets, and parameter restrictions can be specified by the user in a SAS data file. The software will handle continuous as well as categorical covariates. Like Latent GOLD, a Bayesian stabilizing prior can be invoked when sparseness is an issue for parameter estimation (Lanza et al. 2007).

- SAS Proc LTA (`http://methodology.psu.edu/index.php/downloads/proclcalta`). Proc LTA is a companion software package to Proc LCA for latent transition analysis (LTA). LTA, described futher in Chapter 7, is a longitudinal extension of LCA used to estimate baseline (time 1) latent status membership probabilities along with probabilities of transitions in latent status membership over time. Like Proc LCA, Proc LTA reads and writes SAS datasets, and parameter restrictions can be specified by the user in a SAS data file. It has many of the same features as Proc LCA.
- PANMARK (Van de Pol et al. 1991). PANMARK estimates and tests a range of latent class models, although as its name—a combination of panel Markov models—implies, this program was developed to estimate longitudinal latent class models. PANMARK also has a number of features such as its ability to generate multiple sets of starting values, which may be useful in avoiding local maxima, its reporting of asymptotic standard errors for model parameter estimates, and its use of bootstrapping methods for comparing models with different numbers of latent classes.
- GLLAMM (`http://www.gllamm.org/`). GLLAMM is a very general software package that runs under Stata for analyzing *generalized linear latent and mixed models* (GLLAMM), which refers to a wide class of multilevel latent variable models for responses of mixed type, including continuous responses, counts, duration/survival data, and dichotomous, ordered, and unordered categorical responses and rankings. The latent variables can be assumed to be discrete or to have a multivariate–normal distribution. Examples of models in this class are multilevel generalized linear models or generalized linear mixed models, multilevel factor or latent trait models, item response models, latent class models and multilevel structural equation models. Although GLLAMM is free, it requires Stata to run.
- CDAS (`http://faculty.ucr.edu/~hanneman/cdas/general.htm`). The Categorical Data Analysis System (CDAS) is a reincarnation of the now obsolescent MLLSA software. However, CDAS provides a rich environment for many types of categorical data analysis. The current version allows the user to analyze categorical data with loglinear models, logmultiplicative association models with or without loglinear covariates, and latent class models with the EM algorithm.
- WINMIRA (`http://www.scienceplus.nl/winmira`). WINMIRA will estimate and test a large number of discrete mixture models for categorical variables. Both nominal and continuous latent variable models can be estimated. It has an online help system with a detailed description of all software features. The software reads and writes data directly in SPSS file format.

Throughout the remainder of the book, the ℓEM software will be used for fitting latent class and Markov latent class models (see Chapter 7) unless otherwise stated. The ℓEM input statements used in the examples will also be provided to encourage the reader to put the LCA methods and principles to practice. The reader is strongly encouraged to download this free software from the Website given above and use it to replicate the results shown in the examples and illustrations to follow. A copy of the software can also be obtained from the author.

4.2 LATENT CLASS MODELING BASICS

4.2.1 Model Assumptions

In this section, the basic assumptions of the LC model are examined using the notation introduced in Section 3.3.1. Let X denote a latent variable of interest that, in our applications, is assumed to be the true value of the some characteristic to be measured in a survey. To introduce the ideas, we consider three indicators of X denoted by A, B, and C; however, by simple extension of the notation to include more variables, the results can easily be extended for any number of indicators. The assumptions of the standard LC model are as follows:

1. The sample consists of n units sampled without replacement from a large population of N units using simple random sampling.
2. The indicators are locally independent for all values of a, b, c and x:

$$\pi_{abc}^{ABC} = \pi_{a|x}^{A|X}\pi_{b|x}^{B|X}\pi_{c|x}^{C|X}$$

3. The response probabilities, $\pi_{a|x}^{A|X}$, $\pi_{b|x}^{B|X}$ and $\pi_{c|x}^{C|X}$ are homogeneous; that is, the probabilities are the same for any two units selected from the population.
4. The indicators are *univocal*; A, B, and C are all indicators of the same latent variable X.

Assumption 1 means that the sample can be treated as through it were a simple random sample from an infinite population. Thus, the finite population correction factors that generally apply in the estimation of standard errors can be ignored. In addition, the sample is assumed to be unclustered (no stratification or multistage sampling). Methods that are more appropriate for complex survey samples that involve unequal probability sampling with clustering are considered in Section 5.3.

As discussed in the previous chapters, one implication of local independence is that the classification errors associated with A, B, and C are mutually independent. Thus, knowledge of the error in one indicator suggests nothing about the potential errors of the other indicators. For example, if we write

$$\begin{aligned} A &= X + e_A \\ B &= X + e_B \\ C &= X + e_C \end{aligned} \tag{4.2}$$

where e_A, e_B, and e_C are random error terms, then local independence implies that e_A, e_B, and e_C are mutually uncorrelated. If A, B, and C are designed to be indicators of X, then X can be interpreted as the true value and e_A, e_B, and e_C can be interpreted as measurement errors. However, in the traditional LCA setting, X is not necessarily the error-free version of A, B and C (i.e., the true value). It is simply the *common part* of the three manifest variables A, B and C, meaning that the correlation between any two indicators is *Var*(X) or, more generally, a constant times *Var*(X). This more general interpretation was clearly more appropriate for the Stouffer–Toby example of Section 4.1.2 since the construct X was theoretical rather than substantive. In that example, the interpretation of X as well as local independence depended on how the manifest variables were chosen and their relationships with X and each other.

Another interpretation of local independence is often found in the LCA literature [see, e.g., McCutcheon (1987)]. Under this interpretation, *local independence* means that, within the classes of the latent variable X, the variables A, B, and C are independent. In that sense, X *explains* the relationship among these manifest variables. Stated another way, the only thing A, B, and C have in common is X. After accounting for X, there is no association between A, B, and C. The full implications of local independence and how it can be violated in practice are examined in some detail in the next chapter.

Assumption 3 is a generalization of the assumption $\gamma_{\theta\phi} = 0$ in the Bross model of Chapter 3. This assumption is necessary in order to reduce the number of parameters of the model so that the model is identifiable and the parameters can be estimated. Unfortunately, the assumption is seldom satisfied in practice, and, when it is not, the parameter estimates can be severely biased. One remedy is to incorporate grouping variables in the model so that assumption 3 is satisfied *within the levels of the grouping variable*. This approach is discussed later in the chapter.

Finally, assumption 4 is a seemingly simple assumption that can be quite difficult conceptually. To aid the discussion, we provide a formal definition for an "indicator" as follows: *A categorical manifest variable, A, is an indicator of the latent variable X if the correlation of A with X exceeds the correlation of A with Y, where Y is any other latent variable.*

In other words, A is an indicator of X if it measures X better than does any other latent variable that could be measured by A. Note that A may be an indicator of X even though it may be a very poor indicator of X; for example, A may have a very high false-negative probability. Indeed, in some cases, it may be difficult to distinguish between a poor indicator of X and a variable that is not an indicator of X by the definition given above. In most applications, however, it will be clear from the statement of the survey question and the definition of X as to whether the question response can be regarded as an indicator of X.

As an example, suppose that X is an individual's true status on "past-year use of marijuana" and consider the following four potential indicators of X:

A. In the past year, have you smoked marijuana even once?
B. How many times in the past 12 months have you smoked marijuana?
C. In the past year, have you smoked cigarettes?
D. In the past 12 months, have you smoked dope even once?

Obviously, A seems to fit the definition of an indicator of X since it asks specifically about smoking marijuana, even employing the same words used to define X. It is plausible, then, that X has maximal correlation with A among all other latent variables that A could potentially indicate. Likewise, if the response to B is recoded to be a dichotomous variable indicating any (i.e., 1 or more times) past year use, it also satisfies the definition of an indicator of X. Thus, we say that A and B are *univocal indicators of X*.

On the other hand, C is obviously not an indicator of X. Although cigarette smoking may be highly correlated with marijuana smoking in that most marijuana smokers may also smoke cigarettes, the response to question C does not fit the definition of an indicator X; in other words, there exists a latent variable Y for "past-year cigarette smoking" that has a higher correlation with C than X has with C. Since A is an indicator of X and C is an indicator of Y, A, and C are not univocal. Rather, A and C are *bivocal* indicators meaning they measure two different latent variables or constructs. Regarding A and D, the situation is less obvious. Is D an indicator (albeit a poor one) of X, or does it really indicate a different latent variable? Suppose that we define the latent variable Z as an individual's true status regarding "the smoking of any drug such as marijuana, hashish, or cocaine." Now it is obvious that Z and X are not the same latent variables and further that D is an indicator of Z. It therefore cannot be an indicator of X. Therefore, A and Z are also bivocal indicators.

The univocality assumption can be violated by two indicators even if the wording of the questions underlying the indicators is identical. For example, in an interview–reinterview study, the same question that appeared in the original interview is often repeated verbatim in the reinterview. For example, the question may be "Do you currently have a savings account at a bank or credit union?" If the two interviews are months apart, they may not be measuring the same latent variable since an individual's situation may have changed in the interim. To address this issue, a second latent variable may be introduced in the model corresponding to an individual's true status *at the time of the reinterview*. This situation commonly arises in panel surveys where the same or a similar questionnaire may be repeated at different points in time. Models for panel surveys that define a new latent variable at each timepoint are referred to as *Markov latent class models*. These are discussed in some detail in Chapter 7.

There are at least four modeling approaches, or *parameterizations*, for describing the same latent class model. One approach, called *probability model parameterization*, was used in (4.1). The three other modeling approaches are

1. The loglinear model with a latent variable
2. The modified path model, which is a system of logit models with latent and manifest dependent variables
3. The graphical model using path diagrams

The next section covers these approaches and discusses the strengths and weaknesses of each approach.

4.2.2 Probability Model Parameterization of the Standard LC Model

A probability model expresses the likelihood for the observed cross-classification table as a function of the marginal and conditional probabilities associated with each cell of the table. A *latent class probability model* is a probability model that includes one or more latent variables. For example, suppose that the LCA involves three indicators—A, B, and C—of the latent variable X with the number of categories equaling K_X, K_A, K_B, and K_C, respectively. For most examples used in this book, $K_X = K_A = K_B = K_C = K$; however, in general, this will not be the case. A well-specified model for the ABC cross-classification table should unambiguously define the probability that the ith selected unit is classified into cell (a, b, c) of the table in terms the $K_X \times (K_A + K_B + K_C - 2) - 1$ parameters π_x, $x = 1,\ldots, K_X - 1$ and $\pi_{a|x}$, $a = 1,\ldots, K_A - 1$, $\pi_{b|x}$, $b = 1,\ldots, K_B - 1$, and $\pi_{c|x}$, $c = 1,\ldots, K_C - 1$ for $x = 1,\ldots, K_X$. Under the assumptions of the previous section, the probability of cell (a,b,c) is

$$\pi_{abc}^{ABC} = \sum_x \pi_x^X \pi_{a|x}^{A|X} \pi_{b|x}^{B|X} \pi_{c|x}^{C|X} \tag{4.3}$$

In Chapter 3 we referred to the right side of (4.3) as the *kernel* of the likelihood function representation since it specifies that portion of the likelihood that involves only the parameters to be estimated. The full multinomial likelihood for ABC is denoted by $\mathcal{L}(\boldsymbol{\pi}|\mathbf{n})$, where $\boldsymbol{\pi}$ is a vector of parameters and $\mathbf{n}$ is a vector containing the frequencies n_{abc} for cell (a,b,c), for all cells in the ABC table. It can be written as

$$\mathcal{L}(\boldsymbol{\pi} \mid \mathbf{n}) = \frac{n!}{\prod_a \prod_b \prod_c n_{abc}!} \prod_a \prod_b \prod_c \left(\sum_x \pi_x^X \pi_{a|x}^{A|X} \pi_{b|x}^{B|X} \pi_{c|x}^{C|X} \right)^{n_{abc}} \tag{4.4}$$

where n is the sample size; that is, $n = \Sigma_{abc} n_{abc}$. In practice, the logarithm of this likelihood is easier to work with $\log[\mathcal{L}(\boldsymbol{\pi}|\mathbf{n})] = \ell(\boldsymbol{\pi}|\mathbf{n})$, where

$$\ell(\boldsymbol{\pi} \mid \mathbf{n}) = Const + \sum_a \sum_b \sum_c n_{abc} \log\left(\sum_x \pi_x^X \pi_{a|x}^{A|X} \pi_{b|x}^{B|X} \pi_{c|x}^{C|X} \right) \quad (4.5)$$

and where *Const* is the logarithm of the factorial constant in (4.4), which can be ignored in the estimation process. From the definitions of the variables and parameters, it follows that all probabilities will sum to 1 across their mutually exclusive and exhaustive categories:

$$\sum_x \pi_x^X = \sum_a \pi_{a|x}^{A|X} = \sum_b \pi_{b|x}^{B|X} = \sum_c \pi_{c|x}^{CX} = 1 \quad (4.6)$$

This model is often referred to as an *unrestricted* latent class model since only the usual restrictions in (4.6) are imposed on it. *Restricted* LC models impose additional constraints based on theoretical or practical considerations. The two most common types of constraints equate a probability to some constant or two or more probabilities to each other. Structural zero constraints (which are examined in more detail later) should be imposed on cells that have zero probability of being observed. These restricted LC models are discussed later in this chapter.

For more complex modeling involving grouping variables, constraints are imposed on test theories, estimation efficiency, and greater model parsimony. As an example, recall the two-indicator Hui–Walter model imposed these constraints: $\pi_{a|xg}^{A|XG} = \pi_{a|x}^{A|X}$ and $\pi_{b|xg}^{B|XG} = \pi_{b|x}^{B|X}$ for all g, x, a and b; that is, error probabilities were constrained to be equal across the groups defined by G. Inequality constraints are also possible and sometimes desirable when modeling ordinal data, as we shall see in the next chapter. To apply these parameter constraints, the likelihood maximization process must be altered. For example, in the EM algorithm, the parameters must be set to their constrained values at each iteration of the M-step. The likelihood equations associated with the full data likelihood could take a different functional form when constraints are imposed, which will affect the E-step as well. (See, for example the illustration in Section 3.2.2). Further discussions of the methods for implementing the EM algorithm for restricted LC models can be found in Heinen (1996) and Vermunt (1997a).

4.2.3 Estimation of the LC Model Parameters

If X were an observed (manifest) rather than latent variable, the loglikelihood ℓ for arbitrary cell (x,a,b,c) of the XABC table would take the form

$$\ell = \sum_x \sum_a \sum_b \sum_c n_{xabc} \log\left(\pi_{xabc}^{XABC}\right) \quad (4.7)$$

where n_{xabc} is the cell frequency and π_{xabc}^{XABC} follows a model to be specified. Procedures for finding the maximum-likelihood estimates in this situation are available in many standard statistical analysis packages. When X is unobserved, then only frequencies $n_{abc} = \sum_x n_{xabc}$ (i.e., the frequencies of the manifest variables collapsed over X) are known. Let $m_{abc} = E(n_{abc})$, the expected value of n_{abc}, and note that $m_{abc} = n\pi_{abc}^{ABC}$, where $\pi_{abc}^{ABC} = \sum_x \pi_{xabc}^{XABC}$. The loglikelihood takes the form

$$\ell = \sum_a \sum_b \sum_c n_{abc} \log\left(\sum_x \pi_{xabc}^{XABC}\right) \tag{4.8}$$

The summation within the log function in this equation creates problems in the likelihood maximization process, many of which are not yet well-understood [see, e.g., Fienberg et al. (2007)]. The literature on latent class analysis provides a number of algorithms for handling this situation, including the Newton–Raphson algorithm, modified steepest-descent (Habibullah and Katti 1991), Broyden–Fletcher–Goldfarb–Shanno algorithm (Fletcher 1987), Levenberg–Maquardt algorithm (Moré, 1978), and the EM algorithm discussed in Chapter 3.

A number of software programs have implemented at least one of these for estimating LC models, and some of these are discussed later in this chapter. Most software packages implement some version of the EM algorithm (introduced in Section 3.2.2), but may also include options for the Newton–Raphson and other algorithms. Next, we consider how the EM algorithm might be applied to obtain a solution for the LC model with three dichotomous indicators. Extensions to four or more indicators of any dimension will be apparent in the discussion.

EM Algorithm for the Probabilistic Parameterization

Let ABC denote the data table with counts n_{abc} in cell (a,b,c). To simplify the presentation, we will also show that the indicators and the latent variable are binary variables; thus, the model contains seven parameters: $\pi_1^X, \pi_{1|x}^{A|X}, \pi_{1|x}^{B|X}$, and $\pi_{1|x}^{C|X}$ for $x = 1,2$. Recall from Section 3.2.2 that the EM algorithm alternates between the E- and M-steps until the specified convergence criteria are satisfied. Denote the value of the parameter at the tth iteration by the left superscript (t); for example, ${}^{(t)}\pi_1^X$, ${}^{(t)}\pi_{1|x}^{A|X}$, ${}^{(t)}\pi_{1|x}^{B|X}$, and ${}^{(t)}\pi_{1|x}^{C|X}$ for $x = 1,2$. To "prime" the algorithm, we enter a set of starting values of the parameters into E-step. These may be specified by the user or generated randomly by the software. Thus, the first step of the algorithm commences with an initial table ${}^{(1)}$XABC on the basis of these starting values, where the left superscript (1) denotes the first iteration.

In the E-step for the tth iteration, an estimate of the full data cross-classification table denoted by ${}^{(t)}$XABC is computed. In this table, the count in cell (x,a,b,c) denoted ${}^{(t)}n_{xabc}$ is computed as

$$ {}^{(t)}n_{xabc} = {}^{(t)}\pi_{x|abc}^{X|ABC} n_{abc} \tag{4.9}$$

where, by Bayes' rule, we obtain

$$ {}^{(t)}\pi_{x|abc}^{X|ABC} = \frac{{}^{(t)}\pi_{xabc}^{XABC}}{{}^{(t)}\pi_{1abc}^{XABC} + {}^{(t)}\pi_{2abc}^{XABC}} \tag{4.10}$$

Under the standard LC model assumptions, the numerator can be rewritten as ${}^{(t)}\pi_{xabc}^{XABC} = {}^{(t)}\pi_{x}^{X}\, {}^{(t)}\pi_{a|x}^{A|X}\, {}^{(t)}\pi_{b|x}^{B|X}\, {}^{(t)}\pi_{c|x}^{C|X}$. If any fixed constant constraints are to be imposed on the parameters, the parameters would be replaced by their corresponding constrained values at this step.

In the M-step, we use this estimate of table XABC to compute the MLEs of the parameters using the full-data likelihood ($\mathcal{L}_{\text{full}}$). It can be shown that the kernel of the full-data loglikelihood function is given by

$$ \ell_{\text{full}} = \log(\mathcal{L}_{\text{full}}) = \sum_x \sum_a \sum_b \sum_c n_{xabc} \log(\pi_x^X \pi_{a|x}^{A|X} \pi_{b|x}^{B|X} \pi_{c|x}^{C|X}) \tag{4.11}$$

Differentiating this function by each of the seven parameters produces seven likelihood equations. Equating these equations to 0 and solving for the parameters in terms of the estimated quantities in $XABC$ yields the following new estimates of the parameters:

$$ \begin{aligned} {}^{(t+1)}\pi_x^X &= \frac{{}^{(t)}n_{x+++}}{n} \\ {}^{(t+1)}\pi_{a|x}^{A|X} &= \frac{{}^{(t)}n_{xa++}}{{}^{(t)}n_{x+++}} \\ {}^{(t+1)}\pi_{b|x}^{B|X} &= \frac{{}^{(t)}n_{x+b+}}{{}^{(t)}n_{x+++}} \\ {}^{(t+1)}\pi_{c|x}^{C|X} &= \frac{{}^{(t)}n_{x++c}}{{}^{(t)}n_{x+++}} \end{aligned} \tag{4.12}$$

The "+" notation established in the previous chapters is used to denote summation over the subscript that the "+" replaces. For binary variables, these estimates need be computed only for $x = a = b = c = 1$. This completes the M-step and one cycle of the algorithm. To begin the next iteration, these new parameter estimates are used to obtain a new estimate of XABC via the equality in (4.9). The process continues until the absolute difference in the parameter estimates between iterations is negligibly small, say, less than 10^{-8}. At that point, the algorithm is said to have *fully converged.*

The EM algorithm is most advantageous in situations where there is no closed-form solution to the incomplete-data likelihood equations but the *full data likelihood* equations can be easily solved. Thus, it is ideal for many types of LC models. Newton–Raphson-type methods such as Fisher's scoring algorithm require computations of second derivatives of the likelihood equations with respect to the parameters. Besides being more difficult to program, these algorithms tend to be quite slow and highly dependent on starting values for convergence. With the EM algorithm, second derivatives are not needed, which greatly simplifies the algorithm and makes it much faster. But this is also a disadvantage of EM since second partial derivatives are required for model-based estimates of the standard errors of the model estimates. For situtions where standard errors are needed, some efficient options for computing standard errors in EM algorithm applications are described in Louis (1982), Baker (1992), and Jamshidian and Jennrich (2000).

Example: Illustration of the EM Algorithm for the Data in Table 2.5
The past-year marijuana use data in Table 2.5 (of Chapter 2) will be used to illustrate the EM algorithm. The LC model that will be fit to these data assumes the probability for an arbitrary cell in the ABC table, denoted (a,b,c), has the form $\sum_x \pi_x^X \pi_{a|x}^{A|X} \pi_{b|x}^{B|X} \pi_{c|x}^{C|X}$ as in (4.11) where $A = y_1$, $B = y_2$ and $C = y_3$. The ℓEM software will be used with model statement `mod A|X B|X C|X`, which matches the conditional probabilities in the model. A convergence criterion of 10^{-5} was specified. Starting values for the prevalence probability and all six error parameters were set at 0.1. Convergence was obtained in just eight iterations. With random starting values (the ℓEM default option) convergence required 30–40 iterations. Therefore, it is slightly more efficient to provide good starting values if reasonable values are known. However, this is not essential since, in these days of personal computers, computing cycles are fairly inexpensive.[1] Table 4.2 shows the values of the parameter estimates and the loglikelihood for five iterations. The loglikelihood is shown in the last column.

The probabilistic parameterization of the LC model likelihood in (4.4) is quite adequate for many straightforward applications of LCA. This specification has the advantage of being simple to specify, easy to understand for anyone having completed a basic course in probability theory, and compatible with classical true score, true value, and response error theory described in the previous chapters. However, this parameterization has an important limitation in that many constraints that are of interest to survey data analysts are difficult to specify using probability models. An alternative is the loglinear model parameterization. Loglinear and logit (referred to collectively as LL) models are widely used for analyzing cross-classified data and polytomous response variables. Although more complex than probabilistic models, LL

[1]As will be discussed in Chapter 5, some software packages (ℓEM is not one of them) will automatically rerun the model with as many sets of starting as the user specifies. In this way, the user can ensure that the final solution is a global rather than a local MLE (see Section 5.1.6).

Table 4.2 Parameter Estimates by Iteration Using the EM Algorithm

p Parameters	π_1^X	$\pi_{2\|1}^{A\|X}$	$\pi_{1\|2}^{A\|X}$	$\pi_{2\|1}^{B\|X}$	$\pi_{1\|2}^{B\|X}$	$\pi_{2\|1}^{C\|X}$	$\pi_{1\|2}^{C\|X}$	Log($\mathcal{L}$)
Iteration 0	0.1	0.1	0.1	0.1	0.1	0.1	0.1	−11975.455389
Iteration 1	0.0704	0.0494	0.005	0.0109	0.0157	0.0408	0.018	−8083.051911
Iteration 2	0.0764	0.0764	0.0011	0.0113	0.0094	0.0681	0.0141	−7973.840537
Iteration 3	0.0773	0.0837	0.0009	0.0132	0.0081	0.0731	0.0137	−7971.045063
Iteration 4	0.0775	0.0854	0.0008	0.0138	0.0085	0.0746	0.0136	−7970.909115
Iteration 5	0.0775	0.0858	0.0008	0.014	0.0085	0.075	0.0136	−7970.900194
Maximum iteration (8)	0.0775	0.0859	0.0008	0.014	0.0085	0.0751	0.0136	−7970.899527

models have much intuitive appeal and are accessible by a large and growing number of survey statisticians. Models that would be highly complex in a probabilistic framework can be easily specified in the LL model framework. In addition, the more recent software packages for fitting LC models rely to a great extent on the LL model framework. The next section shows how the basic LC model with three indicators can be easily specified using LL models that incorporate a latent variable X denoting the true value of the evaluation variable. As we shall see, LL models with latent variables provide a much higher level of flexibility in the specification of model constraints and have many other advantages over probabilistic models.

4.2.4 Loglinear Model Parameterization

The loglinear model parameterization of the LC model was introduced by Haberman (1979) and Goodman (1973b). This development, like the development of EM algorithms for estimation, greatly increased the applicability and accessibility of LCA. Loglinear models are ideal for exploring the associations between two or more categorical variables, particularly in situations where there is no single dependent variable and all variables are treated as being explanatory of the variation in the cell probabilities. For estimating classification error probabilities, logit models, which are closely related to loglinear models, are often more convenient. Below (in this chapter) and in Appendix B we demonstrate how a loglinear model can be easily transformed into a logit model by including particular effects in the loglinear model. Even more useful in classification error research is the *modified path model*, which is a system of logit models that are simultaneously estimated as a single model. In modified path models, the same variable may be a dependent variable in one submodel and an independent variable in one or more other submodels within the same path model. These are discussed in some detail in Section 4.2.6. Readers who are unfamiliar with loglinear models or who might benefit from a brief review of these models are encouraged to read the primer on loglinear models in Appendix B and to refer to it in the following discussion.

To fix the ideas, we again consider the simple LCA for three indicators given in (4.3). As shown in Appendix B, the logit model for the conditional probability $\pi^{A}_{a|x}$ is given by

$$\pi^{A}_{a|x} = \frac{\exp(u^{A}_{a} + u^{XA}_{xa})}{\Sigma_{a} \exp(u^{A}_{a} + u^{XA}_{xa})} \tag{4.13}$$

where u^{A}_{a} is the main effect and u^{XA}_{xa} is the interaction effect between A and X. Note that, when A is an indicator of X, the interaction term u^{XA}_{xa} will almost always be statistically significant from 0 since A (as well as other indicators) is designed to act as a proxy for the latent variable and should, therefore, be highly correlated with it. The logit models for $\pi^{B}_{b|x}$ and $\pi^{C}_{c|x}$ can be obtained in the same way, replacing A in (4.13) with B and C, respectively. The logit model for π^{X}_{x} is the complex expression

$$\pi^{X}_{x} = \frac{\Sigma_{abc} \exp(u^{X}_{x} + u^{A}_{a} + u^{B}_{b} + u^{C}_{c} + u^{XA}_{xa} + u^{XB}_{xb} + u^{XC}_{xc})}{\Sigma_{xabc} \exp(u^{X}_{x} + u^{A}_{a} + u^{B}_{b} + u^{C}_{c} + u^{XA}_{xa} + u^{XB}_{xb} + u^{XC}_{xc})} \tag{4.14}$$

Note that the numerator of this expression is the sum of the expected cell counts for the X subtable corresponding to $X = x$ and the denominator is the sample size n.

Modified path analysis essentially substitutes the logistic model form of the probabilities in the probabilistic model formulation of the likelihood kernel. It is easy to show the equivalence of the probabilistic model, loglinear model, and modified path model formulations using the exponential forms defined above. The probabilistic form of the expected proportion in cell (x,a,b,c) is

$$\pi^{XABC}_{xabc} = \pi^{X}_{x} \pi^{A|X}_{a|x} \pi^{B|X}_{b|x} \pi^{C|X}_{c|x} \tag{4.15}$$

Now substitute the logistic model representation for each probability on the right side of the equality to obtain

$$\begin{aligned} &\frac{\exp(u^{X}_{x}) \Sigma_{abc} \exp(u^{A}_{a} + u^{B}_{b} + u^{C}_{c} + u^{XA}_{xa} + u^{XB}_{xb} + u^{XC}_{xc})}{\Sigma_{xabc} \exp(u^{X}_{x} + u^{A}_{a} + u^{B}_{b} + u^{C}_{c} + u^{XA}_{xz} + u^{XB}_{xb} + u^{XC}_{xc})} \frac{\exp(u^{A}_{a} + u^{XA}_{xa})}{\Sigma_{a} \exp(u^{A}_{a} + u^{XA}_{xa})} \\ &\times \frac{\exp(u^{B}_{b} + u^{XB}_{xb})}{\Sigma_{b} \exp(u^{B}_{b} + u^{XB}_{xb})} \frac{\exp(u^{C}_{c} + u^{XC}_{xc})}{\Sigma_{c} \exp(u^{C}_{c} + u^{XC}_{xc})} \end{aligned} \tag{4.16}$$

Note that $\sum_{abc} \exp\left(u_a^A + u_b^B + u_c^C + u_{xa}^{XA} + u_{xb}^{XB} + u_{xc}^{XC}\right)$ can be rewritten as

$$\sum_a \exp(u_a^A + u_{xa}^{XA}) \sum_b \exp(u_b^B + u_{xb}^{XB}) \sum_c \exp(u_c^C + u_{xc}^{XC}) \tag{4.17}$$

so from (4.15) and (4.16) we obtain

$$\pi_{xabc}^{XABC} = \frac{\exp(u_x^X + u_a^A + u_b^B + u_c^C + u_{xa}^{XA} + u_{xb}^{XB} + u_{xc}^{XC})}{\sum_{xabc} \exp(u_x^X + u_a^A + u_b^B + u_c^C + u_{xa}^{XA} + u_{xb}^{XB} + u_{xc}^{XC})} \tag{4.18}$$

As noted previously, the denominator is simply the total number of observations, n, while the numerator may be recognized as the loglinear model for the expected cell count $m_{xabc} = n\pi_{xabc}^{XABC}$:

$$m_{xabc} = u_x^X + u_a^A + u_b^B + u_c^C + u_{xa}^{XA} + u_{xb}^{XB} + u_{xc}^{XC} \tag{4.19}$$

Using the shorthand notation for hierarchical models described in Appendix B, this model may be represented concisely as $\{X\, A\, B\, C\, XA\, XB\, XC\}$ or simply as $\{XA\ XB\ XC\}$. Since the models we consider are all hierarchical, only the highest-order interaction terms are needed to represent the model.

The absence of any interactions between A, B, and C specifies that A, B, and C are conditionally independent given X, which is the definition of local independence, a key assumption for the standard LC model. Local dependence can be modeled by adding one or more of the following interactions: XAB, XAC, XBC, AB, AC, or BC. However, in that case, the model would not be identifiable (discussed further in Section 5.1.2). These results provide justifation that the three formulations are equivalent.

In the LL representation, the constraints in (4.6) translate into the usual loglinear modeling restrictions (see Appendix A):

$$\begin{gathered}
\sum_x u_x^X = \sum_a u_a^A = \sum_b u_b^B = \sum_c u_c^C = 0 \\
\sum_x u_{xa}^{XA} = \sum_a u_{xa}^{XA} = \sum_x u_{xb}^{XB} = \sum_b u_{xb}^{XB} = \sum_x u_{xc}^{XC} = \sum_c u_{xc}^{XC} = 0
\end{gathered} \tag{4.20}$$

The EM algorithm can be used to obtain estimates of the u parameters in (4.19). These estimates can then be used to obtain estimates of the parameters π_x^X, $\pi_{a|x}^{A|X}$, $\pi_{b|x}^{B|X}$, and $\pi_{c|x}^{C|X}$ through the logistic equations in (4.13) and (4.14). To see this, suppose that $\hat{u}_x^X$ and $\hat{u}_{a|x}^{A|X}$ denote the MLEs of u_x^X and $u_{a|x}^{A|X}$. Then the MLE of the false-positive probability for A is

$$\hat{\pi}_{1|2}^{A|X} = \frac{\exp(\hat{u}_1^A + \hat{u}_{21}^{XA})}{\exp(\hat{u}_1^A + \hat{u}_{21}^{XA}) + \exp(\hat{u}_2^A + \hat{u}_{22}^{XA})} \tag{4.21}$$

and the MLE of the false-negative probability is

$$\hat{\pi}_{2|1}^{A|X} = \frac{\exp(\hat{u}_2^A + \hat{u}_{12}^{XA})}{\exp(\hat{u}_1^A + \hat{u}_{12}^{XA}) + \exp(\hat{u}_2^A + \hat{u}_{12}^{XA})} \tag{4.22}$$

MLEs for the other error probabilities can be obtained similarly.

Equivalence of the three formulations implies that MLEs of the response (or error) probabilities obtained by maximizing the likelihood corresponding to the model in (4.15) will be equal to the MLEs of the same parameters obtained by maximizing the likelihood in (4.16) and (4.19). Since all three parameterizations will produce the same estimates of the model parameters, one has a choice as to which parameterization to use. Many latent class modelers will use all three, depending on the situation and the constraints imposed on the parameters. However, the LL parameterization does provide some important advantages over its probability model counterpart, due primarily to its generality and the greater flexibility that it affords for specifying many different kinds of classification error models, not just the standard LCA. As an example, latent variable models that include virtually any combination of interactions among the indicators ($AB, XAB, BC,$ etc.) can be explored to test model assumptions such as local independence, univocality, and group homogeneity. These models will be discussed further in Section 5.2. In addition, as we shall see, many theoretical results developed for loglinear and logistic models without latent variables extend readily to loglinear models with latent variables. For that reason, serious students of LCA are advised to become proficient in the theory and application of loglinear models.

4.2.5 Example: Computing Probabilities Using Loglinear Parameters

This example demonstrates how to convert the u-parameter estimates from an LL model to π parameters of the corresponding probability model for dichotomous variables. The same general approach could be used for polytomous data. The standard LC model for three indicators—$\{XA\ XB\ XC\}$—was fit to the data on past-year marijuana use in Table 2.4 using the ℓEM software. The corresponding ℓEM model statement is `mod{XA XB XC}`. Since the data are binary, there are seven parameters plus the main effect, so the model is fully saturated. The estimated u parameters are shown in Table 4.3. Also shown are the corresponding π-parameter estimates for the corresponding probability model. Note that, as a result of the usual LL model constraints, the following relations hold:

$$\begin{gathered} u_2^X = -u_1^X, \quad u_2^A = -u_1^A, \quad u_{12}^{XA} = u_{21}^{XA} = -u_{11}^{XA}, \quad u_{22}^{XA} = u_{11}^{XA}, \\ u_{12}^{XB} = u_{21}^{XB} = -u_{11}^{XB}, \quad u_{22}^{XB} = u_{11}^{XB}, \quad u_{12}^{XC} = u_{21}^{XC} = -u_{11}^{XC}, \quad u_{22}^{XC} = u_{11}^{XC} \end{gathered} \tag{4.23}$$

Table 4.3 Latent Class Model Parameter Estimates for Past-Year Marijuana Use Data in Table 2.4

u Parameters	u_1^X	u_1^A	u_{11}^{XA}	u_1^B	u_{11}^{XB}	u_1^C	u_{11}^{XC}
Estimates	0.44	–1.18	2.37	–0.13	2.25	–0.44	1.70
p Parameters	π_1^X	$\pi_{2\mid 1}^{A\mid X}$	$\pi_{1\mid 2}^{A\mid X}$	$\pi_{2\mid 1}^{B\mid X}$	$\pi_{1\mid 2}^{B\mid X}$	$\pi_{2\mid 1}^{C\mid X}$	$\pi_{1\mid 2}^{C\mid X}$
Estimates	0.078	0.085	0.0008	0.0140	0.0085	0.0751	0.0136

The calculations will be illustrated for $\hat{\pi}_{2\mid 1}^{A\mid X}$ and $\hat{\pi}_{1\mid 2}^{C\mid X}$

$$\pi_{2\mid 1}^{A\mid X} = \frac{\exp(\hat{u}_2^A + \hat{u}_{12}^{XA})}{\exp(\hat{u}_1^A + \hat{u}_{11}^{XA}) + \exp(\hat{u}_2^A + \hat{u}_{12}^{XA})} = \frac{\exp(1.18 + -2.37)}{\exp(-1.18 + 2.37) + \exp(1.18 - 2.37)}$$

$$= \frac{0.3042}{3.2871 + 0.3042} = 0.0847$$

$$\pi_{1\mid 2}^{C\mid X} = \frac{\exp(\hat{u}_1^C + \hat{u}_{21}^{XC})}{\exp(\hat{u}_1^C + \hat{u}_{21}^{XC}) + \exp(\hat{u}_2^C + \hat{u}_{22}^{XC})}$$

$$= \frac{\exp(-0.44 + -1.70)}{\exp(-0.44 + -1.70) + \exp(0.44 + 1.70)} = \frac{0.1175}{0.1175 + 8.510} = 0.0136$$

The model estimate of the true prevalence of past-year marijuana use is 7.8%. The model further suggests that false-negative (i.e., denial of marijuana use) rates are fairly high for all three indicators, but especially for A, which has a false-negative probability of 8.5%. These data are discussed again later in the chapter.

The next section expands the discussion of modified path models to more complex models that take advantage of this formulation. The modified path model combines the intuitive appeal of the probabilistic modeling approach with the power and generality of the logit modeling approach. The idea began with Goodman (1973b), who originally applied the method to manifest variables. Hagenaars (1990) extended the idea to incorporate latent variables that he referred to as the *modified LISREL method*, due to their similarity to structural equation (LISREL) models. In this book these models, whether with or without latent variables, will be referred to simply as *modified path models*.

4.2.6 Modified Path Model Parameterization

It is often more intuitive to specify a probabilistic latent class model rather than a loglinear model when the goal is to estimate the error probabilities for one or more indicators. There are a couple of reasons for this. First, the order in which indicators of X are collected may be important for some types of analysis. Suppose that the indicators A, B, and C are collected in the same interview and in that order. It is possible that obtaining A affected the measurements of B and C. For example, a respondent's memory of the events associated with the latent variable X might improve during the interview with each new question about X. Likewise, the process of obtaining a response to B could influence the accuracy of the response to C. If C is obtained last, then (theoretically) it cannot influence the errors in A or B; nor can B influence the error in A. A modeler wishing to study these local dependences might find it quite intuitive to write the probabilistic model

$$\pi_{xabc}^{XABC} = \pi_x^X \pi_{a|x}^{A|X} \pi_{b|xa}^{B|XA} \pi_{c|xab}^{C|XAB} \tag{4.24}$$

since this specification preserves the chronological ordering of the indicators while reflecting the hypotheses regarding the influence of preceding indicators on posterior ones. Now compare this specification with the equivalent loglinear model for $m_{xab} = n\pi_{xab}^{XAB}$, written as

$$\begin{aligned} m_{xabc} &= u + u_x^X + u_a^A + u_b^B + u_c^C \\ &+ u_{xa}^{XA} + u_{xb}^{XB} + u_{xc}^{XC} + u_{ab}^{AB} + u_{ac}^{AC} + u_{bc}^{BC} \\ &+ u_{abc}^{ABC} + u_{xab}^{XAB} + u_{xac}^{XAC} + u_{xbc}^{XBC} + u_{xabc}^{XABC} \end{aligned} \tag{4.25}$$

which is the model $\{XABC\}$ in shorthand notation. The ℓEM model statement could take a number of equivalent forms. A simple form is `mod X|ABC` or equivalently `mod XABC`.[2] The loglinear specification treats all variables as though they were observed coincidentally. We know from the discussion above that the probabilistic and the loglinear parameterizations will produce the same MLEs for the error probabilities; however, the chronological (and causal) ordering of the variables is obscured in (4.25).

Goodman's modified path model approach (Goodman 1973b) combines the probabilistic and loglinear (or logit) modeling approaches. Like the probabilistic parameterization, the modified path model preserves the causal ordering of the variables by specifying a separate logit model for each probability on the right side of (4.24), similar to what was done for deriving (4.16). However, it has the advantages of the loglinear modeling approach in that any logit or loglinear model can be specified for each conditional probability. For example, suppose that the indicators A, B, and C were observed in the order listed. Then $\pi_{a|x}^{A|X}$ can be replaced by

[2]The order in which the variables are listed does not matter in ℓEM. For example, `mod XABC` is equivalent to `mod ABCX`. In this book, we attempt to consistently list the latent variable first since that is required in the ℓEM label or `lab` statement.

$$\pi_{a|x}^{A|X} = \frac{\exp(u_a^A + u_{xa}^{XA})}{\sum_a \exp(u_a^A + u_{xa}^{XA})} \tag{4.26}$$

Here, the causal ordering is represented by the fact that the indicators B and C do not appear in the model since chronologically they can have no effect on A. Likewise, $\pi_{b|xa}^{B|XA}$ can be replaced by

$$\pi_{b|xa}^{B|XA} = \frac{\exp(u_b^B + u_{xb}^{XB} + u_{ab}^{AB} + u_{xab}^{XAB})}{\sum_b \exp(u_b^B + u_{xb}^{XB} + u_{ab}^{AB} + u_{xab}^{XAB})} \tag{4.27}$$

and $\pi_{c|xab}^{C|XAB}$ by

$$\pi_{c|xab}^{C|XAB} = \frac{\exp(u_c^C + u_{xc}^{XC} + u_{ac}^{AC} + u_{bc}^{BC} + u_{xac}^{XAC} + u_{xbc}^{XBC} + u_{abc}^{ABC} + u_{xabc}^{XABC})}{\sum_b \exp(u_c^C + u_{xc}^{XC} + u_{ac}^{AC} + u_{bc}^{BC} + u_{xac}^{XAC} + u_{xbc}^{XBC} + u_{abc}^{ABC} + u_{xabc}^{XABC})} \tag{4.28}$$

Note that it is possible to specify reduced models for all three conditional probabilities. For example, the XAB term in (4.27) as well as the three and four-way interaction terms in (4.28) can be set to 0 by omitting them from the model to obtain a much more parsimonious specification of local dependence.

Just as the joint probability of π_{xabc}^{XABC} is the product of π_x^X and the three error probabilities in (4.26)–(4.28), the likelihood of the model for π_{xabc}^{XABC} is also the product of the likelihoods associated with the submodels in (4.26)–(4.28). This *product likelihood* can be maximized to obtain estimates of parameters associated with the three submodels. In many cases, the three submodels can be fit as three separate models: one for the XA subtable, one for the XAB subtable, and the last for the $XABC$ subtable. Estimates of π_{xabc}^{XABC} can be computed via (4.24) after replacing each product term by its corresponding MLE. This will be discussed in more detail subsequently.

This model for π_{xabc}^{XABC} is severely overparameterized and not identifiable. It can be made identifiable by leaving out certain higher-order interaction terms, setting other restrictions, or both. In the probabilistic parameterization, the choices are somewhat limited compared with the possibilities using the modified path model parameterization. For example, in the probabilistic parameterization, the term $\pi_{c|xab}^{C|XAB}$ might be replaced by $\pi_{c|q}^{C|Q}$, where the choice Q can be any combination of the variables X, A, and B (XA, AB, etc.). For the logit parameterization, these models can be easily represented, but so can models such as

$$\pi_{c|xab}^{C|XAB} = \frac{\exp(u_c^C + u_{xc}^{XC} + u_{bc}^{BC})}{\sum_b \exp(u_c^C + u_{xc}^{XC} + u_{bc}^{BC})} \tag{4.29}$$

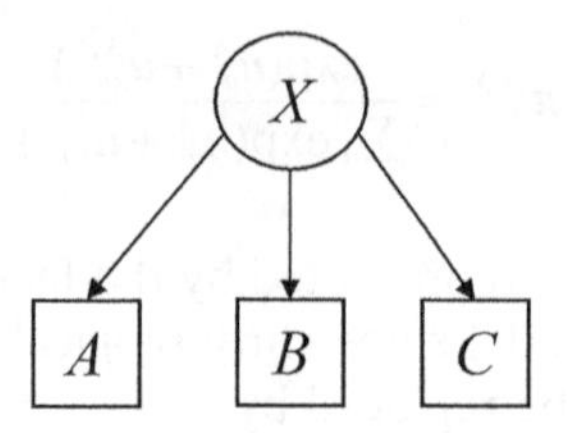

Figure 4.2 Path model diagram for the standard latent class model.

which cannot be easily represented in the probabilistic framework. These models arise quite naturally in the LL model parameterization since they are *nested* within higher-order models; that is, they can be obtained by simply omitting some higher-order terms. Note that, under the probabilistic parameterization, the highest-order interaction that can be formed from the variables constituting Q and C are automatically included in the model. For example, if $Q = XB$, then the u_{xbc}^{XBC} effect is automatically included in the probabilistic specification for $\pi_{c|xb}^{C|XB}$. However, it can be easily excluded from the logit parameterization as shown in (4.29). Thus, a key advantage of the modified path model approach is that models can be more specialized and parsimonious than in the probabilistic parameterization.

The modified path model approach gets its name from the graphical method of displaying the causal relationships among variables via a path model diagram. An example of such a diagram is shown in Figure 4.2, which represents the three-indicator LC model in (4.3). Latent variables are represented by circles and manifest variables, by rectangles. The arrows leading from the latent variable to each indicator shows the direction of causal influence; specifically, X gives rise to A, B, and C. Thus, each arrow represents a conditional probability or interaction of the corresponding indicator with respect to X. Likewise, the absence of a line between two variables represents the conditional independence of the variables. Thus, it is apparent from Figure 4.2 that A, B, and C are mutually locally independent. The order in which A, B, and C appear in the figure can be used to indicate the order the indicators were observed. For example, C is posterior to B and both B and C are posterior to A.

4.2.7 Recruitment Probabilities

In addition to the error probabilities such as $\Pr(A = a|X = x)$, other conditional probabilities are also often of interest in LCA. One such probability is $\Pr(X = x|A = a)$, referred to as the *recruitment probability for the indicator A* (Lazarsfeld and Henry 1968). This is the probability that an individual classified by A into category a is truly in class x. More generally, estimates of recruitment probabilities can used to estimate the true characteristic for individuals having a certain response pattern on the indicator variables.

Using Bayes' rule, the recruitment probability for a single indicator A can be written

$$\Pr(X=x \mid A=a)=\frac{\Pr(A=a \mid X=x)\Pr(X=x)}{\Pr(A=a)}=\pi_{a|x}^{A|X}\frac{\pi_x^X}{\pi_a^A} \tag{4.30}$$

In other words, the probability that an individual in truly in x given he/she is classified into category a is the error probability $\pi_{a|x}^{A|X}$ times the ratio of probability of truly being in x to the probability of being classified in a.

For three indicators, the recruitment probability is an estimate of the true class for an individual with response pattern (a,b,c) for indicators A,B,C. This is computed as

$$\begin{aligned}\pi_{x|abc}^{X|ABC} &= \pi_{abc|x}^{ABC|X}\frac{\pi_x^X}{\pi_{abc}^{ABC}} \\ &= \pi_{a|x}^{A|X}\pi_{b|x}^{B|X}\pi_{c|x}^{C|X}\frac{\pi_x^X}{\pi_{abc}^{ABC}}\end{aligned} \tag{4.31}$$

Thus, the probability that an individual with pattern (a,b,c) truly is in class x is the product of the error probabilities of the indicators times the ratio of the unconditional latent class probability and the (a,b,c) cell probability. The MLEs of the recruitment probabilities are obtained by replacing the parameters in (4.31) by their corresponding MLEs.

Recruitment probabilities can be used to classify each case having a given response pattern into a single latent class using the *modal class* associated with the response pattern. The *modal class* for the response pattern (a,b,c) is the latent class having the highest value of $\pi_{x|abc}^{X|ABC}$, that is, $\max_x(\pi_{x|abc}^{X|ABC})$. As an example, consider the response pattern (1,1,1) for three trichotomous indicators A, B, and C and suppose that the estimates of $\pi_{x|111}^{X|ABC}$ are 0.7, 0.2, and 0.1 for $x=1,2,3$, respectively. The modal class for this response pattern is $X=1$. Nevertheless, if we classified all cases with this response pattern into class 1, we could expect 30% of them to be misclassified since the probability that X is the true class is 0.7.

In general, the probability that classifying cases with a given response pattern by their modal class will be incorrect is $1-\max_x(\pi_{x|abc}^{X|ABC})$. The overall error rate using the modal class classifications is therefore

$$\frac{\sum_{abc} n_{abc}\left[1-\max_x(\pi_{x|abc}^{X|ABC})\right]}{n} \tag{4.32}$$

We may be able to reduce the overall error rate by increasing the modal class probabilities through the addition of grouping variables.

4.2.8 Example: Computing Probabilities Using Modified Path Model Parameters

Here we repeat the analysis above to show how to convert the u-parameter estimates from a modified path model to π parameters of the corresponding probability model. Again, the standard LC model for three indicators was fit to the data on past-year marijuana use in Table 2.4 using the ℓEM software. The model specified was the path model equivalent—$\{X\}\{A|X\}\{B|X\}\{C|X\}$—of the loglinear model $\{XA\ XB\ XC\}$. As before there are seven parameters plus the main effect so the model is fully saturated. The estimated u parameters are shown in Table 4.4. Also shown are the corresponding π-parameter estimates for the corresponding probability model.

The estimates in Table 4.4 are identical to those in Table 4.3 except for u_1^X. This is because in a path model, the parameters of a submodel are computed from the collapsed table involving only the submodel parameters. In the case of u_1^X, the collapsed table X is formed and the parameter u_1^X is obtained by fitting the model

$$\pi_1^X = \frac{\exp(u_1^X)}{\Sigma_x \exp(u_x^X)} \tag{4.33}$$

Substituting the estimate of u_1^X from the table in the formula produces

$$\pi_1^X = \frac{\exp(-1.24)}{\exp(-1.24)+\exp(1.24)} = \frac{0.29}{0.29+3.46} = 0.078 \tag{4.34}$$

To illustrate how conditional probabilities are computed, consider the computation of the false positive probability for A, namely, $\hat{\pi}_{1|2}^{A|X}$. This is computed as follows:

$$\pi_{1|2}^{A|X} = \frac{\exp(u_1^A + u_{21}^{XA})}{\exp(u_1^A + u_{21}^{XA}) + \exp(u_2^A + u_{22}^{XA})} = \frac{\exp(-1.18-2.37)}{\exp(-1.18-2.37)+\exp(1.18+2.37)}$$
$$= 0.000824 \tag{4.35}$$

Table 4.4 Modified Path Model Parameter Estimates for Past-Year Marijuana Use

u Parameters	u_1^X	u_1^A	u_{11}^{XA}	u_1^B	u_{11}^{XB}	u_1^C	u_{11}^{XC}
Estimates	−1.24	−1.18	2.37	−0.13	2.25	−0.44	1.70
π Parameters	π_1^X	$\pi_{2\|1}^{A\|X}$	$\pi_{1\|2}^{A\|X}$	$\pi_{2\|1}^{B\|X}$	$\pi_{1\|2}^{B\|X}$	$\pi_{2\|1}^{C\|X}$	$\pi_{1\|2}^{C\|X}$
Estimates	0.078	0.085	0.0008	0.014	0.0085	0.075	0.014

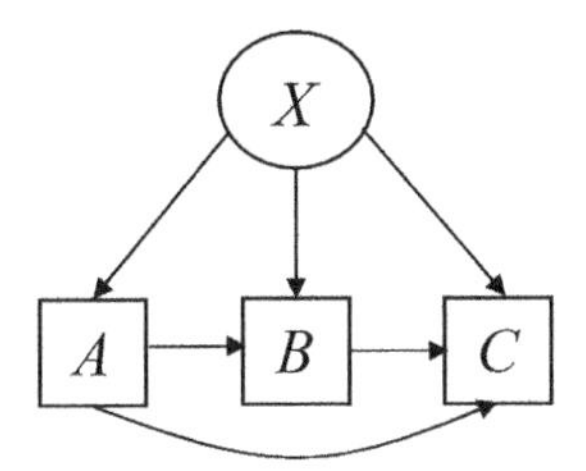

Figure 4.3 Path diagram for a latent class model with direct effects between indicators.

Causal dependencies between the indicators can be represented in various ways in a path diagram. The path diagram in Figure 4.3 shows one possible approach. The arrow pointing from A to B specifies that A directly influences B. Likewise, the arrows pointing from A to C and C to B specify that A directly influences C and B directly influences C, respectively. These so-called *direct effects* (Hagenaars 1988b) are modeled by adding the interaction terms for the pairs of indicators connected by arrows. The word "direct" used in this context means that X is not involved in these interactions. An arrow leading from one indicator to another violates the assumption of local independence. In Section 5.2.4 we show how adding direct effects to the model can address the problems of lack of model fit due to local dependence regardless of the cause of the dependence.

The path diagram in Figure 4.3 shows the terms in the loglinear model. For the $A \rightarrow B$ arrow, the interaction AB is added to the model for $\pi_{b|xa}^{B|XA}$. Note that the XAB is still zero in the model. In ℓEM, the model statement would contain a specification for the submodel $B|XA$ with the added restriction $\{XB\ AB\}$ follows:

```
mod B|XA {XB AB}
```

This is equivalent to specifying the restriction

$$\pi_{b|xa}^{B|XA} = \frac{\exp(u_b^B + u_{xb}^{XB} + u_{ab}^{AB})}{\sum_b \exp(u_b^B + u_{xb}^{XB} + u_{ab}^{AB})} \tag{4.36}$$

for the $\pi_{b|xa}^{B|XA}$ term in the model. Likewise, to add a direct effect between A and C and B and C, the $C|XAB$ submodel, which corresponds to the conditional probability $\pi_{c|xab}^{C|XAB}$ in the product likelihood, would be restricted by $\{XC\ AC\ BC\}$ or in ℓEM code:

```
mod C|XAB {XC AC BC}
```

Either of the two direct effects (AC, BC, or both) can be easily set to 0 by removing the corresponding effect(s) from the model.

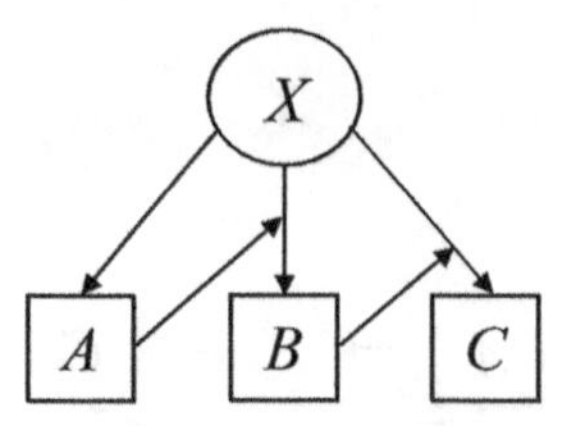

Figure 4.4 Path diagram for a latent class model with XAB and XBC effects.

Unfortunately, direct effects can be quite difficult to interpret in the context of classification error since they are independent of X. Conceptually, how does the direct effect, say, AB, relate to error in B that is modeled by XB? To address this difficulty, the three-way interaction (XAB) could be included in the model, which is more easily interpretable. Moreover, in many cases, local dependence is more effectively modeled by XAB rather than AB (see Section 5.2). One interpretation of XAB is that the error probability for B (i.e., $\pi_{b|x}^{B|X}$) varies according to an individual's response to A. Likewise, B may influence the error in C. The path diagram in Figure 4.4 represents these two situations. To denote the three-way interaction, the arrow is directed at the arrow connecting X and the indicator rather than being directed from indicator to indicator. For example, the arrow leading from A to the $X \rightarrow B$ arrow means that A influences the classification error in B represented by XB. Likewise, the arrow from B to the $X \rightarrow C$ arrow means that B influences the classification error in C represented by XC.

We shall discuss local dependence models in more detail in Section 4.2.8, with a much fuller discussion Section 5.2. The full potential of the modified path modeling will be explored in the next section after grouping variables are incorporated into the model. We shall see that models of increasing complexity are made much simpler by this approach, which allows the analyst to essentially create a submodel for each variable (or subtable) in an analysis. These submodels can be fit simultaneously to their corresponding subtables to yield MLEs for all the parameters in the full table.

4.3 INCORPORATING GROUPING VARIABLES

So far in this chapter, latent class models have consisted of two types of variables: latent variables and indicator variables. A third type of variable that can be incorporated easily into a latent class analysis is the *grouping variable*, which we encountered in describing the Hui–Walter method (Chapter 3). A grouping variable can be any manifest variable that is not intended to be an indicator of X. There are at least three reasons for incorporating grouping variables in the LC model. First, as we saw in the case of the Hui–Walter model, grouping variables can be used to achieve an identifiable model when

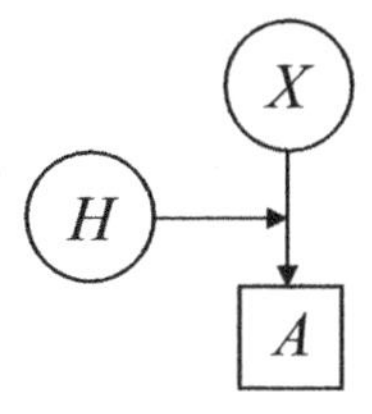

Figure 4.5 Unobserved heterogeneity in the error probabilities for indicator *A*.

the model would otherwise be nonidentifiable. The Hui–Walter approach used a single, dichotomous grouping variable that doubled the number of cell frequencies, thus increasing the degrees of freedom. By placing restrictions on some of the parameters of the fully saturated model for this extended table, we could estimate all the parameters of the restricted model.

Grouping variables can also be added to the model in order to obtain a better-fitting model. For example, a typical model-building approach is to start with the basic LC model that contains only the latent variable and its corresponding indicator variables. Such a simplified model may provide a poor fit to the data. Grouping variables are added that are correlated with the latent variable, the measurement errors in the indicator variables, or both, one by one until the model fit is acceptable. This process, referred to as *stepwise model building*, will be discussed later in the chapter. Thus, grouping variables can be an essential part of the modeling process.

A related reason for using grouping variables is to reduce the effects of *unobserved heterogeneity* in the model. Recall that an essential assumption for an LC model is that the error probabilities are the same for each individual in the population. When this assumption does not hold, the population is said to be *heterogeneous* with respect to the error probabilities. Suppose, however, that there is a categorical variable H such that within the categories of H, the assumption of homogeneity holds, that is, that $\pi_{a|xh}$ is the same for all individuals with the same h. If H is known, then adding the term XHA to the model will achieve *conditional homogeneity* for XA, which is still sufficient for model validity. When H is unknown, the model is said to have *unobserved heterogeneity*. Figure 4.5 is a path diagram that enables us to visualize this effect. In this diagram, H is a latent grouping variable. There are several ways to address this problem, which are discussed further in Section 5.2.1. However, one way is to add one or more grouping variables to the model that are highly correlated with H. If such grouping variables are available, adding them to the model will divide the population into categories or strata that are approximately homogeneous with respect to the error probabilities, thus increasing model validity.

Finally, grouping variables are used to test hypotheses about the causes and correlates of measurement error and variations in the prevalence of the latent variable in the population. For example, the analyst may wish to explore whether the accuracy of an indicator varies by mode of interview, household size, respondent race, type of interviewer, region of the country, and so on. It may also be of interest to test whether the latent prevalence probabilities are the same across age groups once classification errors are removed from the comparisons. These questions and many more can be explored through the use of grouping variables.

Adding grouping variables is quite simple, especially within the loglinear modeling framework. The widely used and standard approaches for building loglinear models can also be applied to latent variable models with a few differences. First, it is important to note that the latent variable loglinear model consists of essentially two model components—the *measurement component*, comprising all main effects and interactions with the indicator variables; and the *structural component*, comprising all other terms, that is, the main effects and interactions among the grouping variables alone as well as the grouping variables with the latent variable(s).

The shorthand notation for representing modified path models that we will use is in the form {structural component}{measurement component}, where each component may be further divided into submodels as needed to provide better structure and clarity. As an example, let F, G, and H denote grouping variables and suppose that we wish to model the data table FGHABC. The structural component is the submodel $\{XFGH\}$, and the measurement component is $\{ABC|XFGH\}$. Thus, the combined model may be expressed in shorthand notation as $\{XFGH\}\{ABC|XFGH\}$, where the symbol "$|XFGH$" denotes that A, B, and C are conditioned on X, F, G, and H. The measurement component can be further divided into submodels corresponding to a logit equation for each indicator as $\{A|XFGH\}\{B|XFGH\}\{C|XFGH\}$. Each of the submodels can be replaced by a simpler form if desired to save degrees of freedom without unduly sacrificing model fit. For example, $\{XFGH\}$ might be reduced to $\{FGH\ XF\ XG\ XH\}$, which can also be expressed as two submodels, $\{FGH\}\{XF\ XG\ XH\}$. Likewise, each submodel constituting the measurement component may also be reduced if desired; for example, $\{A|XFGH\}$ can be reduced to $\{XFA\ XGA\ XHA\}$, $\{B|XFGH\}$ can be reduced to $\{XGB\}$, and $\{C|XFGH\}$ can be reduced to $\{XC\ FC\}$. In this case, the shorthand representation of the full modified path model is

$$\{FGH\ XF\ XG\ XH\}\{XFA\ XGA\ XHA\}\{XGB\}\{XC\ FC\}$$

Separate path diagrams for the structural, measurement and full (combined) model are shown in Figure 4.6. The model is written in modified path model form by writing the probability for cell (f,g,h,a,b,c) of the FGHABC table as

$$\sum_x \pi_{fgh}^{FGH} \pi_{x|fgh}^{X|FGH} \pi_{a|xfgh}^{A|XFGH} \pi_{b|xfgh}^{B|XFGH} \pi_{c|xfgh}^{C|XFGH} \tag{4.37}$$

leaving π_{fgh}^{FGH} unrestricted and imposing the restrictions

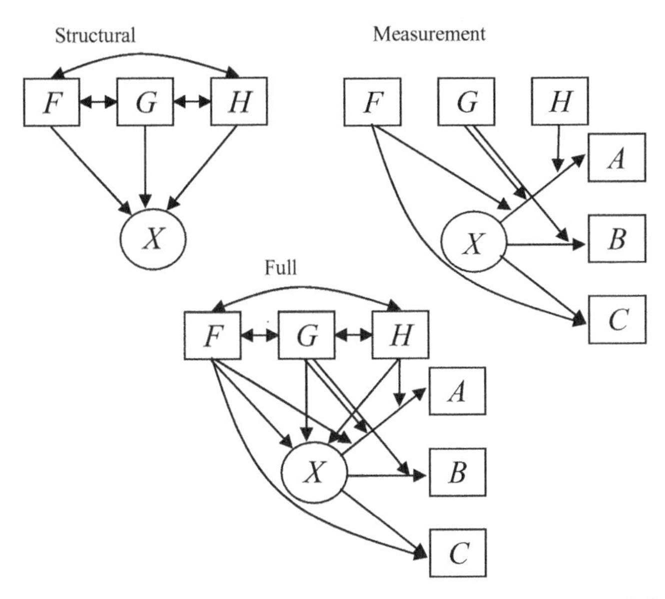

Figure 4.6 Path diagrams for the structural and measurement components and the full model for the model {FGH XF XG XH}{XFA XGA XHA}{XGB}{XC FC}.

$$
\begin{aligned}
\pi_{x|fgh}^{X|FGH} &= \frac{\exp(u_x^X + u_{xf}^{XF} + u_{xg}^{XG} + u_{xh}^{XH})}{\sum_x \exp(u_x^X + u_{xf}^{XF} + u_{xg}^{XG} + u_{xh}^{XH})} \\
\pi_{a|xfgh}^{A|XFGH} &= \frac{\exp(u_a^A + u_{xa}^{XA} + u_{fa}^{FA} + u_{ga}^{GA} + u_{ha}^{HA} + u_{xfa}^{XFA} + u_{xga}^{XGA} + u_{xha}^{XHA})}{\sum_a \exp(u_a^A + u_{xa}^{XA} + u_{fa}^{FA} + u_{ga}^{GA} + u_{ha}^{HA} + u_{xfa}^{XFA} + u_{xga}^{XGA} + u_{xha}^{XHA})} \\
\pi_{b|xfgh}^{B|XFGH} &= \frac{\exp(u_b^B + u_{xb}^{XB} + u_{gb}^{GB} + u_{xgb}^{XGB})}{\sum_b \exp(u_b^B + u_{xb}^{XB} + u_{gb}^{GB} + u_{xgb}^{XGB})} \\
\pi_{c|xfgh}^{C|XFGH} &= \frac{\exp(u_c^C + u_{xc}^{XC} + u_{fc}^{FC})}{\sum_c \exp(u_c^C + u_{xc}^{XC} + u_{fc}^{FC})}
\end{aligned} \tag{4.38}
$$

Two examples in the next section will further exemplify the use of grouping variables.

4.3.1 Example: Loglinear Parameterization of the Hui–Walter Model

The probabilistic parameterization of the Hui–Walter method was discussed in Section 3.3.1. Here we consider the loglinear parameterization of this model. For one grouping variable, G, and two indicators, A and B, the model assumes that the prevalence of the true characteristic varies across the categories of G. In other words, the distribution of X depends on G, which means that the interaction XG must be in the model and moreover, it must be statistically significant. Otherwise, the model may not be identifiable. In addition, the model

```
*  This is the loglinear parameterization
*  G is sex, X = 1 for EMP, 2 for UNEMP and 3 for NLF
*  A, B, C are indicators of X defined analogously
lat 1
man 3
dim 2 2 3 3
lab X G A B
mod {XG XA XB}*Absence of XGA and XGB implies group homogeneity for A and B
probs. dat hw.dat  *data file containing 18 cell frequencies for GAB
*************************************************************************
*  This is the modified path model parameterization
lat 1
man 3
dim 2 3 3 3
lab X G A B
mod X|G        * {XG} is the default model specification for X|G
    A|XG {XA} * The absence of XGA implies group homogeneity for A error probs.
    B|XG {XB} * The absence of XGB implies group homogeneity for B error probs.
dat hw.dat
```

Figure 4.7 ℓEM Input statements for the log linear and modified path model parameterizations of the Hui–Walter model.

assumes that the error probabilities are equal across levels of the grouping variable. This is equivalent to specifying that XA and XB do not depend on G; thus, the XGA and XGB interactions are omitted from the model. Combining these assumptions, we see that the equivalent Hui–Walter loglinear model is $\{XG\ XA\ XB\}$ or, using path model notation, three submodels are specified, represented as $\{XG\}\{XA\}\{XB\}$ or $\{XG\}\{A|X\}\{B|X\}$.

These models were fit using the ℓEM software to the data in Table 3.10. The input statements for both models are shown in Figure 4.7. In this example, the grouping variable is sex and the two indicators represent the labor force classifications from the original survey (A) and the test–retest reinterview survey (B). The latent variable X is interpreted as the true labor force status, where $X = 1$ for employed, $X = 2$ for unemployed, and $X = 3$ for not in the labor force. The indicator categories are defined analogously. The estimated error probabilities for these data were reported in Table 3.11. The corresponding loglinear parameters are shown in Table 4.5. At this point, we are only interested in the columns labeled u and $\exp(u)$. The remaining columns will be discussed subsequently.

The rows of Table 4.5 correspond to effects in the model. As an example, the u parameter for "main" is simply u, the model constant. Likewise, the u parameters for the row labeled X are the parameters u_1^X, u_2^X, and u_3^X. These three parameters are in the rows directly under X labeled 1, 2, and 3, respectively. The other rows are read the same way. For example, row 31 under XA corresponds to the parameter u_{31}^{XA}, and so on. In traditional loglinear analysis, loglinear interaction terms, when they are in the model, are a central focus. However, when the objective of the analysis is classification error evaluation,

Table 4.5 Loglinear Parameter Estimates and Other Statistics for the Hui–Walter Model Applied to the Data in Table 3.10[a]

Effect	u	Std Err	z Value	exp(u)	Wald	df	Prob
Main	0.1395	—	—	1.1496	—	—	—
X	—	—	—	—	—	—	—
1	–4.7114	0.356	–13.236	0.009	—	—	—
2	2.3368	0.2844	8.217	10.3485	—	—	—
3	2.3746	—	—	10.7463	181.5	2	0.000
G	—	—	—	—	—	—	—
1	–0.0194	0.0182	–1.069	0.9807	—	—	—
2	0.0194	—	—	1.0196	1.14	1	0.285
A	—	—	—	—	—	—	—
1	1.1385	0.0794	14.342	3.1222	—	—	—
2	–1.9869	0.2363	–8.41	0.1371	—	—	—
3	0.8484	—	—	2.3358	228.79	2	0.000
B	—	—	—	—	—	—	—
1	2.1397	0.1854	11.541	8.4971	—	—	—
2	–1.8822	0.176	–10.696	0.1523	—	—	—
3	–0.2575	—	—	0.773	579.05	2	0.000
XG	—	—	—	—	—	—	—
11	0.0715	0.0193	3.711	1.0741	—	—	—
12	–0.0715	—	—	0.931	—	—	—
21	0.18	0.0347	5.181	1.1972	—	—	—
22	–0.18	—	—	0.8353	—	—	—
31	–0.2515	—	—	0.7777	—	—	—
32	0.2515	—	—	1.2859	335.57	2	0.000
XA	—	—	—	—	—	—	—
11	3.6329	0.2887	12.582	37.8231	—	—	—
12	–3.5558	0.3544	–10.034	0.0286	—	—	—
13	–0.0771			0.9258	—	—	—
21	–1.6519	0.2364	–6.988	0.1917	—	—	—
22	3.3195	****	****	27.6478	—	—	—
23	–1.6677	—	—	0.1887	—	—	—
31	–1.981	—	—	0.1379	—	—	—
32	0.2363	—	—	1.2665	—	—	—
33	1.7447	—	—	5.7244	226.85	4	0.000
XB	—	—	—	—	—	—	—
11	6.0165	0.3748	16.053	410.1213	—	—	—
12	–3.2919	0.3421	–9.622	0.0372	—	—	—
13	–2.7245	—	—	0.0656	—	—	—
21	–2.892	****	****	0.0555	—	—	—
22	3.1045	****	****	22.2983	—	—	—
23	–0.2125	—	—	0.8085	—	—	—
31	–3.1245	—	—	0.044	—	—	—
32	0.1874	—	—	1.2061	—	—	—
33	2.9371	—	—	18.8603	262.89	4	0.000

[a]*Abbreviations*: Std Err—standard error; df—degrees of freedom; Prob—probability.

these parameters are not particularly useful. For example, in the table, the terms $\exp(u_{xa}^{XA})$ are large whenever $x = a$ as expected since A is an indicator of X. Note that $\exp(u_{11}^{XA})$ and $\exp(u_{22}^{XB})$ are among the largest parameters. These are 37.8 in row 11 under the XA heading and 410.1 in row 22 under XB heading, respectively. These large values portend the results shown in Table 3.11 that EMP is measured with a high degree of accuracy by both indicators. Further, B is more accurate in that regard than A. However, much more information regarding classification errors can be obtained more readily by going straight to the error probabilities themselves which are also provided in the ℓEM output by default. The value of the information in Table 4.5 will become more apparent in the discussion of model building and evaluation below.

4.3.2 Example: Analysis of Past-Year Marijuana Use with Grouping Variables

The National Household Survey on Drug Abuse (NHSDA) design is a stratified, multistage cluster sample of dwelling units selected in approximately 127 primary sampling units (PSUs) in 1994 and 115 PSUs in 1995 and 1996. The PSUs represent geographic areas in the United States; generally defined as counties, groups of counties or metropolitan statistical areas (MSAs). The target population includes persons 12 years old or older who live in households with annual sample sizes in the range of 17,700–18,300 households. Hispanics, blacks, younger persons, and the residents of certain MSAs are oversampled to ensure that the sample sizes are adequate to produce the subpopulation estimates of interest. The response rate is about 80%.

Although field interviewers collect some nonsensitive items in the questionnaire, most of the interview is self-administered. In 2002, the survey was redesigned for audiocomputer-assisted self-interviewing (ACASI) administration and renamed the National Survey on Drug Use and Health (NSDUH). For the years analyzed here, however, the NHSDA used a paper-and-pencil questionnaire. The NHSDA questionnaire was divided into sections corresponding to each substance of interest: tobacco, alcohol, marijuana, cocaine, crack, hallucinogens, inhalants, analgesics, tranquilizers, stimulants, and sedatives. Each section had its own answer sheet to be completed by the respondent. After completing a section, the respondent placed the answer sheet in an envelope that was concealed from the interviewer. Some questions about drug use activities were asked in multiple ways that provided a basis for conducting a repeated measurements analysis of drug usage.

As an example, Biemer and Wiesen (2002) formed three indicators of past-year marijuana use in their analysis from these repeated measurements. The indicators were collected sequentially during the interview with no opportunity for respondents to return to an earlier item to revise their responses. Since the measures follow a chronology, a modified path model approach seems appropriate given that the ordering of the indicators could affect the error probabilities. The first indicator, A, was derived from the recency of use (or

1. How long has it been since you last used marijuana or hashish?
A = "Yes" if either "Within the past 30 days" or "More than 30 days but within past 12 months;" **A = "No" if otherwise.**
2. Now think about the past 12 months from your 12-month reference date through today. On how many days in the past 12 months did you use marijuana or hashish?
B = "Yes" if response is 1 or more days; **B = "No" otherwise.**

Figure 4.8 Two embedded questions on past-year marijuana smoking.

recency) question (Figure 4.8, question 1). The second indicator, B, was derived from the frequency of use (or frequency) question (Figure 4.8, question 2). Both A and B were coded "1" if the corresponding response indicated *past-year* marijuana use and "2" otherwise. The final indicator, C, was derived from a combination of seven questions about past-year marijuana use that were similar to one another although distinct from A and B. These questions appeared on the so-called drug answer sheet at the end of the approximately one-hour interview. A positive response to any of the seven questions was regarded as evidence of past-year use in which case C was coded "1." Otherwise, C was coded "2." Of particular interest in the study was the evaluation of misclassification for A and B. There was less interest in the composite indicator, C, because it as formed primarily as a device for obtaining an identifiable model for the LCA of A and B. Three years of NHSDA data were analyzed: 1994, 1995, and 1996. These data were weighted to account for the unequal probabilities of selection. Sample clustering was also taken into account in computing standard errors using the *design effect* method to be discussed in Section 5.3. Details of the complex sample aspects of this analysis is deferred until then.

The models that Biemer and Wiesen considered were limited to simple extensions of the basic LC model for three measurements incorporating three grouping variables defined by age (G), race (R), and sex (S). The data used in the analysis are provided in Table 4.6. The structural component of the model was fully saturated: $\{XGRS\}$. Since the focus of the investigation was on classification error, no attempt was made to simplify this component by excluding some higher-order interactions. Many alternate forms of the measurement component were considered. For the simplest model considered, the error probabilities $\pi_{a|x}^{A|X}$, $\pi_{b|x}^{B|X}$, and $\pi_{c|x}^{C|X}$ were assumed to be group homogeneous; in other words, the error terms XA, XB, and XC were assumed to not interact with G, R, and S. These assumptions yield the model $\{XGRS\}\{XA\}\{XB\}\{XC\}$ having 54 parameters. Among the locally independent models (i.e., models with no direct effects) that were tested, the most complex model (with 90 parameters) specified that the error probabilities depend on three grouping variables taken one at a time: $\{XGRS\}\{XGA\ XRA\ XSA\}\{XGB\ XRB\ XSB\}\{XGC\ XRC\ XSC\}$. For example, this model assumes that the classification error for an indicator varies by age, race, and

Table 4.6 Selection Probability Weighted Data for Past-Year Marijuana Use

					$R = 1$			$R = 2$			$R = 3$		
G	S	A	B	C	1994	1995	1996	1994	1995	1996	1994	1995	1996
1	1	1	1	1	96.3	101.8	110.1	113.5	112.7	119.8	586.8	584.4	614.6
1	1	1	1	2	4.2	1.1	1.1	5.6	1.3	2.0	27.7	8.8	6.7
1	1	1	2	1	1.0	0.0	0.3	0.2	2.5	0.0	4.2	5.5	3.3
1	1	1	2	2	1.1	1.3	2.5	1.2	0.8	2.9	4.3	3.2	9.3
1	1	2	1	1	0.1	0.1	0.2	1.0	0.1	0.0	0.5	0.0	0.0
1	1	2	1	2	0.4	0.5	0.1	0.0	0.4	0.0	0.4	1.2	1.2
1	1	2	2	1	0.6	1.3	1.3	1.5	1.1	1.2	5.2	3.8	5.9
1	1	2	2	2	13.2	13.4	12.9	10.1	15.7	17.2	68.2	91.7	72.2
1	2	1	1	1	94.5	99.4	103.7	109.8	124.0	121.9	563.8	556.3	583.7
1	2	1	1	2	5.7	1.1	1.1	4.5	0.5	0.4	28.0	6.6	6.6
1	2	1	2	1	0.3	0.8	1.3	0.7	0.9	1.7	2.6	1.2	4.7
1	2	1	2	2	1.3	1.8	0.8	0.7	0.5	2.1	2.4	9.1	4.2
1	2	2	1	1	0.2	0.0	0.0	0.1	0.0	0.3	0.8	0.0	0.0
1	2	2	1	2	0.2	0.2	0.1	0.5	0.7	0.0	0.5	2.2	1.4
1	2	2	2	1	0.9	0.3	0.4	1.3	1.6	1.0	1.5	3.5	7.2
1	2	2	2	2	9.3	11.5	12.4	10.6	7.0	10.0	64.5	81.1	71.5
2	1	1	1	1	121.4	128.3	136.5	101.4	101.7	96.7	577.9	615.9	574.9
2	1	1	1	2	9.5	2.6	4.6	7.2	2.3	3.0	43.3	18.0	15.9
2	1	1	2	1	2.2	2.2	2.7	2.5	4.7	2.3	11.7	23.0	11.6
2	1	1	2	2	1.6	0.7	3.1	2.9	2.8	2.6	24.2	14.1	25.3
2	1	2	1	1	0.0	0.0	0.3	0.5	0.0	0.0	0.0	1.2	1.6
2	1	2	1	2	0.0	0.0	0.2	0.0	0.6	0.7	1.9	0.5	2.1
2	1	2	2	1	2.3	1.8	2.8	2.6	2.2	4.1	13.4	5.9	8.1
2	1	2	2	2	21.5	20.0	26.1	31.6	29.9	40.2	200.4	177.0	218.6
2	2	1	1	1	129.0	134.0	134.7	140.7	148.4	142.3	679.1	656.3	676.9
2	2	1	1	2	5.0	1.9	1.3	5.5	1.4	3.2	44.2	12.1	16.8
2	2	1	2	1	0.5	1.6	1.2	2.1	2.5	3.1	9.1	12.8	15.2
2	2	1	2	2	0.8	2.3	2.2	2.9	1.1	1.5	7.9	12.5	11.1
2	2	2	1	1	0.1	0.0	0.2	0.3	0.3	0.2	0.9	0.0	0.0
2	2	2	1	2	0.5	0.3	0.5	1.0	0.5	0.2	0.0	0.6	0.3
2	2	2	2	1	0.5	0.3	1.0	2.9	1.4	2.7	6.0	6.1	8.2
2	2	2	2	2	11.4	7.6	14.0	16.7	15.9	19.5	136.9	158.5	132.3

sex but not jointly; that is, variation by age is independent of race and sex, variation by race is independent of age and sex, and variation by sex is independent of age and race.

Since A, B, and C were collected in the same interview, the possibility of locally dependent errors was also considered in the analysis. Local dependence was explored and tested using some of the techniques to be described in the next chapter. Among all the models considered in their analysis, the model providing the best compromise between fit and complexity was a locally independent model containing 72 parameters. This model incorporated simple

Table 4.6 ***Continued***

					R = 1			R = 2			R = 3		
G	S	A	B	C	1994	1995	1996	1994	1995	1996	1994	1995	1996
3	1	1	1	1	160.2	166.9	179.6	122.0	129.3	123.4	922.3	909.3	930.6
3	1	1	1	2	11.5	3.6	4.6	11.2	2.4	4.2	65.2	30.2	23.6
3	1	1	2	1	3.4	1.8	3.0	3.3	2.5	3.8	8.4	20.3	13.2
3	1	1	2	2	1.4	1.6	1.1	4.4	2.7	1.5	9.5	10.6	14.4
3	1	2	1	1	0.0	0.0	0.0	0.0	0.0	1.1	0.0	0.0	0.6
3	1	2	1	2	0.0	0.0	0.0	0.0	0.9	1.7	1.2	0.4	1.2
3	1	2	2	1	0.5	1.2	1.5	2.5	1.4	2.6	7.7	9.4	2.4
3	1	2	2	2	14.3	17.0	12.8	18.9	18.3	20.5	154.2	146.0	133.7
3	2	1	1	1	160.3	172.0	173.9	181.8	182.0	185.1	1049.9	1050.3	1040.8
3	2	1	1	2	6.4	1.3	0.3	7.2	2.8	3.1	55.4	12.7	15.5
3	2	1	2	1	1.5	0.6	1.3	5.1	3.6	2.7	9.5	10.9	9.5
3	2	1	2	2	1.5	0.7	0.6	0.5	2.1	2.2	9.8	6.3	4.2
3	2	2	1	1	0.0	0.1	0.3	0.1	0.2	0.0	0.6	0.0	1.0
3	2	2	1	2	0.3	0.0	0.0	0.3	0.0	0.5	2.6	0.9	0.0
3	2	2	2	1	0.5	0.0	0.3	0.7	2.1	2.1	3.6	6.9	10.9
3	2	2	2	2	6.6	3.0	5.2	12.6	10.9	11.1	72.5	72.7	76.4
4	1	1	1	1	321.8	342.7	381.5	385.9	409.7	423.4	3699.3	3891.8	3952.4
4	1	1	1	2	8.1	2.5	3.4	17.5	6.1	8.6	152.0	42.3	63.9
4	1	1	2	1	3.3	2.4	1.6	9.3	8.5	7.8	17.4	9.2	39.4
4	1	1	2	2	2.1	1.9	1.6	0.7	6.5	2.8	10.1	19.0	5.2
4	1	2	1	1	0.0	0.5	0.0	0.0	0.0	2.6	0.0	0.0	5.1
4	1	2	1	2	0.0	0.0	0.0	0.0	0.6	0.6	4.4	0.0	5.3
4	1	2	2	1	1.4	0.4	0.7	2.6	5.8	10.0	15.6	8.7	19.0
4	1	2	2	2	7.1	6.5	4.4	26.4	18.1	14.3	201.1	143.9	181.7
4	2	1	1	1	363.2	383.6	414.3	556.9	569.1	580.1	4391.6	4493.6	4669.1
4	2	1	1	2	6.8	3.3	0.6	13.1	2.3	8.3	119.2	23.7	30.2
4	2	1	2	1	2.2	1.8	0.4	3.1	5.0	4.8	17.3	10.8	14.9
4	2	1	2	2	0.3	0.0	0.0	1.2	1.3	5.3	6.8	11.2	6.5
4	2	2	1	1	0.2	0.0	1.2	0.0	0.0	0.0	1.2	0.0	0.0
4	2	2	1	2	0.0	0.0	0.0	0.0	0.0	0.5	0.0	0.0	0.0
4	2	2	2	1	1.1	0.3	0.0	3.1	0.8	2.5	6.0	5.9	0.0
4	2	2	2	2	4.2	1.6	2.7	14.7	6.2	12.9	63.7	83.9	58.1

two-way interaction terms between each grouping variable and indicator. More specifically, the best measurement model was

$$\{XGRS\}\{XA\ GA\ RA\ SA\}\{XB\ GB\ RB\ SB\}\{XC\ GC\ RC\ SC\}$$

using the modified path model notation. This model specifies that classification varies by grouping variable independently of an individual's true classification. The ℓEM model statement follows immediately from this specification of the model:

```
mod    GRSX
       A|XGRS      {XA  GA  RA  SA}
       B|XRS       {XB  GB  RB  SB}
       C|XGRS      {XC  GC  RC  SC}
```

Interestingly, the simple relationship between the indicators and grouping variables was preferred over the more complex but perhaps more easily interpreted forms for the interactions. For example, $\{XGA\ XRA\ XSA\}$ has an easier interpretation than $\{XA\ GA\ RA\ SA\}$ since the former model (call it the *three-way* model) specifies that the error probabilities vary across the levels of each of the grouping variables age, race, and sex. By contrast, the latter model (i.e., the *two-way* model) reflects dependence of misclassification on age, race, and sex but not the full dependence represented in the three-way model. Biemer and Wiesen explored the interpretation generically for the simple form $\{XA\ FA\}$, where A is dichotomous and F is a polytomous grouping variable with categories $f = 1,\ldots, K_F$. They found that this model holds only if the product $\lambda_{+f}\lambda_{-f}$ is the same for all f where

$$\lambda_{+f} = \frac{\Pr(A = 1 \mid X = 2)}{\Pr(A = 2 \mid X = 2)} \tag{4.39}$$

is the odds of a false-positive error and

$$\lambda_{-f} = \frac{\Pr(A = 2 \mid X = 1)}{\Pr(A = 1 \mid X = 1)} \tag{4.40}$$

is the odds of a false-negative error.

As an example, suppose that the false-positive error is near 0 in all groups; that is, the probability an individual is classified as a past-year marijuana user when he/she truly is not a user is approximately 0. Then $\lambda_{+f} = \lambda_{+f}\lambda_{-f} = 0$ for all f for any value of λ_{-f} and the condition is satisfied. On the other hand, if both λ_{+f} and λ_{-f} are positive and both vary across groups, then they must exhibit a negative correlation across the levels of F in order for their product to remain approximately constant. In other words, if the odds of a false-positive error are higher than the average odds for some group, then the odds of a false-negative must be lower than the average odds for that group in order that the product $\lambda_{+f}\lambda_{-f}$ to remain constant.

Examples of such an indicator are not uncommon in practice [see, e.g., Biemer and Wiesen (2002)]. Items with high sensitivity (i.e., low false-negative error) for some groups, may often exhibit low specificity (high false-positive error) for the same groups. Consider the question "Do you use marijuana?" Some individuals may interpret the term "use" as "routinely or often use." Individuals who seldom or only occasionally use marijuana would respond

Table 4.7 Observed Inconsistencies in the Three Indicators of Marijuana Use

	1994		1995		1996	
Indicator	n	%	n	%	n	%
A versus B	241	1.35	263	1.48	293	1.61
A versus C	854	4.80	380	2.14	452	2.48
B versus C	883	4.96	409	2.31	491	2.69
A versus B versus C	989	5.55	526	2.96	618	3.39

negatively (high false-negative error). On the other hand, this somewhat extreme wording is not likely to elicit many false positives. In fact, the false-positive error rate is likely to be near zero except for inadvertent or random positive responses (such as coding errors). Therefore, the model $\{XA\ FA\}$ may apply to items that have (1) zero or near-zero false-positive or false-negative probabilities in all groups or (2) sensitivities and specificities that vary with negative correlation across groups.

One important motivation for the LCA of these data was the apparent inconsistencies among the various questions about drug use in the survey. To illustrate, consider Table 4.7, which shows the rate of disagreement among all combinations of the indicators A, B, and C defined above. This rate is computed as the unweighted proportion of observations in the off-diagonal cells of the cross-classification table for the variables in each combination. The disagreement rate for A versus B varies around 1.50%, while the rate for C versus A or B is considerably higher, particularly for the year 1994, where the disagreement rate among all three measurements is 5.55%. Since C is a composite of many questions, a plausible hypothesis for the high rate of inconsistency with C is that C is the more accurate indicator, in which case disagreement with C indicates potential misclassification error by the other two measurements. In support of this argument, estimates of past-year marijuana use are highest for C, which suggests better accuracy since marijuana use tends to be underreported in the NHSDA [see, e.g., Turner et al. (1992) and Mieczkowski (1991)]. This is but one of the hypotheses that can be explored in the LCA.

Before continuing with this analysis, we will discuss the tools needed for estimating and evaluating alternative models for describing the error in the three indicators. The next section considers the EM algorithm under the log-linear parameterization and some methods for building models and testing their fit to the data.

4.4 MODEL ESTIMATION AND EVALUATION

In this section, we consider some of the best practices from the LCA literature for building parsimonious models, determining model fit, testing model significance, and various other hypotheses of interest for the model parameters.

These results will then be applied to the NHSDA marijuana data described in the previous section as a demonstration of the methodology. To facilitate this discussion, we first examine some aspects of the EM algorithm for the LL parameterization of the LC model.

4.4.1 EM Algorithm for the LL Parameterization

In Section 4.2.3 the EM algorithm was discussed in some detail for the probabilistic parameterization. In this section, the EM algorithm is applied for the loglinear and logit parameterizations of the LC model. As in Section 4.2.3, the focus here is on the basic ideas involved illustrated using simple models that avoid many of the complexities encountered in the implementation of EM in general latent variable LL models. Readers interested in exploring these issues in more detail should read Goodman (1974b), Mooihaart and van der Heijden (1992), and Vermunt (1997a). The approach is quite similar to that described in the probabilistic framework. The essential idea involves obtaining a preliminary estimate of the full-data table, say, XABC, using starting values for the joint distribution of X and the indicators. This constitutes an initial (iteration 1) of the E-step. Applying the specified model, the likelihood of the full-data table is then computed. The MLE of the model parameters can then be obtained by maximizing the full-data likelihood using methods for LL without latent variables; for example, iterative proportional fitting (IPF) is often used at this step. This constitutes the initial M-step. Using these new parameter estimates, a new (estimated) full-data table XABC is obtained (another E-step). This is followed by another M-step and so on until all parameter estimates from the M-step satisfy user-specified convergence criteria.

To illustrate, consider the standard LC model for three indicators in Figure 4.2. For generality, we allow the number of categories to differ for latent and manifest variables. In order to begin the E-step, starting values for all the parameters are needed. These are usually specified as error probabilities, namely, ${}^{(1)}\pi_x$, ${}^{(1)}\pi_{a|x}$, ${}^{(1)}\pi_{b|x}$ and ${}^{(1)}\pi_{c|x}$ for $x = 1,\ldots, K_X$, $a = 1,\ldots, K_A$, $b = 1,\ldots, K_B$ and $c = 1,\ldots, K_C$. The starting values must satisfy all the constraints of the model; in particular, those in (4.6) requiring error probabilities for an indicator and prevalence probabilities for the latent classes to sum to 1. The starting values can be uniform random numbers in the interval (0,1), although it may be more advantageous to specify plausible values the basis on of prior knowledge of their probable sizes or from an analysis of the observed. As an example of the latter approach, starting values for π_x, $\pi_{b|x}$, and $\pi_{c|x}$ can be computed by letting one of the indicators, say, A, play the role of X, as a gold standard measurement. Then the parameter estimates can be computed using the techniques for known true values described in Section 2.5. Random starting values are usually adequate if the model is well identified. If the model is weakly identifiable, good starting values can help to avoid false solutions arising from local maxima. The ideas are discussed further in the next chapter, where the topic of identifiability is discussed in more detail. Starting values can also be provided for the u parameters of the loglinear model. This may be more convenient in some cases, although it may be less intuitive and more difficult to arrive at plausible values. Essentially any starting values satisfying the model constraints can be used to form the initial estimate of the XABC table.

For the three indicator LC model, the E-step computes

$$^{(1)}n_{xabc} = n_{abc}\frac{^{(1)}\pi_{xabc}}{^{(1)}\pi_{+abc}} \tag{4.41}$$

where, according to the model, we obtain

$$^{(1)}\pi_{xabc} = {}^{(1)}\pi_x\,{}^{(1)}\pi_{a|x}\,{}^{(1)}\pi_{b|x}\,{}^{(1)}\pi_{c|x} \tag{4.42}$$

or, if starting values of the u parameters are provided, we can equivalently compute

$$^{(1)}n_{xabc} = n_{abc}\frac{\exp(^{(1)}u_x^X + {}^{(1)}u_{xa}^{XA} + {}^{(1)}u_{xb}^{XB} + {}^{(1)}u_{xc}^{XC})}{\Sigma_x \exp(^{(1)}u_x^X + {}^{(1)}u_{xa}^{XA} + {}^{(1)}u_{xb}^{XB} + {}^{(1)}u_{xc}^{XC})} \tag{4.43}$$

Note that in either case, $\Sigma_x {}^{(1)}n_{xabc} \equiv n_{abc}$; that is, the original observed frequencies are always preserved at this step. Next, the M-step computes new parameter estimates as shown in (4.12). For more complex models, the IPF algorithm can be used to obtain the estimates iteratively. Subsequent iterations essentially repeat these two steps until convergence is reached.

Estimation via the EM algorithm proceeds similarly for modified path models. Initial starting values for the parameters of each submodel must be specified. The cell probabilities for the complete data table can be computed as in (4.42), where now each component in the product on the left-hand side is replaced by its corresponding logit model representation. As noted above, this is equivalent to a standard loglinear model with the same parameters. For example, in the present case, this substitution yields (4.43). The corresponding full-data likelihood would be formed according to this logit model. The likelihood can then be maximized using standard methods such as IPF to obtain new parameter estimates.

A number of improved EM algorithms have been discussed in the literature offering faster convergence. For example, ℓEM uses a modification of the EM algorithm referred to as the *EM1 algorithm* (Rai and Matthews 1993). Since it uses only one IPF iteration at the M-step rather than iterating the IPF to convergence, it can be much faster. The EM1 is a special case of the ECM algorithm developed by Meng and Rubin (1993) that is based on conditional maximization. This approach strategically fixes some parameters to their previous values while updating the others. A more detailed discussion of these and other algorithms can be found in Vermunt (1997a, pp. 64–69).

When latent variables are added to loglinear models, the observed and expected values of the second derivatives of the loglikelihood are no longer identical [see Appendix A in Heinen (1996) for a proof of this)]. This leads to two ways of obtaining model-based estimates of standard errors and parameter covariances. The usual method is to estimate the information matrix **I**, defined as the negative expected value of the matrix of second partial derivatives of the loglikelihood. However, another possibility is to use the inverse of the matrix of second partial derivatives of the loglikelihood to the parameters multiplied by –1. There are situations when the latter is preferred over the former [see, e.g., Efron and Hinkley (1978)].

Let $\ell(\boldsymbol{\pi})$ denote the loglikelihood where $\boldsymbol{\pi} = [\pi_1, \pi_2, \ldots, \pi_p]'$ is the $p \times 1$ column vector of LC model parameters. For example, when X, A, B, and C are dichotomous, the parameter vector for the model in Figure 4.2 is $\boldsymbol{\pi} = [\pi_1^X, \pi_{2|1}^{A|X}, \pi_{1|2}^{A|X}, \pi_{2|1}^{B|X}, \pi_{1|2}^{B|X}, \pi_{2|1}^{C|X}, \pi_{1|2}^{C|X}]'$. Alternatively, the parameters can be defined in terms of the u parameters of the loglinear model. In that case, $\boldsymbol{\pi} = [u_1^X, u_1^A, u_1^B, u_1^C, u_{11}^{XA}, u_{11}^{XB}, u_{11}^{XC}]'$. In either case, the information is defined as

$$\mathbf{I} = \mathbf{D}^{-1} \quad \text{where} \quad \mathbf{D} = [d_{ij}] \quad \text{and} \quad d_{ij} = -\frac{\partial^2 \ell(\boldsymbol{\pi})}{\partial \pi_i \, \partial \pi_j} \tag{4.44}$$

The standard error of $\hat{\pi}_i$ is the square root of the ith diagonal element of MLE of **I** for $i = 1, \ldots, p$.

Bartholomew and Knott (1999, p. 140) note that (4.44) does not perform well for small sample sizes or when the parameter estimates are close to their boundary values.[3] The bootstrap method can be applied and is preferred in these cases [see, e.g., De Menezes (1999)]. This method constructs multiple simulated datasets (called *pseudosamples*) using the expected frequencies estimated from the model. It then fits a latent class model to each pseudosample. The variation of the parameter estimates across pseudosamples provides empirical estimates of their standard errors.

4.4.2 Assessing Model Fit

Identifiable models can be tested using the traditional methods of categorical data analysis. For example, model fit can be assessed using chi-square and F distributions, model effects can be tested using Wald statistics, and penalty-based indices such as the (Bayesian information criterion) (BIC) and the (Akaike information criterion) (AIC) can be used to compare nonnested models. In this section, we review these and other fit statistics and demonstrate their use for latent class models.

[3]Boundary values and other estimation problems are discussed in more detail in the next section.

Chi-Square Goodness-of-Fit Tests

Latent class models, or more generally, loglinear models with latent variables, can be tested using chi-square goodness-of-fit statistics in the same fashion as traditional loglinear models. Let n_{abc} denote the observed frequency for cell (a,b,c) of the ABC table, and let $\hat{m}_{abc}$ denote the estimated frequency under the hypothesized model. The *Pearson chi-square* statistic, given by

$$X^2 = \sum_{abc} \frac{(n_{abc} - \hat{m}_{abc})^2}{\hat{m}_{abc}} \tag{4.45}$$

and the *likelihood ratio chi-square* statistic, given by

$$L^2 = 2\sum_{abc} n_{abc} \log \frac{n_{abc}}{\hat{m}_{abc}} \tag{4.46}$$

are useful statistics for assessing the fit of the model for $\hat{m}_{abc}$. If the model fits the data perfectly, both X^2 and L^2 will be 0; but, in general, X^2 and L^2 will be greater than 0, indicating some lack of fit. If either statistic is too large (statistically significant), then the model should be rejected. Otherwise, the fit is deemed acceptable. This process is sometimes difficult to grasp for LCA novices since it seems to run counter to traditional hypothesis testing. Traditionally, a statistically insignificant test result suggests no evidence to support the null hypothesis. For model testing, an insignificant test suggests *support* of the hypothesized model; that is, the model is said to provide an acceptable fit if the value of the chi-square statistic is small compared to the appropriate percentile of a chi-square distribution.

Thus, to determine whether a model is acceptable, we compare the computed values of X^2, L^2, or both to $(1 - \alpha)100$ percentile of a χ^2_{df} distribution. Here degrees of freedom (df) is the model degrees of freedom computed as

$$\text{df} = k - \text{npar} - 1 \tag{4.47}$$

where k is the number of cells (with positive expected frequency) and npar is the number of free (i.e., unconstrained) u or π parameters in the model not including boundary parameters. Obviously, if df = 0, no test of model fit can be performed and the model is fully saturated. If df < 0, the model is nonidentifiable because there are too many parameters. In that case, the number of parameters must be reduced or the number of cell frequencies must be increased, for example, by adding grouping variables. Adding grouping variables increases k and npar simultaneously. However, npar can be reduced to some extent by imposing parameter restrictions. In this way, otherwise nonidentifiable models can be transformed into identifiable models.

To illustrate, when A, B, C, and X are all trichotomous variables, then $k = 3^3 = 27$ and npar = 20. Therefore, according to (4.47), the X^2 and L^2 should be compared to an appropriate critical value from the χ^2_6 distribution. This is usually done using the *p value* of the chi-square statistic, which is defined for X^2 as $\Pr(\chi_{df} > X^2 | H_0)$ with a similar definition for L^2. In words, the p value is the probability of observing a value of X^2 as large as the one computed under the model when the model is true. One might interpret the p value as the probability that the model is correct. The usual approach is to reject the model if $p \leq 0.05$; otherwise, the model is deemed acceptable.

A few cautions and exceptions should be observed when applying the chi-square criteria. Extremely large p values (say, 0.7 or higher) could signal that the model is *overfitted*; that is, sampling and spurious errors are being modeled along with the phenomenon that actually generated the data. If there are many parameters in the model, the model may not be a good representation of the true underlying phenomenon. In that case, variables that cannot be supported by the substantive theory or strongly justified by theoretical interest should be removed from the model to lower the p value. A model that is overfitted would be expected to provide a much poorer fit to a new set of data generated by the same process under identical conditions. Therefore, the estimates should not be trusted. Some analysts recommend using cross-validation methods as insurance against overfitting. For further guidance on cross-validation methods, see Mosteller and Tukey (1968) or more recently, Collins and Lanza (2010).

Conversely, models with extremely small p values should not always be rejected. For large datasets (say, $n > 5000$), chi-square goodness-of-fit tests can have very high power and very small Type I error, resulting in ultraconservative model testing. This is apparent when the differences between the observed and predicted cell frequencies seem quite small yet the model p value is 0.001 or less. Strict adherence to the chi-square goodness-of-fit criterion would reject the model. However, if the estimated and observed cell frequencies are still close, the model could still be considered a good-fitting model despite the small chi-square p value. In fact, as noted previously, adding additional parameters and variables to the model in order to increase the p value could result in overfitting. In these situations, the index of dissimilarity is often used to assess the adequacy of the model.

The *index of dissimilarity* is defined as

$$D = \frac{\sum_{abc} |n_{abc} - \hat{m}_{abc}|}{2n} \tag{4.48}$$

This index is always positive and less than 1. It can be interpreted as the smallest proportion of observations that would need to move to other cells in order for the model to fit the data perfectly. For example, a D of 0.10 means that the model and observed cell counts differ by a total of $0.1n$ observations (i.e., 10% of the observations). The model and observed data would agree perfectly if not for these observations. Models with dissimilarity indices less than 0.05 can be regarded as fitting the data adequately in most situations. However, this is only a rule of thumb. For some applications, it may be desirable to require a somewhat smaller D (say, 0.01 or less).

4.4.3 Model Selection

A number of methods are available for determining which model among a number of alternatives is the best one. Ideally, one should have good theoretical notions regarding the causes of measurement error and the indicators that should perform best in the situation at hand. The plausibility of the model and the model estimates are essential considerations in the model selection process. If the model does not make theoretical sense, or if the estimates are implausible, the model should be questioned even when it provides a reasonable fit to the data. In most cases, using a purely statistical–mechanical approach for model selection will not produce satisfactory results.

An important strategy for model selection is the comparison of competing models to choose the best one. The same methods used in traditional loglinear modeling can also be applied for LCA. As an example, two nested latent variable loglinear models can be compared using the so-called conditional L^2-test statistic. The model M_2 is said to be *nested* within the model M_1 if M_2 can be obtained from M_1 by restricting some of the M_1 parameters. As an example, the model $\{XA\ GA\ XGB\ XGC\}$ is nested within $\{XGA\ XGB\ XGC\}$ since the former model can be obtained from the latter one by setting XGA to zero. Restricting the false-positive probabilities for A and B to be equal is another way that one model (the restricted one) can be nested in another (the unrestricted one). Note that the models $\{XGA\ XB\ XC\}$ and $\{XA\ XGB\ XC\}$ are not nested since one cannot be obtained by adding parameters to the other.

Let $L^2(M_1)$ and $L^2(M_2)$ denote the likelihood ratio chi-square statistics for M_1 and M_2, respectively. Then it is well known that

$$L^2(M_2 \mid M_1) = L^2(M_2) - L^2(M_1) \tag{4.49}$$

is distributed as a chi-square with $\mathrm{df}_{2|1} = \mathrm{df}_2 - \mathrm{df}_1$ degrees of freedom under the null hypothesis that M_2 is the correct model. If the null hypothesis is true, $L^2(M_2|M_1)$ will be small compared to the appropriate $\chi^2_{\mathrm{df}_{2|1}}$ critical value. The p value for $L^2(M_2|M_1)$ can be used to determine whether M_1 can be reduced to M_2 without significant loss of fit. If $p \le 0.05$, then M_2 is said to provide an acceptable fit; otherwise, M_2 should be rejected in favor of M_1.

Another useful approach for comparing alternative models is to use the information-theoretic measures AIC and BIC, especially for models that are not nested. For a model M, define

$$\begin{aligned} \mathrm{AIC}(\mathcal{L}) &= -2\log \mathcal{L}_{\max} + 2 \times \mathrm{npar} \\ \mathrm{BIC}(\mathcal{L}) &= -2\log \mathcal{L}_{\max} + \mathrm{npar} \times \log(n) \end{aligned} \tag{4.50}$$

where $\mathcal{L}_{\max}$ is the maximized likelihood value for model M. The model having the lowest $\mathrm{AIC}(\mathcal{L})$ [or $\mathrm{BIC}(\mathcal{L})$] is preferred under this criterion since it provides the best balance between two factors: model fit (reflected in the value of $\log \mathcal{L}_{\max}$) and model parsimony [represented by either 2npar or npar $\times \log(n)$]. To see this, note that, for the $\mathrm{AIC}(\mathcal{L})$, 2npar increases as the number of parameters increases. However, $-2\log\mathcal{L}_{\max}$ becomes more negative as more parameters are added since the likelihood in an increasing function of npar. Thus npar and $-2\log\mathcal{L}_{\max}$ counterbalance one another with the npar term acting as a sort of "penalty" for adding more parameters and with it, more model complexity. The same can be said of the BIC criterion, except that, for moderate to large sample sizes, the penalty term for model complexity [viz., npar $\times$ $\log(n)$] is larger. Thus, AIC($\mathcal{L}$) and BIC($\mathcal{L}$) are useful to identify the model that strikes the best balance between model fit and model parsimony. Unfortunately, the theory does not provide a statistical test for determining whether an AIC($\mathcal{L}$) or BIC($\mathcal{L}$) for one model is significantly lower than the corresponding measure for another model. Rather, the measures are meant to be used as indices of the comparative fit and somewhat of a rough guide for model selection.

Because of the huge penalty it extracts for adding new parameters, BIC($\mathcal{L}$) tends to favor less complex models than AIC($\mathcal{L}$). Results of empirical investigations of the AIC($\mathcal{L}$) and BIC($\mathcal{L}$) for use in LC models (Lin and Dayton 1997) suggest that AIC($\mathcal{L}$) should be preferred for sample sizes of about 2000 cases or less. BIC($\mathcal{L}$) is preferred for larger sample sizes and for models containing relatively few parameters. In practice, the two indices select the same or very similar models. Still, both are usually computed and considered in the model selection process.

For loglinear models, an alternative formulation of the AIC($\mathcal{L}$) and BIC($\mathcal{L}$) statistics is computed denoted by AIC(L^2) and BIC(L^2) and given by

$$\begin{aligned} \mathrm{AIC}(L^2) &= L^2 - 2\times\mathrm{df} \\ \mathrm{BIC}(L^2) &= L^2 - \mathrm{df}\times\log(n) \end{aligned} \tag{4.51}$$

[see, e.g., Vermunt (1997b)]. As Vermunt points out, these indices essentially compare the model of interest with a fully saturated model that fits the data perfectly. To see this, note that the loglikelihood kernel for the perfectly fitting model is $\mathcal{L} = \sum_{abc} n_{abc} \log n_{abc}/n$. If $\mathcal{L}_{\max} = \sum_{abc} n_{abc} \log\hat{\pi}_{abc}$ is the likelihood for the model of interest, the different $\mathcal{L}_{\max} - \mathcal{L}$ is just L^2 for the model of interest. Likewise, npar for the fully saturated model is equal to k, the number of cells in the table. Therefore, k minus npar for the fully saturated model is the model df. It therefore follows that subtracting AIC for the model of interest from the AIC of the perfectly fitting model produces AIC(L^2). Again, by the differing signs associated with the measure of model fit and model degrees of freedom, these statistics also represent a compromise model fit and parsimony. As in the case of AIC($\mathcal{L}$) and BIC($\mathcal{L}$), we wish to minimize these statistics in the model-fitting process. Unless otherwise specified, the BIC(L^2) criteria will be used for examples and illustrations in this book and will be denoted simply as BIC.

When using either AIC or BIC to determine the best model, it is important that the models be fit to the same data or cross-classification table. A common mistake in using these measures is to fit a model to, say, table GHABC and another to table GABC (where the manifest variable indicator H has been dropped from the input table). The information theoretic measures for the two models will not be computed correctly since the two likelihood functions, $\mathcal{L}_1$ and $\mathcal{L}_2$, say, will not pertain to the same data table and will consequently not be comparable. As an example, suppose that we wish to compare the BICs for the models $\{XGH\}\{XGA\}\{XB\}\{XC\}$ and $\{XG\}\{XA\}\{XB\}\{XGC\}$. Although the latter model does not include the grouping variable H, both models should be fit to the *GHABC* table in order the BICs (or AICs) to be comparable.

4.4.4 Model-Building Strategies

Determining the Number of Latent Classes

In general applications of LCA, the number of latent classes specified for X need not agree with the number of categories of the indicators. In these applications, the indicators themselves may differ with respect to their numbers of categories. Recall from the Stouffer–Toby example in Section 4.1.2 that the precise meaning of X was deduced from LCA through the estimated error probabilities. In these applications, the first decision that an analyst encounters in building an LC model is the number of latent classes that should be specified for X. The usual approach is to fit models with varying numbers of latent classes and then choose the model with the number of classes that provides the best fit. This choice can be difficult because the usual chi-square goodness-of-fit tests are of little use. The reason is that reducing the number of latent classes creates boundary constraints that violate chi-square asymptotics. For example, a three-class model can be reduced to a two-class model by setting the size of one class of the former model to 0 (e.g., $\pi_3^X = 0$) and, hence, the boundary constraint. Typically BIC or AIC is used to identify the best fitting model; but, more recently, parametric bootstrapping methods (McLachlan and Peel 2000) and Markov chain Monte Carlo methods (Bensmail et al. 1997) have also been used successfully. Nevertheless, determining the dimension of X can be quite challenging, especially in complex models. One issue that must be addressed is whether the class definitions change across population subgroups. In the end, of course, the latent classes must still have theoretically plausible interpretations.

Fortunately, as noted in Section 4.1, the issue of the number of latent classes is not particularly relevant for classification error evaluation applications. In these applications, X is a well-defined, measurable characteristic representing the error-free version of the indicator variables. As such, the number of classes for X is given by the definition of X. Typically, although there are a few important exceptions that we will encounter later, the number of categories of the indicator variables and the number of classes of X are the same and have exactly the same definitions. Since the focus of this book is on error evaluation, how to choose the number of latent classes for general LCA applications is not addressed. Some useful references for exploring that question are Collins and Lanza (2010), Heinen (1996), and McCutcheon (1987).

Selection of Covariates

The process for selecting covariates (specifically, grouping variables) for latent class models parallels that used in standard loglinear modeling. The key difference is that identifiability can become an issue when latent variables are being modeled. In addition, as noted previously, both the structural and the measurement components of the latent variable loglinear model need to be considered when determining what variables may be important in the analysis. As an example, for the analysis of past-year marijuana use described in Section 4.3.2, the grouping variables age, race, and sex were selected primarily because they were believed to be correlated with marijuana use. As Biemer and Wiesen (2002) show, younger age groups tend to use marijuana more than do older adults, blacks more than whites and males more than females. However, these variables may be only weakly correlated with classification errors. Their analysis showed that older adults tend to deny their use of marijuana only slightly more than do younger persons. Females and blacks also tend to underreport more, but the differences between males and nonwhites are rather small.

Ideally, other variables that are believed to be highly correlated with classification error should also be included in the LCA. For example, persons having higher incomes may be more inclined to underreport their use of illegal drugs since they have more to loose by such admissions. For surveys employing multiple interview modes, the mode of interview is often related to measurement error with more private modes tending to elicit more accurate responses, all else being equal. However, the interview mode is not likely to be correlated with drug use itself, so, although it may help the measurement component, it may not improve the fit of the structural component of the model. Characteristics of the interviewer such as experience, skill, and demographic characteristics may also affect misclassification. These variables may not be important for the structural component, however.

Stepwise selection methods are commonly employed for model building. This involves either *backward elimination* or *forward selection* methods. Backward elimination begins with a very complex model—one that includes almost all of the grouping variables of interest as well as interaction terms (up to, say, third or fourth order). Assuming that this model fits the data well and is identifiable and theoretically acceptable, the process of eliminating insignificant terms begins, often using the *Wald statistic* to identify model effects to retain or delete.

The Wald statistic provides a convenient way to assess the statistical significance of the set of parameter estimates for a given model effect (e.g., XGA). Let $\hat{\mathbf{u}}$ denote the column vector of parameters for some set of parameters (e.g., the set of parameters comprising an interaction term u_{xag}^{XAG} for all values of x, a, and g). Let $\widehat{\mathbf{Cov}}(\hat{\mathbf{u}})$ denote the variance–covariance matrix for $\hat{\mathbf{u}}$. Then, under the null hypothesis (i.e., assuming $E(\hat{\mathbf{u}}) = \mathbf{0}$),

$$W^2 = \hat{\mathbf{u}}'\widehat{\mathbf{Cov}(\hat{\mathbf{u}})}\hat{\mathbf{u}} \tag{4.52}$$

is distributed approximately and asymptotically as a chi-square with degrees of freedom (df) equal to the number of independent parameters of $\hat{\mathbf{u}}$. For example, suppose that X, A, and G each have three categories. Then the term XGA contains eight independent parameters. Thus there are 8 degrees of freedom for the Wald chi-square test. The Wald statistic tests the hypothesis that $XGA = 0$; that is, all six parameters are 0. When $\hat{\mathbf{u}}$ consists of one independent parameter, then W is equal to the Student's t statistic, that is, the ratio of the parameter to its standard error.

The last three columns of Table 4.5 list the Wald statistics, the degrees of freedom, and the p value, respectively, corresponding to each term (main effect or interaction) for the Hui–Walter model. Terms that have small p values (say, 0.15 or smaller) should be retained in the model. Terms with larger p values are prime candidates for deletion at this step of the stepwise regression. Usually, higher-order interactions are deleted, if possible, before lower-order effects. At the same time, the model should remain hierarchical. For example, in Table 4.5, only one effect (G), having a p value of 0.285, is a candidate for removal. However, it should still be retained in the model because one or more higher-order interactions involving G are significant.

According to Hauck and Donner (1977), the Wald test is not as powerful for testing a reduced model as is using the conditional chi-square test. For model selection, however, it is quite useful as a quick method for identifying terms that are insignificant and should be removed from the model during the backward elimination process. The conditional likelihood ratio chi-square can be used to verify the final model.

The backward elimination process continues by removing an insignificant term based on the Wald test, rerunning the model, removing an additional term based on the Wald test, again rerunning the model, and so on until all remaining terms in the model are significant at some appropriate level (e.g., $\alpha = 0.05$ or, more conservatively, $\alpha = 0.15$ can be used). In Table 4.5, all the interaction terms are significant, so none should be removed and the backward elimination process is terminated.

One problem that arises with backward elimination is that the initial model may be so complex and highly parameterized that the EM algorithm fails to converge properly owing a preponderance of 0 or very small cells (*sparseness*). In that case, some other method should be used (e.g., the forward selection method) to identify a much smaller pool of candidate grouping variables. Then the backward elimination method can proceed with this much reduced set of variables in the model.

The *forward selection method* described here is a process building a model that successively considers more and more complex models while employing backward elimination to keep the model from growing too large. It begins with a very parsimonious model; for example, the three-indicator latent class model $\{XA\ XB\ XC\}$, is a good starting point. It then draws a grouping variable from the candidate variable pool and adds it to the model. Usually this is done by including the interactions of this variable with a one or more error terms. For example, to add the grouping variable *F*, the model {*XFA XFB XFC*} is specified. The fit of this model can be compared with {*XGA XGB XGC*}, {*XHA XHB XHC*}, and so on to identify the single grouping variable having the greatest improvement on the basis of the AIC or BIC criterion. Each time a grouping variable is added, the model can be reduced using backward elimination before proceeding to more complex models. This helps to keep the number of parameters in the model manageable.

Suppose that the best variable from this step of the process is *F*. To determine whether model fit can be further improved through the addition of terms, a second grouping variable can be selected by considering the models {*XFGA XFGB XFGC*}, {*XFHA XFHB XFHC*}, and so forth using the BIC (or AIC) criterion to identify the best variable. As before, backward elimination can be applied to reduce the number of parameters before moving forward to consider a third grouping variable, a fourth grouping variable, and so on.

To illustrate, consider the model {*XFGHA XFGHB XFGHC*}, which is quite complex because it essentially specifies the LC model {*XA XB XC*} within each combination of the grouping variables, *FGH*. A more parsimonious model might be found by applying the backward elimination process to eliminate higher-order interaction terms. For example, it may be possible to replace the term *XFGHA* by all possible second-order interactions of the grouping variables with the error term: {*XFGA XFHA XGHA*}. If the loss in fit is insignificant (the basis of the conditional chi-square test), this reduced model can replace the more complex one with no significant loss in fit. The process can then be repeated with *XFGHB* and *XFGHC*. The backward elimination process continues in this fashion for the highest-order interaction terms in the model, replacing each high-order interaction by its lower-order subordinates.

The single grouping variable model can be compared with the best double, triple, and higher grouping variable models to determine the final model. The best model is the one that (a) fits the data (i.e., has a model p value of 0.05 or more) and (2) has the smallest BIC (or AIC, if that is preferred).

4.4.5 Model Restrictions

Model restrictions can be introduced to reduce the complexity of the models and test hypotheses about effects that constitute an LC model. By imposing restrictions on the model parameters, one can perform detailed multigroup comparisons or test a wide range of special models that may be of interest to the analyst. When the probability model parameterization is used, the restrictions take the form

$$\pi_v^V = \upsilon \tag{4.53}$$

where V is a some combination of indicators and grouping variables either with or without the latent variable and v is either a constant or another probability, say, $\pi_{v'}^{V'}$. Some examples of choices for V or V' are $A|X$, XGB, X, and GX. Quite often v is a boundary value (0 or 1). Inequality constraints are also possible and these will be addressed in the next chapter.

As an example, for sensitive dichotomous items such as drug use or undergoing an abortion, for example, it might be reasonable to assume that the probability of a false-positive report (e.g., falsely reporting the use of an illicit drug or an abortion) is 0 for some indicator, say, A. In that case, constraints of the form $\pi_{1|2}^{A|X} = 0$ or equivalently $\pi_{2|2}^{A|X} = 1$ can be imposed on the model. In other situations, the assumption of parallel measurements may be plausible for two or more indicators. Thus, constraints such as $\pi_{b|x}^{B|X} = \pi_{c|x}^{C|X}$ for $b = c$ might be imposed. Since this restriction nests the restricted model within the unrestricted one, the conditional chi-square test can be used to test the hypothesis of parallel indicators. It is also conceivable that a priori information may be available, suggesting that one or more probabilities should be fixed at some constant other than 0 or 1. For example, from a prior study, the false-negative rate for C may be known to be 0.1 which is imposed by setting $\pi_{2|1}^{C|X} = 0.1$. Such situations are quite rare, however.

For the loglinear and modified path model parameterizations, restrictions of the form in (4.53) can still be used as they can be easily converted to loglinear restrictions via the mechanism in (4.13). In fact, probability constraints can be more intuitive than loglinear constraints in many situations. However, both loglinear and probability constraints can be used within the same model if desired. One type of restriction that we have previously discussed is the usual restrictions for loglinear models in (4.20). But these restrictions are automatically imposed on the estimation by default in all software packages. Another way to impose the zero-equality constraint on a model effect is to simply omit the term from the model. For example, leaving the term XGA out of the submodel $A|XG$ is equivalent to setting $\pi_{a|xg}^{A|XG} = \pi_{a|xg'}^{A|XG}$ for all g and g'.

Restrictions must also be used to properly model *structural zero cells*. A *zero cell* is an empty cell, that is, with a cell frequency of 0. This can occur for tables of higher dimension or when the overall sample size is small. When the zero cell has a nonzero expected frequency, it is called a *sampling zero cell*. However, when the cell has a zero expected frequency (i.e., a zero probability of observation), it is called a *structural zero cell*. This topic is discussed in more detail in the next chapter. It is mentioned here because it is another important use of model restrictions—namely, to set the cell probabilities associated with the structural zero cells to 0.

Model restrictions alter the degrees of freedom of the model. For every parameter that is restricted, one degree of freedom should be added to the model degrees of freedom. For every structural zero cell that is restricted, one degree of freedom should be subtracted from the model degrees of freedom. As an example, the unrestricted model $\{XA\ XB\ XC\}$ with binary indicators has seven parameters plus the main effect. Therefore the model degrees of freedom is 0 (i.e., the model is fully saturated). However, suppose that we restrict A, B, and C to be parallel indicators: $\pi_{1|2}^{A|X} = \pi_{1|2}^{B|X} = \pi_{1|2}^{C|X} = \pi_{1|2}$, say and $\pi_{2|1}^{A|X} = \pi_{2|1}^{B|X} = \pi_{2|1}^{C|X} = \pi_{2|1}$, say. Since four parameters are restricted by these constraints, the model degrees of freedom is 0 + 4 or 4 degrees of freedom Alternatively, one can simply add up the number of parameters to be estimated and subtract that from 8, the number of cells with positive expected frequencies. Thus, two error parameters and one latent class parameter make three parameters in total to be estimated. Therefore, the model degrees of freedom is 8 − 3 − 1 (for main effect) = 4. Note that if cell (2,2,2) were a structural zero cell, the model degrees of freedom would be one less (i.e., 3) since the number of cells with positive expectation is 7, not 8.

Depending on the software used for estimation, it may be important to verify the model degrees of freedom reported by the software using the formula in (4.47) since the reported degrees of freedom can be in error in some cases. For example, in most software packages, specifying 0 or 1 starting values for some probabilities will fix those probabilities to 0. Yet, the software may not adjust the degrees of freedom appropriately. (For example, this is the case for ℓEM in some situations.) Likewise, the software may allow the user to set certain parameters to a fixed constant other than 0 or specify structural zero cells, but these, too, may not be properly handled when computing the model degrees of freedom. Some software packages will add one degree of freedom for every observed or estimated zero cell when the program cannot distinguish between the structural zero and the estimated 0 parameters. For a structural zero, the degrees of freedom should be reduced by 1 while for estimated 0s it should be increased by 1. The responsibility for verifying the degrees of freedom lies with the user. The computation of degrees of freedom in these complex situations is considered in the next chapter.

4.4.6 Example: Continuation of Marijuana Use Analysis

To illustrate these concepts, we continue the example from Section 4.3.2. Recall that in this study, the objective was to evaluate the classification error in three indicators of past-year marijuana use. A secondary objective was to obtain an estimate of the prevalence of use that was corrected for classification error bias.

Selecting the Best Model

Since there are 3 years of data, the decision as to whether to pool the data into one large dataset and perform a combined analysis or to perform a separate analysis for each year was considered. The latter approach was taken for several reasons:

1. It was a way of cross-validating the model selection results. Since there were no design changes during 1994–1996, the error structure was thought to be the same for the 3 years. Thus, a model that was optimized for the 1994 data could then be fitted to the 1995 and 1996 data. Failure

of the model to fit one or both of these years could be evidence of overfitting the 1994 data or other model failure.

2. Separate analyses by year also provided greater opportunities for comparisons of the LCA prevalence estimates to the official NHSDA estimates, which are also computed separately for each year.
3. Separate analysis reduced the complexity of the models, which improved the estimation process and aided the interpretability of the results.

A wide range of latent class models was explored in order to identify the best model for producing estimates of classification error. The best model was defined as the one satisfying two criteria: (1) the likelihood ratio chi-square p value for the model should be greater than 0.01, indicating that the model fits the data reasonably well; and (2) the BIC should be the smallest among all models satisfying criterion 1.

Since the sample is quite large, the p value criterion is more liberal than the minimum 0.05 p value typically used for model selection. An alternative to criterion 1 is to require that the dissimilarity index be no greater than 0.05. This might be preferable in some situations; for example, for sparse datasets that render chi-square statistics invalid. Criteron 2 follows the Lin and Dayton (1997) recommendation for comparing complex models with large sample sizes as in the present case. Identifiability of each model satisfying these criteria was verified, when possible, using the rank of the information matrix (described in the next chapter). When boundary values were encountered, the identifiability was established by using multiple starting values.

The models considered were simple extensions of the basic modified path model $\{XGRS\}\{XA\}\{XB\}\{XC\}$ described in Section 4.3.2, which is referred to as model 0. As previously noted, all the models contain the term $GRSX$, which postulates that the prevalence of past-year use of marijuana, π_x, differs by each combination of age, race and sex. Four other models were defined as follows:

Model 1 Adds the first-order interaction terms to reflect variation in the indicator error rates by age, race, and sex $\{XGRS\}\{XA\ GA\ RA\ SA\}\{XB\ GB\ RB\ SB\}\{XC\ GC\ RC\ SC\}$.

Model 2 Interactions between the indicators are added to model 1 to check the local independence assumption. This produced Model 1 + $\{AB\}(BC\}$. The AC interaction term was not added due to problems of identifiability. This topic is discussed in some detail in Section 5.1.

Model 3 Extends Model 1 by adding second-order interactions between the grouping variables and the error terms: $\{XGRS\}\{XGA\ XRA\ XSA\}\{XGB\ XRB\ XSB\}\{XGC\ XRC\ XSC\}$.

Model 4 Analogous to Model 2, Model 4 adds the terms AB and BC to model 3 to test local dependence.

The fit statistics for these five models by year are provided in Table 4.8. Also provided in this table are the fit statistics for the same models to a revised

Table 4.8 Model Diagnostics for Alternative Classification Error Models by Year[a]

Model	df	npar	L^2	p Value	BIC
1994					
Model 0: $\{XGRS\}\{XA\}\{XB\}\{XC\}$	138	54	199.24	0.0005	−1151
Model 1: $\{XGRS\}\{XA\ GA\ RA\ SA\}$ $\{XB\ GB\ RB\ SB\}\{XC\ GC\ RC\ SC\}$	120	72	82.96	0.9960	−1092
Model 2: model 1 + $\{AB\}\{BC\}$	118	74	72.11	0.9997	−1083
Model 3: $\{XGRS\}\{XGA\ XRA\ XSA\}$ $\{XGB\ XRB\ XSB\}\{XGC\ XRC\ XSC\}$	102	90	63.61	0.9990	−935
Model 4: model 3 + $\{AB\}\{BC\}$	100	92	57.25	0.9998	−921
1995					
Model 0: $\{XGRS\}\{XA\}\{XB\}\{XC\}$	138	54	252.20	0.0000	−1098
Model 1: $\{XGRS\}\{XA\ GA\ RA\ SA\}$ $\{XB\ GB\ RB\ SB\}\{XC\ GC\ RC\ SC\}$	120	72	90.38	0.9799	−1084
Model 2: model 1 + $\{AB\}\{BC\}$	118	74	89.77	0.9753	−1065
Model 3: $\{XGRS\}\{XGA\ XRA\ XSA\}$ $\{XGB\ XRB\ XSB\}\{XGC\ XRC\ XSC\}$	102	90	51.42	1.0000	−947
Model 4: model 3 + $\{AB\}\{BC\}$	100	92	51.22	1.0000	−927
1996					
Model 0: $\{XGRS\}\{XA\}\{XB\}\{XC\}$	138	54	244.46	0.0000	−1110
Model 1: $\{XGRS\}\{XA\ GA\ RA\ SA\}$ $\{XB\ GB\ RB\ SB\}\{XC\ GC\ RC\ SC\}$	120	72	155.46	0.0163	−1022
Model 2: model 1 + $\{AB\}\{BC\}$	118	74	139.83	0.0831	−1018
Model 3: $\{XGRS\}\{XGA\ XRA\ XSA\}$ $\{XGB\ XRB\ XSB\}\{XGC\ XRC\ XSC\}$	102	90	111.95	0.2354	−889
Model 4: model 3 + $\{AB\}\{BC\}$	100	92	107.19	0.2933	−874
1994 Revised					
Model 0: $\{XGRS\}\{XA\}\{XB\}\{XC\}$	138	54	212.08	0.0000	−1139
Model 1: $\{XGRS\}\{XA\ GA\ RA\ SA\}$ $\{XB\ GB\ RB\ SB\}\{XC\ GC\ RC\ SC\}$	120	72	105.00	0.8336	−1069
Model 2: model 1 + $\{AB\}\{BC\}$	118	74	96.08	0.9308	−1059
Model 3: $\{XGRS\}\{XGA\ XRA\ XSA\}$ $\{XGB\ XRB\ XSB\}\{XGC\ XRC\ XSC\}$	102	90	75.27	0.9782	−923
Model 4: model 3 + $\{AB\}\{BC\}$	100	92	64.95	0.9974	−914

[a]Abbreviations: df—degrees of freedom; npar—number of free (unconstrained) parameters, being fit by the model.
Source: Biemer and Wiesen (2002).

version of the 1994 data that will be explained subsequently. The results in Table 4.8 indicate that Model 0 does not describe the data for any year. Model 1 fits the data well, satisfies criterion 2, and is identifiable for all 3 years. Models 2 and 4, which add the correlated error terms, also fit the data very well; however, the models do not satisfy criterion 2. Moreover, the conditional L^2 test comparing Models 1 and 2 is significant for 1994 and 1996 but not for 1995. This seems implausible since the surveys were essentially the same for all 3

```
* LEM Input Statements  for Past-Year Marijuana Use Example
*
man 6
lat 1
dim 2 3 4 2 2 2 2
lab X R G S A B C
mod XGRS    {XGRS}
    A|XGRS {XA GA RA SA}
    B|XGRS {XB GB RB SB}
    C|XGRS {XC GC RC SC}
rec 192
rco
sta A|X [.9 .1 .1 .9]
sta B|X [.9 .1 .1 .9]
sta C|X [.9 .1 .1 .9]
wma XA xa.wma      * Write marginal XA to compute A error probabilities
wma XB xb.wma      * Write marginal XB to compute B error probabilities
wma XC xc.wma      * Write marginal XC to compute C error probabilities
dat mrj94wtd.txt  * File containing 1994 weighted data
* dat mrj95wtd.txt  File containing 1995 weighted data
* dat mrj96wtd.txt  File containing 1996 weighted data
```

Figure 4.9 ℓEM input statements for fitting model 1 to the data in Table 4.6.

years. Another important consideration is that interpreting the differences in the estimates among the years would be made easier by using the same model for all 3 years. Therefore, Model 1 was accepted for all 3 years, and this model was selected for the subsequent analysis. The ℓEM input statements for fitting this model to the data in Table 4.6 is provided in Figure 4.9.

The local dependence models did not fit the data as well as did those postulating locally independent error. This was somewhat surprising given prior concerns that respondents may attempt to conceal their use of marijuana during the interview and thereby generate systematic errors in their responses to repeated questions about this in the same interview. Instead, response errors for the indicators appear to behave independently. This suggests that response errors are more likely to arise from such random factors as confusion generated by question wording or differences in the interpretation of the questions. This latter explanation also may explain the presence of the indicator by grouping variable interaction terms in the best model (e.g., GA, RA, SA). As noted in Section 4.3.2, these second-order interactions imply that false-negative and false-positive error rates vary inversely across groups. For example, groups that adopt a broader interpretation of the question may tend to respond positively, and thus have a higher false-positive error rate, while groups adopting a narrower interpretation may tend to respond negatively, resulting in a higher false-negative error rate.

Response error correlations may be quite different for questions about more stigmatized and strictly outlawed drugs such as cocaine and heroin or for questions regarding current use of illegal drugs rather than past-year use. For these the tendency to deliberately deny use may be greater, which will induce more local dependence among the indicators. Model 2 or 4 may emerge as the best model for describing the error in more sensitive questions on drug use.

Results

This section presents the Model 1 estimates of the classification error rates for each indicator, first for the total population and then by the age, race, and sex. We compared these estimates with estimates derived from Models 2, 3, and 4 and found that they differ somewhat from the estimates we report, particularly at the group level. However, the major findings that are reported below do not depend on the choice of model. Estimates of the classification error rates for all three indicators of past-year drug use were derived from the best model. Table 4.9 shows the estimated classification error rates (expressed as percentages) for the total population, for all 3 years, including a revised 1994 dataset, denoted by 1994′. This dataset is identical to the 1994 dataset for indicators A and B, but differs importantly for indicator C as described below. Standard errors, which are provided in parentheses, assume simple random sampling and do not take into account the unequal probability cluster design of the NHSDA. Consequently, they may be understated.

A number of points can be made from these results. First, note that the false-positive rates for all 3 indicators are very small across all 3 years except for indicator C in 1994, where it is 4.07%, more than 4 times that of the other two measurements. This was investigated further by comparing the questions C for all 3 years. The analysis revealed that, prior to 1995, one of the questions used for constructing C seemed quite complicated and potentially confusing to many respondents. However, after 1995, this question was eliminated. Therefore, a plausible hypothesis for the high false-positive rate for C in 1994 is the presence of this highly complex and confusing question.

Table 4.9 Comparison of Estimated Percent Classification Error by Indicator[a]

True Classification	Indicator of Past Year Use	1994	1994′	1995	1996
Yes ($X = 1$)	Recency = no ($A = 2$)	7.29 (0.75)	6.93 (0.72)	8.96 (0.80)	8.60 (0.79)
	Direct = no ($B = 2$)	1.17 (0.31)	1.18 (0.31)	0.90 (0.28)	1.39 (0.34)
	Composite = no ($C = 2$)	6.60 (0.70)	7.18 (0.72)	5.99 (0.67)	7.59 (0.74)
No ($X = 2$)	Recency = yes ($A = 1$)	0.03 (0.02)	0.03 (0.02)	0.01 (0.01)	0.08 (0.02)
	Direct = yes ($B = 1$)	0.73 (0.07)	0.76 (0.07)	0.78 (0.07)	0.84 (0.07)
	Composite = yes ($C = 1$)	4.07 (0.15)	1.23 (0.09)	1.17 (0.08)	1.36 (0.09)

[a]Standard errors are shown in parentheses.
Source: Biemer and Wiesen (2002).

To test this hypothesis, a new indicator was created by deleting the problematic question from indicator C and rerunning the model. The model fitting process was repeated for these data and Model 1 again emerged as the best model using criteria 1 and 2 above. Those results, and the error parameter estimates for these data, are shown in Table 4.9 in the column labeled 1994′. Note that for the revised indicator C, the false-positive rate for C dropped from 4.07% to only 1.23%, consistent with the other two indicators. Thus, it appears that LCA successfully uncovered an important problem with the composite question.

A second finding of interest from Table 4.9 focused on the false-negative error probability estimates. It is curious that A and C have much higher false-negative rates than B for all 3 years. Biemer and Wiesen (2002) examined the wording of the survey questions underlying A and B for clues (see Figure 4.8). Since question 1 asks how long ago marijuana was used and question 2 asks how many times marijuana was used within the last year, one hypothesis is that very infrequent users of marijuana may disagree that they are "users" of the substance, for example, to avoid being labeled as "marijuana users." These infrequent users might respond "no" to question 1 while responding "yes" to question 2. To test this hypothesis, responses for the frequency question were cross-classified by the A classification. The hypothesis would be confirmed if a disproportionate number of respondents who were classified as "No use in the past 12 months" by indicator A and "Yes, use in the past 12 months" by indicator B, are light or infrequent users who responded "1 to 2 days" to the frequency question.

The results of this analysis are reported in Table 4.10. Consistent with the hypothesis, more than 58% of respondents who indicated "No past-year use" for A and "Past-year use" for B were classified as infrequent users ("1 to 2 days") by the frequency question. Compare this with only about 16% in the

Table 4.10 Distribution of Reported Days of Use in Past Year by Recency of Use in Past Year: 1994–1996 NHSDAs[a]

Number of Days Used Marijuana in Past Year from Frequency Question	Percent Reporting No Past-Year Use on Recency Question	Percent Reporting Past-Year Use on Recency Question
>300	5.84	10.00
201–300	0.96	5.54
101–200	0.93	9.06
51–100	1.45	10.01
25–50	2.96	10.50
12–24	4.76	11.95
6–11	6.06	11.76
3–5	18.41	15.51
1–2	58.63	15.67
Total	*100.00*	*100.00*

[a]Based on 53,715 responses.
Source: Biemer and Wiesen (2002).

"1 to 2 days" category among persons consistently classified consistently as users by A and B. The analysis demonstrates the utility of latent class analysis for identifying questionnaire problems and, to some extent, provide evidence of the validity of the method.

The LC model estimates of classification were examined for differences by grouping variable. Estimates by sex are presented in Table 4.11, race/Hispanicity in Table 4.12, and age in Table 4.13. The entries in the tables are false positive (corresponding to the rows "true" = "yes," "observed" = "no") and false negative (corresponding to the rows "true" = "no," "observed" = "yes") probability estimates with their simple random sampling standard errors. For A, Biemer and Wiesen found very few significant differences by age, race, or sex. One exception is 1995, where the false-negative error rate for the 35+ age group was significantly higher ($p < 0.05$) than the other age groups. There were no important differences by sex for B and C as well. However, black respondents tend to have a higher false-positive error rate for indicator B than do whites or Hispanics. The oldest age group has a higher false negative error rate for indicator C than do other age groups. For both B and C, the false-positive error rates were higher for members of the 18–25 and the 26–35 age groups, and the false-negative error rates for these two age groups are lower than for the other two age groups.

Since we would not expect a large difference in classification error by sex, the null findings for this variable for all three indicators is plausible. The findings of the higher false-negative and lower false-negative error rates for the 35+ age group, particularly for indicator C, fit the pattern hypothesized in the previous discussion of the interpretation of group by indicator interaction for model 1. It is plausible that the 35+ group could have very different interpretations of some drug use questions than the younger age groups have. For example, a person who smoked marijuana in the past year infrequently might indicate no past-year marijuana use. A greater tendency to narrowly interpret the drug use questions in this manner for the 35+ age group would lead to the differences by age shown in the table.

Finally, three types of prevalence estimates of past-year marijuana were compared for all three years produced by Model 1. Table 4.14 provides the results for 1996; the results for 1994 and 1995 are similar. The first column of estimates in the table are based on the recency question, the second column of estimates (column 3) is based on Model 1, and the third set of estimates (in column 4) corresponds to the official NHSDA estimates. The NHSDA estimate essentially classifies a respondent as a user if either A or B is positive. The tables contain an overall estimate (row 1) as well as estimates by race/ethnicity, age, and sex domains.

Because the recency question is biased downward by a substantial false-negative error, we expect the estimates based on A to be uniformly lower than the model-based estimates. Further, since the NHSDA estimates tend to be biased upward due to the false-positive error in A and B, we expect those estimates to be uniformly higher than the model estimates. This ordering of

Table 4.11 Estimated Classification Probabilities and Standard Errors[a] by Sex for Indicators A, B, and C

Classification			Indicator A				Indicator B				Indicator C			
True	Observed	Group	1994	1994′	1995	1996	1994	1994′	1995	1996	1994	1994′	1995	1996
Yes ($X = 1$)	No ($A = 2$)	Males	7.28	6.69	8.35	7.96	1.03	1.01	0.68	1.17	6.04	6.54	4.98	6.39
			(0.94)	(0.90)	(1.05)	(0.95)	(0.30)	(0.29)	(0.23)	(0.31)	(0.67)	(0.74)	(0.67)	(0.72)
		Females	7.29	7.37	9.85	9.75	1.42	1.49	1.22	1.78	7.61	8.31	7.47	9.77
			(1.29)	(1.28)	(1.39)	(1.44)	(0.44)	(0.46)	(0.42)	(0.48)	(0.90)	(1.05)	(1.02)	(1.14)
No ($X = 2$)	Yes ($A = 1$)	Males	0.04	0.04	0.02	0.09	0.87	0.94	1.03	1.04	4.71	1.46	1.50	1.72
			(0.02)	(0.02)	(0.01)	(0.03)	(0.11)	(0.11)	(0.12)	(0.12)	(0.24)	(0.13)	(0.13)	(0.14)
		Females	0.04	0.03	0.01	0.07	0.61	0.60	0.56	0.67	3.51	1.04	0.88	1.06
			(0.02)	(0.02)	(0.01)	(0.02)	(0.08)	(0.08)	(0.08)	(0.19)	(0.19)	(0.10)	(0.09)	(0.10)

[a]Standard error values shown in parentheses.

Source: Biemer and Wiesen (2002).

Table 4.12 Estimated Classification Probabilities and Standard Errors[a] by Race for Indicators A, B, and C

Classification			Indicator A				Indicator B				Indicator C			
True	Observed	Group	1994	1994′	1995	1996	1994	1994′	1995	1996	1994	1994′	1995	1996
Yes ($X = 1$)	No ($A = 2$)	Hispanic	9.34	8.77	11.08	10.84	1.02	0.99	1.20	1.89	7.44	8.08	6.28	9.65
			(3.02)	(2.91)	(3.32)	(3.16)	(0.35)	(0.35)	(0.48)	(0.72)	(1.14)	(1.67)	(1.49)	(1.91)
		Black	7.53	6.40	12.58	10.16	0.61	0.59	0.51	1.06	7.54	7.51	8.73	9.01
			(2.19)	(2.00)	(2.83)	(2.31)	(0.21)	(0.19)	(0.20)	(0.35)	(1.07)	(1.35)	(1.80)	(1.54)
		White/other	7.05	6.84	8.23	8.09	1.27	1.28	0.92	1.40	6.37	7.03	5.55	7.14
			(0.84)	(0.83)	(0.93)	(0.89)	(0.36)	(0.36)	(0.32)	(0.37)	(0.71)	(0.77)	(0.73)	(0.78)
No ($X = 2$)	Yes ($A = 1$)	Hispanic	0.03	0.02	0.01	0.06	0.90	0.94	0.64	0.67	3.75	1.15	1.08	1.09
			(0.02)	(0.02)	(0.01)	(0.03)	(0.25)	(0.25)	(0.21)	(0.20)	(0.47)	(0.23)	(0.24)	(0.22)
		Black	0.03	0.04	0.01	0.06	1.40	1.52	1.33	1.18	3.74	1.27	0.77	1.18
			(0.02)	(0.02)	(0.01)	(0.03)	(0.28)	(0.29)	(0.29)	(0.26)	(0.42)	(0.22)	(0.17)	(0.21)
		White/other	0.04	0.04	0.01	0.08	0.62	0.63	0.72	0.82	4.15	1.25	1.24	1.42
			(0.02)	(0.02)	(0.01)	(0.02)	(0.07)	(0.07)	(0.08)	(0.08)	(0.17)	(0.09)	(0.09)	(0.10)

[a]Standard error values shown in parentheses.
Source: Biemer and Wiesen (2002).

Table 4.13 Estimated Classification Probabilities and Standard Errors[a] by Age for Indicators A, B, and C

Classification		Age in	Indicator A				Indicator B				Indicator C			
True	Observed	Years	1994	1994′	1995	1996	1994	1994′	1995	1996	1994	1994′	1995	1996
Yes ($X = 1$)	No ($A = 2$)	12–17	4.91	4.53	7.23	10.57	1.76	1.74	1.52	2.27	6.88	7.70	6.31	10.02
			(1.62)	(1.52)	(1.69)	(2.02)	(0.62)	(0.59)	(0.55)	(0.66)	(0.93)	(1.31)	(1.10)	(1.48)
		18–25	8.19	7.64	7.14	8.88	0.71	0.68	0.29	0.74	5.23	5.20	4.11	5.26
			(1.31)	(1.25)	(1.25)	(1.28)	(0.22)	(0.21)	(0.10)	(0.23)	(0.67)	(0.77)	(0.66)	(0.74)
		26–35	8.29	8.15	7.88	8.00	0.90	0.93	0.44	1.13	5.37	4.74	4.78	6.62
			(1.62)	(1.57)	(1.58)	(1.59)	(0.29)	(0.29)	(0.17)	(0.36)	(0.69)	(0.72)	(0.79)	(0.97)
		>35	6.51	6.24	13.92	7.34	1.69	1.74	1.71	2.00	9.31	11.61	9.55	10.44
			(1.40)	(1.36)	(2.02)	(1.53)	(0.52)	(0.52)	(0.60)	(0.61)	(1.06)	(1.36)	(1.28)	(1.29)
No ($X = 2$)	Yes ($A = 1$)	12–17	0.05	0.05	0.02	0.06	0.59	0.64	0.49	0.64	4.71	1.40	1.39	1.30
			(0.03)	(0.03)	(0.02)	(0.02)	(0.18)	(0.18)	(0.16)	(0.19)	(0.49)	(0.24)	(0.24)	(0.21)
		18–25	0.03	0.03	0.02	0.07	1.42	1.56	2.41	1.82	5.88	1.98	2.09	2.34
			(0.02)	(0.02)	(0.02)	(0.03)	(0.28)	(0.29)	(0.36)	(0.32)	(0.51)	(0.27)	(0.29)	(0.30)
		26–35	0.03	0.03	0.02	0.08	1.08	1.09	1.41	1.17	5.70	2.16	1.72	1.85
			(0.02)	(0.01)	(0.01)	(0.03)	(0.20)	(0.20)	(0.23)	(0.21)	(0.43)	(0.26)	(0.23)	(0.23)
		>35	0.04	0.03	0.01	0.09	0.53	0.54	0.37	0.62	3.18	0.81	0.83	1.08
			(0.02)	(0.02)	(0.01)	(0.03)	(0.07)	(0.07)	(0.06)	(0.08)	(0.17)	(0.08)	(0.08)	(0.09)

[a]Standard error values shown in parentheses.
Source: Biemer and Wiesen (2002).

Table 4.14 Comparison of Recency, Latent Class Model, and NHSDA Prevalence Rates (in percent) for Past-Year Marijuana Use, 1996[a]

Domain	Recency	LC Model	NHSDA
Race/ethnicity			
Hispanic	5.72 (0.55)	6.36 (0.59)	7.04 (0.44)
Black	8.79 (0.62)	9.72 (0.66)	11.09 (0.65)
White/other	7.11 (0.21)	7.65 (0.22)	8.43 (0.45)
Age in years			
12–17	11.27 (0.72)	12.54 (0.76)	12.99 (0.78)
18–25	20.40 (0.83)	22.34 (0.86)	23.85 (1.08)
26–34	9.44 (0.53)	10.19 (0.55)	11.33 (0.54)
35+	2.94 (0.16)	3.08 (0.17)	3.76 (0.34)
Sex			
Male	9.55 (0.31)	10.29 (0.33)	11.38 (0.58)
Female	4.95 (0.22)	5.41 (0.23)	6.02 (0.31)
Total	*7.16 (0.19)*	*7.75 (0.20)*	*8.60 (0.35)*

[a]Standard error values shown in parentheses.
Source: Biemer and Wiesen (2002).

the three sets of estimates is apparent in the 1996 results as well as the other 2 years not shown in the table.

This analysis indicates that the population-level estimates from the NHSDA are, on average, between 0.7 and 0.9 percentage points higher than the corresponding model-based estimates. However, for some subgroups, the difference may be as high as 1.5 percentage points. The national-level estimates based on the single recency question are between 0.5 and 0.6 percentage points lower than the corresponding model-based estimates. At the subgroup level, the difference may be as large as 1.9 percentage points. Thus, the consequences of assuming no false-negative responses (e.g., for the recency prevalence estimator) or no false-positive responses (e.g., for the NHSDA estimator) can be substantial, particularly for domain estimates.

One reason the NHSDA estimator may be preferred over the model-based estimator despite its upward bias is that it may at least partially compensate for false-negative errors not accounted for by the latent class analysis. For example, respondents who use marijuana yet deny with probability 1 each time the question is asked in the survey (so-called *consistent falsifiers*) are not accounted for in the false-negative probability estimates. Our limited simulation studies to date indicate that the bias in the false-negative error estimates may be substantial when the proportion of users that are consistent falsifiers exceeds 15%; however, more study is needed to fully understand this effect. Still, it is possible that the NHSDA estimator is less biased than the model-based estimator when the consistent falsifier bias is considered.

It is also possible that the patterns of false-positive bias for population subgroups of interest are very different than the patterns of consistent falsifier bias for these subgroups. For example, if one assumes that false-positive error occurs as a result of questionnaire complexity or difficulty, then the population subgroups most prone to false-positive bias are low-literacy subgroups and individuals who are careless completing the NHSDA answer sheets regardless of age, race, sex, or other criteria. By using the false-positive error to compensate for the consistent falsifiers, the tacit assumption is that the same population subgroups that inadvertently answer they used marijuana when they did not are the same groups that deny their use consistently. While it may be true that marijuana use is higher for some groups who have low literacy, to base a bias adjustment on the assumption that false-positive and consistent denial patterns are similar is highly questionable.

Discussion

The example illustrates that latent class analysis methods have considerable potential for providing more valid estimates of self-reported drug use; however, there may be even greater potential for exploiting these methods for identifying problems in questionnaire design and question wording. The questionnaire problems discovered in the Biemer and Wiesen's analysis would have been much more difficult to discover using other means of analysis. For example, the inconsistency analysis of Table 4.6 suggests that, for the 1994 NHSDA, indicator C is quite inconsistent with indicators A and B, considerably more so than indicator A and B are with each other. However, it is not apparent that the problem is due to false-positive error in indicator C and or false-negative error in the other two indicators. The LC model estimates of false-positive and false-negative error for the three indicators quickly and unequivocally determined the problem to be false-positive errors in C.

Similarly, an analysis of the inconsistencies for indicators A and B clearly demonstrates a problem for one of the two indicators, but it is not apparent that the source of the inconsistency was the response of low-frequency users to the recency question. LCA quickly led us to suspect indicator A was the source of the inconsistency. This led to further investigation using more traditional analysis, which uncovered the source of the problem. These analyses are illustrative of the utility of LCA for pointing out problems in the execution of surveys.

Because the three indicators in our analysis are all obtained in one interview, correlated errors among the indicators was suspected. The LCA provided evidence to the contrary, which is useful in that it suggests the absence of social desirability influences on responses to past-year marijuana use questions in the NHSDA. However, as we shall see in other examples, local dependence models can be important for the analysis of embedded indicators for more stigmatized behaviors.

Finally, these results were useful to questionnaire designers as they implemented a redesign of the NHSDA and its conversion to computer-assisted self-interviewing (CASI). Questions on drug use that were in use prior to 1999 were revised in the CASI version, so the problems uncovered for the recency question and the potential biases associated with the former NHSDA estimation approaches are no longer a threat to the survey. In the CASI implementation, the marijuana sequence uses a "gate" question that asks respondents whether they have ever used marijuana in their entire lifetimes. Only those who respond positively to this gate question are asked more detailed questions about their marijuana use and recency. Latent class analysis has not yet been implemented for this new design to determine whether the gains in accuracy expected from CASI administration were realized.

CHAPTER 5

Further Aspects of Latent Class Modeling

The previous chapters introduced the basic concepts of latent class modeling as they apply to survey measurement error evaluations. In this chapter, we expand on these ideas and introduce some additional applications of LCA. We begin with a discussion of a few important estimation issues of which every latent class analyst should be aware.

5.1 PARAMETER ESTIMATION

Many analysts find the application of LC models difficult in practice because of a number of estimation problems that can produce unexpected results in the estimation of the model parameters or the interpretation of the LCA results. The problems include nonidentifiability, data sparseness, boundary values/improper solutions, local maxima, and local dependence. All of these issues were mentioned in previous chapters. Some problems, like local independence and nonidentifiability, are idiosyncratic of latent variable modeling and are not encountered in most other types of analysis. Other problems, like data sparseness and boundary solutions, are quite common in categorical data analysis but may introduce new complications for LC models. This chapter discusses these problems in more detail. We first begin with a discussion of two analytical techniques that are quite useful for investigating many estimation issues encountered in LCA applications: simulation and expeculation.

5.1.1 Simulation and "Expeculation"

For investigating issues in model estimation, it is often useful to generate an artificial population that follows a specified model with known parameters.

Latent Class Analysis of Survey Error By Paul P. Biemer

Once constructed, such a population can be sampled repeatedly to simulate a sampling process for a real population. By studying the behavior of estimators computed from these repeated, simulated samples, information about the sampling distributions of the estimators can be obtained, including the variance and bias of the estimators, confidence interval coverage probabilities, issues of identifiability, and local maxima. Indeed, for many complex LC models, this may be the only way to study the estimation issues since analytical closed form expressions of the estimators and their distributions cannot be obtained.

As an example, suppose that we wish to study the effect that violating the assumption of local independence has on estimates from a standard LC model. We begin by generating a simulated population of some size N using a model that violates the local independence assumption in a known way. Now suppose that a single random sample of size n is selected from this population using simple random sampling without replacement. Fitting the standard LC model to these data, we obtain parameter estimates that are biased as a result of model misspecification. The estimates are also subject to sampling variance unless $n = N$. However, let us assume that n is quite small compared to N. Now imagine repeating the sampling and estimation process for this population many (say, 1000) times. The average bias over these repetitions is an estimate of the model misspecification bias, and the standard deviation of the bias estimates is an estimate of the standard error of the bias estimate. In addition, the standard deviation of the parameter estimates over repeated samples is also an estimator of their standard errors. In this way, the effect of violating the local independence assumption on the bias and standard errors of the estimates can be assessed empirically for a particular population. This technique is often referred to as *Monte Carlo simulation* [see, e.g., Robert and Casella (2004)].

A related method that is used extensively in LC model methodology research does not require repeatedly drawing random samples. Instead, one simply creates a table with frequencies that are exactly equal to the expected cell frequencies under the population model. This technique is appropriate if interest is only on the expected values or biases of the estimators rather than their standard errors or sampling distributions. Since the literature apparently has not invented a name for this technique, we shall refer to it as "expeculation" in this book since it is related to simulation but replaces sample frequencies with their expected values (*expec*ted value sim*ulation*).

A key difference between expeculation and simulation is that for the latter, the observed cell frequencies, say, $\{n_j, i = 1,\ldots,k\}$, where k is the number of cells, can vary substantially from sample to sample, especially for small n. With expeculation, the cell frequencies are exactly equal to $m_j = n\pi_j$, where π_j is the cell probability computed under the assumed population model and specified population parameters and m_j is the expected frequency for cell j under this model. In other words, $n_j = m_j$ for $j = 1,\ldots, k$. If the same model used to generate $\{n_j, j = 1,\ldots, k\}$ is fitted to this table, the model MLEs should agree with the model parameters used to generate the data, apart from any computer roundoff errors. In other words, $\hat{m}_j$ should equal m_j. A difference between the model parameter and its estimate $(\hat{m}_j - m_j)$ can be interpreted as model misspecification bias. This bias estimator is theoretically identical to the bias that would be estimated by a simulation with an infinite number of resamples from the model population. However, expeculation provides no information on the standard errors of the model parameter estimates for the misspecified model. In this and remaining chapters, we will make use of expeculation and simulation for various design and model scenarios in order to assess the bias under an assumed model for populations that depart from the model. An illustration of expeculation for checking model identifiability is provided in the next section.

5.1.2 Model Identifiability

As discussed previously, a necessary condition for the parameters of the LC model to be uniquely estimated is that the model be identifiable. An *identifiable* model is one whose likelihood has one and only one maximum. A necessary (but not sufficient) condition for identifiability that applies to all loglinear models is that the number of parameters to be estimated not exceed the number of observed frequencies or, alternatively, that the model degrees of freedom be nonnegative. As an example, the Hui–Walter model for dichotomous manifest variables described in the previous chapter is just identified because the model degrees of freedom is 0.

The three-indicator LC model for dichotomous indicators and two latent classes is also just identified since it allows up to seven parameters (plus 1 for the overall mean). The seven parameters of this model include the latent proportion and two error parameters (false positive and false negative) for each indicator for a total of six error parameters. In general, the three-indicator LC model K categories corresponding to K latent classes has $K \times K \times K$ frequencies and thus provides sufficient degrees of freedom for estimating the overall mean and $K^3 - 1$ additional parameters. For K-dimensional X, there are $K - 1$ latent class size parameters and $K(K - 1)$ error probabilities to be estimated for each indicator. Thus, the total number of parameters in the model is $(3K + 1)(K - 1)$. The model degrees of freedom, which is computed as the difference in the number of cells and the number of parameters, is $(K^3 - 1) - (3K + 1)(K - 1) = K^3 - 3K^2 + 2K$. Note that when $K = 2$, the model degrees of freedom is 0 and the model is fully saturated. For $K > 2$, the model degrees of freedom is positive and chi-square goodness-of-fit tests can be performed. If the model is nonidentifiable as a result of *overparameterization* (i.e., negative degrees of freedom), parameter restrictions can be imposed so that the number of parameters to be estimated is reduced.

It is important to realize that models with nonnegative degrees of freedom may still not be identifiable. For example, as seen in Chapter 3, the binary two-indicator model contains four frequencies and three parameters (viz.,π, θ,

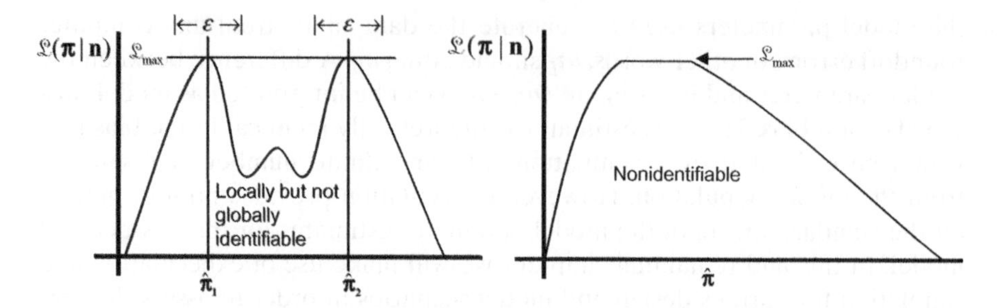

Figure 5.1 Locally identifiable and nonidentifiable likelihoods.

and ϕ) but is not identifiable. Nonidentifiable models are more common when the dimension of X is greater than the dimension of the indicators. For example, the model with four binary indicators (A,B,C,D) is not identifiable when X has three classes even though the model degrees of freedom is 1 (Goodman 1974a). If a model is not identifiable, the model estimates are not unique and are, therefore, invalid. For example, repeated maximizations of the model likelihood will produce different solutions. Therefore, verifying that a model is identifiable is an important step in an LCA.

Model identifiability can be difficult to prove. In most situations, the best that can be done is to verify that the model is *locally identifiable*. According to Goodman (1974a), a model is *locally identifiable* if the maximum-likelihood solution is unique within some small interval (i.e., an ε-neighborhood) around the solution. This is less restrictive than *global identifiability*, in which the MLE must be unique throughout the entire parameter space. These concepts are illustrated graphically in Figure 5.1. Goodman further shows that a necessary and sufficient condition for *local identifiability* is that the information matrix be positive definite [see also Dayton and MacReady (1980)]. Formann (1985) provided a proof that the likelihood is locally identifiable if the eigenvalues of the information matrix are positive. This condition is easily checked in the estimation process when using methods that rely on second partial derivatives such as Fisher's scoring method or the Newton–Raphson algorithm. If the EM algorithm is used, the check is not done automatically because second derivatives are needed for parameter estimation. Nevertheless, software packages employing EM such as ℓEM, MPlus, and Latent GOLD still provide this check.

Tests for identifiability based on the rank of the information matrix typically do not include boundary parameters (e.g., probabilities estimated to 0 or 1). Therefore, this identification test does not check as to whether boundary parameters are identified. Identifiability of these parameters can be tested by other means. For example, zero-parameter estimates due to structural zeroes in the data and boundary parameters with large first-order derivatives may both be considered identifiable. However, parameters that are estimated 0 or

1 with first-order derivatives approximately equal to 0 may not be identifiable. In that case, the method of multiple starting values, discussed next, should be used to test whether these boundary parameters are identifiable.

A common way of checking local identifiability when using the EM algorithm is to estimate the model multiple times using different starting values. If the solutions from these multiple maximizations are identical for the same maximum value of the likelihood, the model is probably identifiable. But should multiple runs with different starting values produce quite different parameter estimates for the same maximum-likelihood value, the model is definitely not identifiable (Hagenaars 1990, p. 112).

Bartholomew and Knott (1999, p. 151) discuss the relationship between identifiability and precision. If the likelihood is fairly flat in the neighborhood of an estimate, then the estimate will have a very large variance even though the likelihood is identifiable. As the likelihood approaches absolute flatness (as in Figure 5.1), the estimate becomes nonidentifiable. Therefore, estimates having very large variances could be *nearly* or *weakly identifiable*. One measure of the strength of model identification is the condition number, defined by the ratio of the largest to the smallest eigenvalues of the information matrix. As a general rule of thumb, a ratio that exceeds 5,000 is considered an indication of weak identification. In that case, standard errors of parameters will be large and some parameters may be highly correlated. To achieve stronger identification, one or more of these parameters should be fixed or set equal to other parameters. Identifiability is not likely to be a problem for models that can be estimated with reasonable precision. Bartholomew and Knott suggest performing LCA only when the sample size is sufficiently large and the number of latent classes to be estimated is four or fewer (i.e., $K \leq 4$). However, such advice may be unrealistic in many applications.

Expeculation can be used is to check model identifiability for simple models. Let $\{n_j\}$ denote the cell frequencies obtained under the model M by expeculation (i.e., $n_j = m_j$, $j = 1,\ldots,k$), and assume that we wish to check whether M is identifiable. If M is identifiable, then fitting the model M to $\{n_j\}$ should produce MLEs of the parameters that are equal to the parameter values assumed for M, apart from roundoff error. If they differ appreciably, then M is nonidentifiable. This is explained further in illustrations to follow.

5.1.3 Checking Identifiability with Expeculation

Expeculation can be used to check the identifiability of a model. For example, suppose that we wish to determine whether the model $\{XA\ XB\ XC\}$ with equal error probabilities constraints (i.e., $\pi_{a|x}^{A|X} = \pi_{b|x}^{B|X} = \pi_{c|x}^{C|X}$) imposed is identifiable for trichotomous variables. First we generate a dataset using expectulation that exactly fits this model. Then we backfit the model to the expeculated dataset and observe the parameter estimates. If the model is identifiable, the estimates should agree exactly with the parameters used to expeculate the dataset.

Table 5.1 Model Parameters Assumed for the Expeculated Data in Table 5.2

$X = x$	$\pi_{1\|x}^{A\|X}$	$\pi_{2\|x}^{A\|X}$	$\pi_{3\|x}^{A\|X}$	π_x^X
1	0.70	0.20	0.10	0.5
2	0.10	0.80	0.10	0.3
3	0.25	0.10	0.65	0.2

To illustrate, let us expeculate a dataset with $n = 10{,}000$ cases from this model with the parameters shown in Table 5.1. Since equality constraints were imposed for the error probabilities for all three indicators, only the conditional probabilities for A are shown in the table. So, for example, $\pi_{a|1}^{A|X}$ for $a = 1, 2,$ and 3 are 0.7, 0.2, and 0.1, respectively, and $\pi_1^X = 0.5$.

The (1,1,1) cell probability is computed as

$$\begin{aligned}\pi_{111}^{ABC} &= \pi_1^X \pi_{1|1}^{A|X} \pi_{1|1}^{B|X} \pi_{1|1}^{C|X} + \pi_2^X \pi_{1|2}^{A|X} \pi_{1|2}^{B|X} \pi_{1|2}^{C|X} + \pi_3^X \pi_{1|3}^{A|X} \pi_{1|3}^{B|X} \pi_{1|3}^{C|X} \\ &= 0.5(0.70)^3 + 0.3(0.10)^3 + 0.2(0.25)^3 \\ &= 0.174925\end{aligned} \tag{5.1}$$

and after multiplying by 10,000, the cell frequency is 1749.25. Likewise for cell (3,2,3), the cell probability is

$$\begin{aligned}\pi_{323}^{ABC} &= \pi_1^X \pi_{3|1}^{A|X} \pi_{2|1}^{B|X} \pi_{3|1}^{C|X} + \pi_2^X \pi_{3|2}^{A|X} \pi_{2|2}^{B|X} \pi_{3|2}^{C|X} + \pi_3^X \pi_{3|3}^{A|X} \pi_{2|3}^{B|X} \pi_{3|3}^{C|X} \\ &= 0.5(0.10)(0.20)(0.10) + 0.3(0.10)(0.80)(0.10) + 0.2(0.65)(0.10)(0.65) \\ &= 0.01185\end{aligned} \tag{5.2}$$

and, multiplying by 10,000 the cell frequency is 1185, and so on. The values for all 27 cells are shown in Table 5.2.

Next, the model {XA XB XC} with equal error probability constraints for A, B, and C was fit to these data. The ℓEM input instructions for this is shown in Figure 5.2. The reader may verify that the parameters in Table 5.1 are reproduced to within $\pm 10^{-5}$ when this model is fit by running the ℓEM statements in Figure 5.2. Thus, there is no evidence of nonidentifiability for this model. If more evidence is desired, this process can be repeated by expeculating other data sets for various sets of model parameters.

Next we illustrate expeculation for a nonidentifiable model. Suppose that the indicator C for this population is unobserved. To form this dataset, the counts in Table 5.2 are collapsed over C to obtain nine counts [2605, 990, 705, 990, 2140, 470, 705, 470, 925] corresponding to cells (a,b) ordered as [11, 12, 13, 21,...], respectively. As before, we set $XA = XB$ so that the model contains eight parameters: $\pi_x^X, x = 1, 2,$ $\pi_{a|1}^{A|X}, a = 1, 2, \pi_{a|2}^{A|X}, a = 1, 2,$ and $\pi_{a|3}^{A|X}, a = 1, 2.$

Table 5.2 An Expeculated Dataset for Model {XA XB XC} with $XA = XB = XC$

Pattern ABC for $A = 1$	Expeculated Cell Count	Pattern ABC for $A = 2$	Expeculated Cell Count	Pattern ABC for $A = 3$	Expeculated Cell Count
111	1749.25	211	526.5	311	329.25
112	526.5	212	337	312	126.5
113	329.25	213	126.5	313	249.25
121	526.5	221	337	321	126.5
122	337	222	1578	322	225
123	126.5	223	225	323	118.5
131	329.25	231	126.5	331	249.25
132	126.5	232	225	332	118.5
133	249.25	233	118.5	333	557.25

```
* LEM Code for {XA XB XC} with Constraints XA=XB=XC
lat 1
man 3
dim 3 3 3 3
lab X A B C
mod X
    A|X
    B|X eq1 A|X    * equates error probabilities for A and B
    C|X eq1 A|X    * equates error probabilities for A and C
dat [1749.25 526.5  329.25 526.5  337    126.5  329.25 126.5  249.25
     526.5   337    126.5  337    1578   225    126.5  225    118.5
     329.25  126.5  249.25 126.5  225    118.5  249.25 118.5  557.25 ]
```

Figure 5.2 ℓEM input statements to fit {XA XB XC} with equality constraints.

Note that the model {XA XB} is fully saturated but nonidentifiable. Fitting this model to the above mentioned data with ℓEM confirmed that the parameter estimates differ considerably from the parameters in Table 5.1. For example, one run produced 0.2337, 0.4013, and 0.3650 for π_x^X, $x = 1, 2, 3$, respectively, with a maximum value of the loglikelihood of –20198.22570. Another run produced 0.3020, 0.4048, and 0.2931 for π_x^X, respectively, with the same maximum value of the loglikelihood. The other parameter estimates also differed considerably from those reported in the first run. A check for local maxima was performed by running the model multiple times. Each run produced the same maximum value of the loglikelihood, but the parameter estimates varied considerably, which clearly demonstrates that the model is not identifiable. If desired, alternate datasets could be generated using different parameters than those in Table 5.1 to check this result. Although this is no proof of the general nonidentifiability of {XA XB} with $XA = XB$ for trichotomous indicators, it provides ample evidence that, for all the datasets considered, the model is not identifiable.

5.1.4 Data Sparseness

In fitting LC models with multiple grouping variables and interactions or models that involve variables that have many categories, the data table can become *sparse*—that is, contain many cells with 0 or very small counts. Sparseness is also a problem in less complex models when n is small. One general measure of data sparseness used throughout the loglinear modeling literature is the ratio n/k, where k is the number of cells in the table. A sparse data table is one with a low value of this ratio, say, less than 5. Sparse tables can cause a number of problems in an LCA, as will be discussed in this chapter.

Fortunately, sparse data do not cause severe problems for parameter estimation (Langeheine et al. 1996; Collins et al. 1993). The EM algorithm is still effective for obtaining the MLE sparse tables, although some parameters may not be identifiable if the sparseness results in *zero fitted cells* and *margins* (described below). However, sparseness, especially zero-frequency cells, are particularly problematic for the chi-square tests of model fit since the true distribution of the test statistic is no longer well approximated by the theoretical chi-square distribution.

Cells with zero frequencies can be divided into two types: (1) *sampling* zero and (2) *structural* zero cells. Sampling zero cells have small, but nonzero, probabilities of being observed; in other words, their expected frequencies are positive. When the sample size is small, such cells will contain no observations by chance. However, as the sample size increases, expected frequencies increase and these zero cells can become populated. As an example, in a survey of drug use, the expected frequency of recent college graduates addicted to heroin may be quite small. Although such persons may exist in the population, they are unlikely to be sampled unless the sample size is quite large.

Structural zeroes are cells that are theoretically impossible to observe. They either do not exist in the population or exist but have a zero probability of selection by the sample method. As an example, for evaluating the accuracy of a census, a postenumeration survey (PES) is conducted. Persons in the PES that are missed by the census are the basis for an estimate of the census miss rate or *undercount*. Persons in the PES and the census can be cross-classified in a 2×2 table as either "in the census" or "not in the census" and "in the PES" and "not in the PES." However, some persons are missed by both the census and the PES (the so-called *fourth cell*). Obviously, this cell cannot be observed and is, therefore, a structural zero cell. Note, however, that such persons exist in the population; they nevertheless have a zero probability of observation. As another example, consider the cross-classification of persons by sex and their reason for hospitalization. Some reasons (e.g., pregnancy) are sex-specific, causing certain cells to have zero probability of being observed.

Sampling zero cells are considered part of the observed data and require no special constraints in the modeling. By contrast, structural zero cells are not part of the data. Proper modeling of these cells requires constraints that fix the corresponding cell probabilities to be zero. Otherwise, the model will treat these cells as sampling zeros with possibly important consequences for parameter estimation, model fitting, and so forth. In addition, the degrees of freedom for the model should be adjusted to reflect the structural zero constraints imposed on the model. A general formula for adjusting the degrees of freedom is the following:

$$\mathrm{df} = (k - z) - (\mathrm{npar} - \mathrm{npar}_0) \tag{5.3}$$

where z is the number of structural zero cells, npar is the number of parameters in the model, and npar_0 is the number of parameters that cannot be estimated because of zero cells. Note that when $\mathrm{npar}_0 > 1$, the model is not identifiable. The methods described in the last section for obtaining an identifiable model (e.g., imposing constraints, reducing model complexity, and collapsing cells) should be applied in this situation.

Another problem that may arise when one or more cells have an estimated zero expected count (*zero fitted cells*), that is, when $\hat{m}_j$ is 0. In some cases, a marginal frequency is estimated to be zero (*zero fitted marginal*). The ℓEM software provides an error message notifying the user of either of these situations. When either of these situations arises, then npar should be increased by the number of zero fitted cells and npar_0 should be increased by the number of parameters that cannot be estimated because there are zero fitted marginal totals.

An important issue in dealing with sampling zero cells or, more generally, sparse tables, is the validity of the chi-square test statistics. Particularly for LC models, sparseness can invalidate tests of model fit using L^2 and X^2 because they are no longer chi-square distributed. Haberman (1977) and Agresti and Yang (1987) have shown that for testing of nested models in loglinear analysis, the conditional likelihood test using L^2 is much less affected by data sparseness. Therefore, although it may be difficult to ascertain model fit, model selection can still proceed normally. For example, for determining whether XGA can be removed from the model, comparing the L^2 statistics from the models containing XGA versus the same model replacing XGA with GA XG XA can still be considered a valid test only when n/k is quite small, provided all other assumptions of the conditional chi-square test are met.

Because both X^2 and L^2 provide a chi-square test of model fit, the question of which statistic is better arises. In most situations, the two measures will provide very nearly the same value, so the question is moot. However, in sparse tables, they will differ as they are no longer chi-square or even identically distributed. If they differ considerably, then the validity of either chi-square goodness-of-fit test is questionable. Indeed, an important reason to compute both X^2 and L^2 is so that this comparison can be made. Many data analysis software packages do this automatically. Studies have shown that X^2 is often closer to chi-square than L^2 in sparseness situations (Agresti 1990, p. 247). However, Collins et al. (1993) warn that the variation in X^2 can be still be quite large and the test can be unreliable. Still, the literature seems to favor X^2 as long as $n/k > 1$ as it tends to be more robust to data sparseness.

One method for dealing with model testing for sparse tables is to reduce the number of grouping variables in the model or the categories of the manifest variables to reduce the number of cells, k. Reducing k can also restore identifiability if the reason for nonidentifiability is empty cells or table sparseness. Another approach is to empirically estimate the exact distribution of either X^2 or for a particular dataset and model using bootstrapping methods (Efron 1979). Both the Latent GOLD and PANMARK (Van de Pol et al. 1998) software have implemented this approach for latent class models following on the work of Aitken et al. (1981), Noreen (1989), and Collins et al. (1993). (This feature is not available in ℓEM.) Alternatively, lower-order marginal distributions of the items can be used instead of the full-data table. Feiser and Lin (1999) have a test based on first- and second-order marginal distributions of the items. Even when the full-data table is sparse, the first and second marginal distributions may not be, and their test is more valid.

Another option is to use a measure of model fit that is more robust to small cell sizes. One such measure is the *power divergence* statistic developed by Cressie and Read (1984), denoted by CR. As a measure of model deviance, CR is often less susceptible to data sparseness than is either X^2 or L^2. It is implemented in most software packages with its power parameter (λ) equal to $\frac{2}{3}$, which then takes the form

$$\mathrm{CR} = \frac{9}{5}\sum_{abc} n_{abc}\left[\left(\frac{n_{abc}}{\hat{m}_{abc}}\right)^{2/3} - 1\right] \tag{5.4}$$

where n_{abc} is the count in cell (a,b,c) and $\hat{m}_{abc}$ is the model estimate of $m_{abc} = E(n_{abc})$ (see Section 4.2.3). When L^2 (or X^2) and CR diverge, CR is likely to more closely approximate a chi-square random variable. However, one should use caution since CR will also be invalid if sparseness is severe. An alternative to model deviance measures is the dissimilarity index defined in (4.48). This measure is often employed to gauge model fit when the chi-square statistics are not valid. Unfortunately, a statistical test of model fit using D is not available. In addition, D is not appropriate for testing and comparing nested models.

As noted in Chapter 4, AIC and BIC defined in (4.51) are useful for comparing alternative models, especially models that are not nested. The model having the lowest AIC or BIC is preferred under this criterion because it provides the best balance between two factors: model fit (reflected in the value of L^2) and model parsimony (represented by df $\times \log n$ for BIC and 2df for AIC). Like D, there is no statistical test for determining whether the BIC for one model is significantly lower than the corresponding measure for another model. Rather, the measures are meant to serve as indices of the comparative fit and somewhat of a rough guide for model selection.

If a problem occurs in the algorithm for computing MLE estimates as a result of sampling zero cells or very small cells, the model parameters may not converge, and the software output will typically provide a warning such as "Information matrix is not of full rank" or "Boundary or nonidentifiable parameters." In addition, one or more of the error probabilities will be at or very near their boundary values (i.e., 1 or 0), which corresponds to a loglinear parameter of $\pm\infty$. The estimated standard errors for these parameters will be quite large in comparison to the remaining estimates. One method for solving this problem is to add a small constant (say, 0.001) to each cell of the observed frequency table. However, Agresti (1990, p. 250) advises caution in applying this method. He recommends rerunning the model with various additive constants (10^{-3}, 10^{-5}, 10^{-8}, etc.) to determine the sensitivity of the estimates to the choice. The smallest constant that will alleviate the computational problems without unduly influencing the key parameter estimates should be used.

5.1.5 Boundary Estimates

It is not uncommon for one or more model probability estimates to be either 0 or 1. Boundary estimates (sometimes referred to as *improper solutions*) occur for a number of reasons:

1. The probability may have been estimated to be very small, say, 10^{-6}, which is essentially zero at the level of precision of the printed output.
2. The optimum value of the probability estimate is actually negative; however, the maximization algorithm stopped moving in the direction of the optimum once the estimate reached 0 and remained there until convergence.
3. The data may contain zero cells, resulting in a u-parameter estimate of $\pm\infty$; in other words, the parameter could not be estimated. Consequently, the corresponding error probability estimate is zero.
4. The probability estimate may have been constrained to be zero.

Situation 1 poses no problem for model testing because the estimate is optimal and the standard error exists and is approximately zero as well. Situation 2 occurs when the sample size is too small or if the model is misspecified. In the latter case, appropriately dropping one or more interaction terms from the model will alleviate the problem. Likewise, condition 3 may be a consequence of inadequate sample or model complexity. It can be handled using the methods described in the last section for data sparseness.

Finally, situation 4 is quite common when dealing with highly stigmatized questions such as drug use, abortions, and sexual perverseness. For these issues, it is highly unlikely that respondents will respond in error in socially undesirable ways. For example, if A is an indicator denoting the response to a question as to whether the respondent has ever had an abortion, it may be reasonable to assume that the probability of a false positive is zero, so we constrain $\pi_{1|2}^{A|X}$ to be zero. Another situation that arises in the general application of latent class models is the testing of the number of classes for X. For example, in typology analysis, an analyst may wish to compare the fit of a two-class model with that of a three-class LC model. It can be shown that the two-class model fit can be obtained by setting $\pi_x^X = 0$ for either $x = 1, 2$, or 3 in the three-class model.

The boundary value estimates under conditions 2 and 3 above can pose problems for model testing since the information matrix is not of full rank and therefore cannot be inverted. This implies that the standard errors of the LC model parameters cannot be computed in the usual way. Nor can interval estimates or significance tests be obtained by standard loglinear modeling procedures. One possible remedy is to constrain the LC model probabilities that are estimated on the boundary to either 1 or 0 and rerun the model. The degrees of freedom should also be adjusted as discussed in Section 4.4.5 to account for the fixed constraints (for, e.g., adding one model degree of freedom for each constrained parameter). The standard errors of the unconstrained parameters so obtained are valid if one is willing to assume that these were a priori constrained based on theoretical considerations as in situation 4. The resulting set of parameter estimates is referred to as a "terminal" solution; they are correct provided the true parameter value of the boundary estimate is actually zero (Goodman 1974a).

De Menezes (1999) proposed obtaining standard errors using the bootstrap variance estimation approach [see, e.g., Efron and Tibshirani (1993)]. This is done by generating a large number of samples of size n using the estimated LCM of interest and then reestimating the model for each synthetic sample. The square root of the variance of an estimate computed over these synthetic samples provides an estimator of the standard error of the estimate. Another approach is to impose a Bayesian prior distribution on the model parameters, referred to as "Bayesian posterior mode estimation" by Galindo-Garre and Vermunt (2006). This latter approach has been implemented in the Latent GOLD software (Vermunt and Magidson 2005a).

5.1.6 Local Maxima

The EM algorithm will sometimes converge to a local maximum rather than a global maximum. The usual strategy for avoiding this problem is to rerun the EM algorithm some number of times, say, $J > 1$ times, each time using a different set of starting values for the parameters. Suppose that the maximum-likelihood value for the jth run is $\mathcal{L}_{\max,j}$. Then the global maximum value of the likelihood, say, $\mathcal{L}_G$ will satisfy

$$\mathcal{L}_G \geq \max_j \{\mathcal{L}_{\max,j}\} \tag{5.5}$$

A set of starting values that produces the global maximum, denoted by $\arg\max_j\{\mathcal{L}_{\max,j}\}$, is called an *optimal set* of starting values. There may be many optimal sets; that is, $\arg\max_j\{\mathcal{L}_{\max,j}\}$ may not be unique. If the parameter esti mates associated with an optimal set are not the same across all optimal sets, then the model is nonidentifiable. However, if the model is identifiable, then the unique set of estimates produced by each $\arg\max_j\{\mathcal{L}_{\max,j}\}$ is the model (local) MLE. In general, using the starting values associated with $\max_j\{\mathcal{L}_{\max,j}\}$ solution is no guarantee of the global maximum; however, the probability that it is the optimal set increases as J increases. Depending on the complexity of the model and the sample size, J between 20 and 100 is usually sufficient to produce good results. However, as mentioned earlier, if the indicators are not sufficiently correlated with X, even $J = 1000$ may not find the global maximum (Berzofsky 2009).

Small values of J can produce good results if good starting values are selected. For example, constraining the starting values of the error parameters to the interval [0.1,0.5] will usually require fewer runs than will the range (0,1) provided the optimal set is in the restricted range. However, Wang and Zhang (2006) showed that the probability that a randomly generated set of starting values is optimal is relatively high. Therefore, in most cases, the EM algorithm may have to be rerun only a dozen times or so to identify the global maximum. Some programs (e.g., PANMARK) do this automatically for user-specified J. For these programs, it is a good idea to specify an J of 100 or more. In other programs (like ℓEM), rerunning the algorithm is a manual process but it is easy to write a customized macro in a general software package (such as SAS or SPSS) to run the LCA software the desired number of times. Almost all LCA programs will list the seed value[1] used for each run, which can be saved by the custom macro. However, if the LCA software is rerun manually, usually 5–10 times is sufficient if each set of starting values produces the same set of estimates. This solution can be accepted as the global MLE. Otherwise, the algorithm should be rerun another 5 or more times, again with different starting values. If new maxima arise, the program should be rerun until one has confidence that the largest of the maxima has been found.

[1]As noted previously, unless the user prespecifies a starting value for a parameter, LC software programs will generate a starting value at random using a pseudorandom-number generator. For the same set of starting values, the EM algorithm will produce the same set of model parameter estimates. The *seed value* (which can either be specified by the user or randomly generated by the software) is an integer used to initialize the pseudorandom-number generator. Since the same sequence of random numbers will be generated from the same seed value, the user can reproduce a particular set of ML estimates by specifying the seed value used to produce it.

For very complex models, rerunning the algorithm to full convergence can be time-consuming. It is well known that the EM algorithm increases the likelihood quickly in its early stage and slows as it approaches convergence. Wang and Zhang (2006) suggest rerunning the EM with just a few (e.g., 50–100) iterations in the process of identifying an optimal set of starting values. Then the algorithm can be rerun one final time with the optimal set to full convergence. This will greatly reduce the time required to find a global maximum. Wang and Zhang also note that the probability of finding the global maximum increases, for any J, as the sample size increases and as the correlation between the latent variable and the indicators increases. On the other hand, the probability of a global maximum decreases with the amount of extreme probability values (i.e., boundary or near-boundary values) and as the number of latent classes increases. To avoid local maximum solutions is it a good idea to keep the number of latent classes small, say, five or fewer, if possible.

5.1.7 Latent Class Flippage

Another problem that often arises in fitting latent class models is a misalignment of the categories of the latent variable X and the categories of the indicators, A, B,…, referred to here as *latent class flippage*. For example, consider the artificial data in Table 5.1. The proportion of positive values for A, B, and C are 0.27, 0.36, and 0.17, respectively. These should compare well with the prevalence probability of X. However, fitting the standard LCA in ℓEM yields an estimated value of the prevalence of X, namely, π_1^X, 0.9 of which is implausible, especially considering that these data were generated using the value of π_1^X. It appears that the software is confusing the positive and negative categories of X. In other words, the categories $X = 1$ and $X = 2$ are flipped in the estimation process. Another consequence of latent class flippage is that the error probabilities for the indicators are also flipped. The model applied for generating the data in the table used the parameter values $\pi_{1|2}^{A|X} = 0.2$ and $\pi_{2|1}^{A|X} = 0.1$. However, the software returned the implausible estimates of 0.9 and 0.8, respectively, for these two parameters. The estimates of the false-positive and false-negative probabilities for B and C were also implausibly large. After rerunning the model, a new solution emerged with estimates appropriately aligned to the data generating model, that is, with $\pi_1^X = 0.1$, $\pi_{1|2}^{A|X} = 0.1$ and $\pi_{2|1}^{A|X} = 0.2$. Apparently, which category of X is associated with the positive categories of A, B, and C depends on the set of starting values used to generate the solution.

Latent class flippage occurs as a result of the symmetry in the likelihood function for a latent class model. The data in Table 5.3 were generated by the parameters in the first row of Table 5.4. However, the same cell frequencies can be produced using the estimates in the second row of Table 5.4. Both sets of parameters produce the same data and the same value of the likelihood function. Depending on where the likelihood maximization algorithm starts, it may find either solution. Particularly if the starting values are generated at

Table 5.3 Artificial Data for Three Indicators

Pattern	111	112	121	122	211	212	221	222
Count	67	64	23	116	28	196	52	454

Table 5.4 Two Sets of Parameters that Will Generate the Data in Table 5.1

Model	π_1^X	$\pi_{1\|2}^{A\|X}$	$\pi_{2\|1}^{A\|X}$	$\pi_{1\|2}^{B\|X}$	$\pi_{2\|1}^{B\|X}$	$\pi_{1\|2}^{C\|X}$	$\pi_{2\|1}^{C\|X}$
Data-generating model	0.1	0.2	0.1	0.3	0.15	0.1	0.2
Equivalent model	0.9	0.9	0.8	0.85	0.7	0.8	0.9

random, a given software package may be just as likely to return the estimates in the first row as the second. It is up to the data user to determine whether latent class flippage has occurred and, if so, to realign the estimates in the output appropriately.

For simple models, flippage is fairly easy to identify. For example, for dichotomous indicators, an error rate that is higher than the accuracy rate will usually suggest that the positive and negative latent classes have been flipped. In that case, the estimate for $\pi_{1|2}^{A|X}$ is reported as for $\pi_{1|1}^{A|X}$ and vice versa. However, in complex models (e.g., in models with polytomous indicators that interact with two or more grouping variables), latent class flippage can be very difficult to discern. It is not uncommon in fitting these models for latent classes to flip inconsistently from group to group. It is therefore important to carefully examine the latent class and error probability estimates for each group to ensure that they are plausible since this is not checked by the LCA software. Comparing the estimates over many runs and starting values is one way to detect the problem.

One way to avoid latent class flippage is to simply supply a set of plausible starting values for at least one set of error probabilities (e.g., $\pi_{a|x}^{A|X}$). In this way, the first category of A is associated with the first category of X, the second category of A with the second category of X, and so on. This aligns the indicators with latent variable as the iterative algorithm is initiated. The alignment will then be maintained throughout the algorithm to the final solution. As an example, for the data in Table 5.3, supplying the starting values $\pi_{1|2}^{A|X} = 0.1$ and $\pi_{2|1}^{A|X} = 0.1$ consistently produced the solution in the first row of Table 5.4 over 100 runs.

Supplying plausible starting values is not an absolute safeguard against latent class flippage since the iterative algorithm can still venture off into an implausible area of the parameter space during the likelihood maximization process. In some cases it may be necessary to supply starting values for all indicators as well as their interactions with the grouping variables (e.g., starting values for $\pi_{a|xgh}^{A|XGH}$ for grouping variables G and H). Note, however, that these starting values should be removed when checking the models for identifiability.

5.2 LOCAL DEPENDENCE MODELS

As noted in the Section 4.2.1, a key assumption of LCA is that of local independence. Recall that three indicators, A, B, and C, are locally independent if and only if

$$\pi_{abc|x}^{ABC|X} = \pi_{a|x}^{A|X} \pi_{b|x}^{B|X} \pi_{c|x}^{C|X} \tag{5.6}$$

Failure of (5.6) to hold implies that the indicators are *locally dependent.* If local dependence is not appropriately considered in the modeling process, latent class models will exhibit poor fit (i.e., L^2 will be large), the estimates of the model parameters will be biased, sometimes substantially so, depending on the severity of the violation, and their standard errors will be too large (Sepulveda et al. 2008). One measure of this local dependence severity is the LD index, defined by

$$\text{LD}_{abc|x} = \frac{\pi_{abc|x}^{ABC|X}}{\pi_{a|x}^{A|X} \pi_{b|x}^{B|X} \pi_{c|x}^{C|X}} \tag{5.7}$$

for all a, b, c, and x, provided the conditional probabilities in the denominator are all positive. A necessary and sufficient condition for local independence is that LD = 1. For dichotomous outcome measures, (5.7) simplifies to

$$\text{LD}_{111|x} = \frac{\pi_{111|x}^{ABC|X}}{\pi_{1|x}^{A|X} \pi_{1|x}^{B|X} \pi_{1|x}^{C|X}} = 1 \tag{5.8}$$

for $x = 1,2$. Because local dependence usually results in better agreement among the indicators than would be explained by X alone, $\text{LD}_{111|x}$ is usually at least 1. The more the LD index deviates from 1 the greater the violation of the local independence assumption.

To illustrate the potential effects of local dependence when a locally independent model is fit, a simple dataset was expeculated using the parameters in the first row of Table 5.5 with $n = 5000$. For these parameters, the pairs A,C and B,C are locally independent while the pair A,B is not. Thus, $\text{LD}_{111|x}$ for $x = 1,2$ is

$$\begin{aligned} \text{LD}_{11|1} &= \frac{\pi_{11|1}^{AB|X}}{\pi_{1|1}^{A|X} \pi_{1|1}^{B|X}} = \frac{\pi_{1|1}^{A|X} \pi_{1|11}^{B|XA}}{\pi_{1|1}^{A|X} \pi_{1|1}^{B|X}} = \frac{\pi_{1|11}^{B|XA}}{\pi_{1|1}^{B|X}} = \frac{(1-0.130)}{(1-0.150)} = 1.02 \\ \text{LD}_{11|2} &= \frac{\pi_{11|2}^{AB|X}}{\pi_{1|2}^{A|X} \pi_{1|2}^{B|X}} = \frac{\pi_{1|2}^{A|X} \pi_{1|21}^{B|XA}}{\pi_{1|2}^{A|X} \pi_{1|2}^{B|X}} = \frac{\pi_{1|21}^{B|XA}}{\pi_{1|2}^{B|X}} = \frac{(1-0.280)}{0.104} = 6.92 \end{aligned} \tag{5.9}$$

Table 5.5 Illustration of Local Dependence between A and B

Parameter	π_1^X	$\pi_{1\|2}^{A\|X}$	$\pi_{2\|1}^{A\|X}$	$\pi_{1\|2}^{B\|X}$	$\pi_{2\|1}^{B\|X}$	$\pi_{1\|2}^{C\|X}$	$\pi_{2\|1}^{C\|X}$	$\pi_{2\|11}^{B\|AX}$	$\pi_{2\|21}^{B\|AX}$	$\pi_{1\|12}^{B\|AX}$	$\pi_{1\|22}^{B\|AX}$
True	0.10	0.050	0.100	0.104	0.150	0.100	0.200	0.130	0.280	0.360	0.090
Estimated	0.12	0.082	0.033	0.129	0.089	0.304	0.102	0.089	0.089	0.129	0.129

The value of $LD_{11|2}$ suggests a high level of local dependence. The expeculated cell frequencies for cells (1,1,1), (1,1,2), (1,2,1), (1,2,2), and so on of the ABC table are

$$[321.3 \quad 151.2 \quad 61.2 \quad 141.3 \quad 67.275 \quad 353.475 \quad 400.225 \quad 3504.025]$$

The model $\{XA\ XB\ XC\}$ was fit to these data (the ℓEM model statement is `mod A|X B|X C|X`). The correct model, namely, $\{XA\ XAB\ XC\}$, cannot be fit to these data because it is nonidentifiable. The ℓEM estimates appear in the last row of Table 5.5. Note that all the estimates are biased, not just the estimates for parameters involving A and B.

As noted previously, there are essentially three possible causes of local dependence: (1) unexplained heterogeneity, (2) correlated error, and (3) bivocality. Determining the cause of the local independence is important in some situations, because it can provide clues as to how to repair the model failure. The next section discusses these three causes and their remedies.

5.2.1 Unexplained Heterogeneity

As noted in Chapter 4, a key assumption of the LC model is that the error parameters $\pi_{a|x}^{A|X}, \pi_{b|x}^{B|X}, \ldots$ are homogeneous in the population. However, this is seldom true for survey data analysis because the probabilities of correctly classifying individuals often depend on individual characteristics such as age, race, sex, education, and interest in the survey. If these variables are known for each respondent, this so-called *unexplained* heterogeneity can be modeled by including grouping variables, as suggested in Section 4.3.

As an example, if the error probability $\pi_{a|x}^{A|X}$ depends on some grouping variable H, replacing $\pi_{a|x}^{A|X}$ by $\pi_{a|xh}^{A|XH}$ in the model will correct for this heterogeneity. (*Note*: This is equivalent to adding the XHA interaction to the model.) If the probabilities $\pi_{a|x}^{A|X}$ are the same for all individuals in group H, then H is said to explain the heterogeneity, and the assumption of homogeneous error probabilities for $\pi_{a|x}^{A|X}$ will hold within each level of H.

In some cases, the heterogeneity cannot be explained exactly by the grouping variables available for modeling. This situation is often referred to as *unobserved* heterogeneity. It still may be possible to explain most of the heterogeneity using known grouping variables. To illustrate, suppose that H is unobserved but another variable G that is observed is highly correlated with H. Then, replacing $\pi_{a|x}^{A|X}$ by $\pi_{a|xg}^{A|XG}$ in the model (i.e., adding the XGA

interaction to the model) will approximately account for the heterogeneity in $\pi_{a|x}^{A|X}$; that is, the variation in $\pi_{a|x}^{A|X}$ will be *nearly* zero within the levels of G. If the correlation between G and H is weak, this approach will not be effective (Berzofsky 2009). In some cases, several grouping variables may be required to approximately explain the heterogeneity in the error probabilities.

A consequence of unexplained heterogeneity is local dependence. To see this, suppose that A and B satisfy conditional local independence given H

$$\pi_{ab|xh}^{AB|XH} = \pi_{a|xh}^{A|XH}\pi_{b|xh}^{B|XH} \tag{5.10}$$

or, in words, the classification errors for A and B are independent within the groups defined by H. In order for unconditional local independence to hold, we must have

$$\text{LD}_{abx} = \frac{\pi_{ab|x}^{AB|X}}{\pi_{a|x}^{A|X}\pi_{b|x}^{B|X}} = \sum_h \pi_h^H \frac{\pi_{a|xh}^{A|XH}}{\pi_{a|x}^{A|X}}\frac{\pi_{b|xh}^{B|XH}}{\pi_{b|x}^{B|X}} = 1 \tag{5.11}$$

which, in general, is not true unless $\pi_{a|x}^{A|X} = \pi_{a|xh}^{A|XH}$ and $\pi_{b|x}^{B|X} = \pi_{b|xh}^{B|XH}$ for all h; that is, the error probabilities are homogeneous across the levels of H.

As an example, let $G = 1$ for males and $G = 2$ for females and assume that males and females are equally divided in the population (i.e., $\pi_1^G = \pi_2^G = 0.5$). Suppose, for measuring cocaine prevalence, that the false-negative probabilities for males and females are 0.4 and 0.05, respectively, for both A and B (equal error probabilities for both indicators), specifically, $\pi_{2|11}^{A|XG} = 0.4$ and $\pi_{2|12}^{A|XG} = 0.05$. Further assume that the false-positive probabilities are zero for both A and B. The marginal accuracy probabilities are then $\pi_{1|1}^{A|X} = \pi_{1|1}^{B|X} = 0.5(1-0.4)+0.5(1-0.05) = 0.775$. Since A and B are locally independent conditional on G, we can write

$$\begin{aligned}\text{LD}_{11|1} &= \frac{\pi_{11|1}^{AB|X}}{\pi_{1|1}^{A|X}\pi_{1|1}^{B|X}} \\ &= \frac{\pi_1^G\pi_{1|11}^{A|XG}\pi_{1|11}^{B|XG} + \pi_2^G\pi_{1|12}^{A|XG}\pi_{1|12}^{B|XG}}{\pi_{1|1}^{A|X}\pi_{1|1}^{B|X}} \\ &= \frac{0.5(1-0.4)^2 + 0.5(1-0.05)^2}{0.775^2} = 1.05\end{aligned} \tag{5.12}$$

which, since it deviates from 1, indicates local dependence. If the variable sex (G) is added to the model via the terms $\pi_{a|xg}^{A|XG}$ and $\pi_{b|xg}^{B|XG}$, the model will be correctly specified and this source of heterogeneity will be eliminated along with its consequential bias.

The best strategy for eliminating local dependence due to unexplained heterogeneity is to add grouping variables to the model that explain the variation in the error probabilities. Error probabilities usually vary by demographic variables such as age, race, sex, and education, which are also important for describing the variation in the structural components of the model. In addition, interviewer characteristics such as experience, age, sex, and, for some items, race, should also be considered for the measurement components. The usual model selection approaches [see, e.g., Burnham and Anderson (2002)] can be applied to achieve model parsimony.

5.2.2 Correlated Errors

When the questions for assessing X are asked in the same interview, respondents may provide erroneous responses to each question either deliberately or inadvertently causing errors to be correlated across indicators. For example, one question may ask "Have you used marijuana in the past 30 days?" and another question later in the questionnaire may ask "In the past 30 days, have you smoked marijuana at least once?" If there is a tendency for respondents who falsely responded negatively to the first question to also respond negatively to the second question, a positive correlation in the false-negative errors will result. Such correlation is sometimes called *behavioral correlation.*

Even if the replicate measurements are separated by longer periods of time (e.g., days or weeks), correlated errors may still be a concern if respondents tend to use the same response process to arrive at their erroneous responses. Likewise, comprehension errors caused by inappropriate wording of questions may cause errors that are correlated across the indicators. Figures 4.3 and 4.4 (in Chapter 4) are examples of path diagrams depicting correlated errors between the indicators.

As an example, the analysis by Biemer and Wiesen (2002) described in Section 4.4.6 found that respondents tended to underreport infrequent marijuana use for the question "Have you used marijuana in the past 30 days?" They speculated that the reason may be the word "use," which could imply more frequent use as in the term "marijuana user." This effect is unlikely to depend on their memory of responses given in previous posing of the same question. Therefore, if this question is asked in the same way in separate interviews occurring weeks apart, errors may still be correlated even though respondents may have no memory of their previous responses.

Errors will be correlated if respondents who provided a false-negative response for indicator A have a higher probability of false-negatively reporting on indicator B than will respondents who correctly responded positively to A. Mathematically, this is expressed as $\pi_{2|12}^{B|XA} > \pi_{2|11}^{B|XA}$ or, equivalently, as

$$\pi_{1|11}^{B|XA} > \pi_{1|12}^{B|XA} \tag{5.13}$$

which indicates a tendency for respondents who answer A correctly to also answer B correctly. Also, note that $\pi_{1|1}^{B|X} = \pi_{1|1}^{A|X}\pi_{1|11}^{B|XA} + \pi_{2|1}^{A|X}\pi_{1|12}^{B|XA} < \pi_{1|1}^{A|X}\pi_{1|11}^{B|XA} + \pi_{2|1}^{A|X}\pi_{1|11}^{B|XA}$ from (5.13). It follows that A and B are locally dependent because

$$
\begin{aligned}
\mathrm{LD}_{11|1}^{AB|X} &= \frac{\pi_{11|1}^{AB|X}}{\pi_{1|1}^{A|X}\pi_{1|1}^{B|X}} \\
&= \frac{\pi_{1|1}^{A|X}\pi_{1|11}^{B|XA}}{\pi_{1|1}^{A|X}(\pi_{1|1}^{A|X}\pi_{1|11}^{B|XA}+\pi_{2|1}^{A|X}\pi_{1|12}^{B|XA})} \\
&= \frac{\pi_{1|11}^{B|AX}}{\pi_{1|1}^{A|X}\pi_{1|11}^{B|XA}+\pi_{2|1}^{A|X}\pi_{1|12}^{B|XA}} \\
&> \frac{\pi_{1|11}^{B|AX}}{\pi_{1|1}^{A|X}\pi_{1|11}^{B|XA}+\pi_{2|1}^{A|X}\pi_{1|11}^{B|XA}} = 1
\end{aligned}
\tag{5.14}
$$

Likewise, the correlated errors can occur for false positives responses as well (i.e., $\pi_{1|21}^{B|XA} > \pi_{1|22}^{B|XA}$), with similar consequences for local dependence.

For modeling correlated errors between a pair of indicators, A and B, the *joint-item method* can be used. For this method, the locally dependent pair of items is simply replaced in the model by a joint item, AB, having K^2 categories corresponding to the cells in the $K \times K$ cross-classification of A and B. In the model, the probabilities $\pi_{a|x}^{A|X}$ and $\pi_{b|x}^{B|X}$ are replaced by $\pi_{ab|x}^{AB|X}$. For example, for dichotomous indicators ($K = 2$), AB has four categories corresponding to $ab = 11,12,21,22$. Thus, the probability $\pi_{ab|x}^{AB|X}$ represents six parameters: $\pi_{11|x}^{AB|X}, \pi_{12|x}^{AB|X}$ and $\pi_{21|x}^{AB|X}$ for $x = 1,2$. This is two more parameters than were required for the terms $\pi_{a|x}^{A|X}$ and $\pi_{b|x}^{B|X}$. With only three indicators, the unrestricted model with the joint item is nonidentifiable since the model degrees of freedom is –2. However, in many situations, the model can be made identifiable through the addition of grouping variables with suitable parameter restrictions. This can be a trial-and-error modeling exercise where identifiability is checked for each restriction using the methods discussed in Section 5.1.2.

Note that the joint-item method is equivalent to adding the term XAB to the submodel for $\pi_{b|xa}^{B|XA}$ in the modified path model formulation. A more parsimonious solution that is just as effective for modeling behavioral local dependence in many applications is the *direct-effect* method. For this method, XAB is replaced by AB in the submodel. Removing X from the AB interaction tacitly assumes that the correlation between the errors for A and B is the same in each latent class.

5.2.3 Bivocality

A third source of local dependence is *bivocality*, sometimes called *multidimensionality* or *multiple factors*. Suppose that A is an indicator of X and B is an indicator of Y. If $\pi_{y|x}^{Y|X} < 1$ for $x = y$ (i.e., X and Y can disagree), A and B are said to be *bivocal* (Alwin 2007). Bivocal indicators jointly measure two different latent constructs rather than one. As an example, suppose that A asks "Have you ever been raped?" and B asks "Have you ever been sexually assaulted?" Clearly, A and B are not indicators of the same latent variable since sexual assault (Y) is a broader concept than rape (X). Therefore, $\pi_{1|1}^{Y|X} < 1$, and a person who has been sexually assaulted but not raped may answer "no" to A and "yes" to B without error. Two bivocal indicators may perfectly measure their respective constructs, yet a model that assumes the indicators to be univocal may erroneously attribute classification error to either or both indicators. An appropriate model for this situation specifies two latent variables that, depending on the correlation between X and Y and other indicators in the model, may or may not be identifiable.

Suppose that there are three indicators A, B, and C, where A and B are indicators of X and C is an indicator of Y (see Figure 5.3). In this situation, the model $\{XA\ XB\ YC\}$ is not identifiable, so the analyst must resort to fitting the model $\{XA\ XB\ XC\}$. In that case, the indicator C will exhibit very high error probabilities relative to A and B. In fact, it can be shown that the error probabilities for C satisfy

$$\pi_{c|x}^{C|X} = \sum_{y} \pi_{c|y}^{C|Y} \pi_{y|x}^{Y|X} \tag{5.15}$$

Fortunately, A, B, and C are still locally independent and hence all the parameters of the model, including $\pi_{c|x}$, can still be estimated unbiasedly. To see this, note that the $\text{LD}_{abc|x}$ is 1 because $\pi_{abc|x}^{ABC|X}$ can be written as $\sum_y \pi_{abc|xy}^{ABC|XY} \pi_{y|x}^{Y|X}$. Note further that $\pi_{a|xy}^{A|XY} = \pi_{a|x}^{A|X}$ and $\pi_{b|xy}^{B|XY} = \pi_{b|x}^{B|X}$. Therefore

$$\begin{aligned}
\text{LD}_{abc|x} &= \frac{\sum_y \pi_{abc|xy}^{ABC|XY} \pi_{y|x}^{Y|X}}{\pi_{a|x}^{A|X} \pi_{b|x}^{B|X} \pi_{c|x}^{C|X}} \\
&= \frac{\sum_y \pi_{a|x}^{A|X} \pi_{b|x}^{B|X} \pi_{c|y}^{C|Y} \pi_{y|x}^{Y|X}}{\pi_{a|x}^{A|X} \pi_{b|x}^{B|X} \pi_{c|x}^{C|X}} \\
&= \frac{\pi_{a|x}^{A|X} \pi_{b|x}^{B|X} \sum_y \pi_{c|y}^{C|Y} \pi_{y|x}^{Y|X}}{\pi_{a|x}^{A|X} \pi_{b|x}^{B|X} \pi_{c|x}^{C|X}} \\
&= \frac{\pi_{a|x}^{A|X} \pi_{b|x}^{B|X} \pi_{c|x}^{C|X}}{\pi_{a|x}^{A|X} \pi_{b|x}^{B|X} \pi_{c|x}^{C|X}} = 1
\end{aligned} \tag{5.16}$$

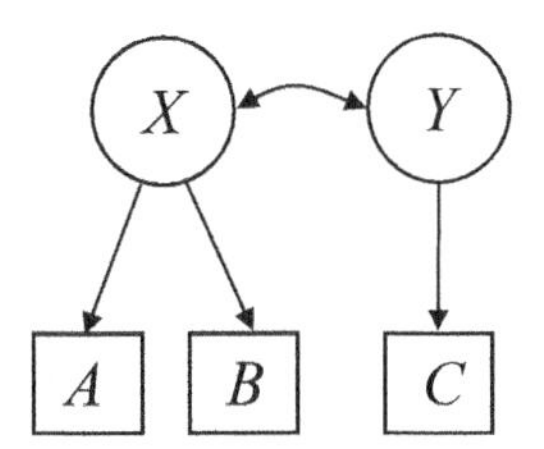

Figure 5.3 Bivocality with three indications.

This situation is analogous to that considered by Biemer and Wiesen (2002) (see Section 4.4.6). Indicator C in that analysis was found to be bivocal relative to A and B. However, replacing C by C' (a univocal indicator of X) did not change the estimates for A and B. This suggests that, for both models, the local independence assumption was satisfied.

Now suppose that there are four indicators: A, B, C, and D. Further assume that items underlying C and D were poorly constructed and, rather than measuring X, they measure another (potentially closely related) latent variable, Y. Thus, A, B, C, and D are bivocal indicators of X. Bivocality is a common problem when multiple indicators of some latent construct are embedded in the same interview and the questionnaire designer tries to conceal the repetition from the survey respondent. If the same question wording is used to create the multiple indicators, respondents may become confused or irritated by the repetition if they become aware of it. Even if they do not perceive the repetition, using the same question wording may induce correlated errors. For this reason, questions are usually reworded and/or reformatted to obscure their repetitiveness. But this is very difficult to do for more than a few indicators without changing the meaning of the questions and creating bivocal indicators.

For the situation illustrated in Figure 5.4, bivocality will induce local dependence—sometimes referred to as *causal* local dependence—since C and D are locally dependent with respect to X even though they may be locally *independent* with respect to Y. To see this, suppose that $\pi_{cd|y}^{CD|Y} = \pi_{c|y}^{C|Y}\pi_{d|y}^{D|Y}$. Thus, we can write $\pi_{cd|x}^{CD|X}$ in terms of Y as $\sum_y \pi_{c|y}^{C|Y}\pi_{d|y}^{D|Y}\pi_{y|x}^{Y|X}$ with similar expressions for $\pi_{c|x}^{C|X}$ and $\pi_{d|x}^{D|X}$. Now, using these results, we write

$$\mathrm{LD}_{cd|x} = \frac{\pi_{cd|x}^{CD|X}}{\pi_{c|x}^{C|X}\pi_{d|x}^{D|X}} = \frac{\sum_y \pi_{c|y}^{C|Y}\pi_{d|y}^{D|Y}\pi_{y|x}^{Y|X}}{\pi_{c|x}^{C|X}\pi_{d|x}^{D|X}} = \frac{\sum_y \pi_{c|y}^{C|Y}\pi_{d|y}^{D|Y}\pi_{y|x}^{Y|X}}{\left(\sum_y \pi_{c|y}^{C|Y}\pi_{y|x}^{Y|X}\right)\left(\sum_y \pi_{d|y}^{D|Y}\pi_{y|x}^{Y|X}\right)} \quad (5.17)$$

From this expression, it is clear that $\mathrm{LD}_{cd|x}$ is not 1 unless $\pi_{y|x}^{Y|X} = 1$ when $y = x$ (i.e., the indicators are univocal).

Determining whether the indicators in an analysis are bivocal is, to a large extent, subjective since a rigorous statistical test for bivocality has yet to be

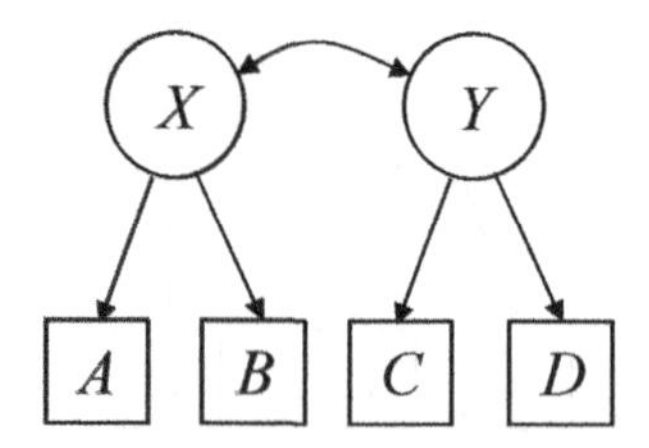

Figure 5.4 Bivocality with four indicators.

developed. In order for a manifest variable A to be an indicator of the latent variable X, the correlation of A with X must exceed the correlation of A with any other potential latent variable. In other words, A is an indicator of X if it measures X better than any other latent variable *that could be* measured by A. Thus, the indicators in an LCA are univocal if and only if this is true for all the indicators. It is possible that $\pi_{a|x}^{A|X}$ for $a \neq x$ is very large yet A is still an indicator of X by the definition given in Section 4.2.1. For this reason, it may be difficult to distinguish between a poor indicator of X and a bivocal indicator other than subjectively, using substantive theory to justify the assumptions.

As mentioned previously, a bivocal model can be used to address causal local dependence in situations where there are at least two indicators of each latent variable such as the model shown in Figure 5.4. The appropriate model is $\{XY\ XA\ XB\ YC\ YD\}$, which is identifiable as long as X and Y are sufficiently correlated. Using simulated data, Berzofsky (2009) showed that, as the correlation between X and Y decreases, the probability of finding a global maximum approaches zero. This is an example of *weak identifiability* referred to in Section 5.1.2.

5.2.4 A Strategy for Modeling Local Dependence

The possibility of local dependence from all three sources should be explicitly considered in any LCA. Fortunately, as we have seen, methods are available for modeling and remedying all three types of local dependence provided identifiable models can be found. The model-fitting strategy can be extremely important since identifiability is a critical issue in modeling local dependence. A model-building strategy we have used with some success is to consider the possibility of bivocality first. As noted previously, causal local dependence is not an issue when only one indicator is bivocal. When there are four or more indicators, bivocality should be considered by subjectively assessing whether each measure is an indicator of the same latent variable. If bivocality is suspected, a second latent variable should be added to the model to account for it as was done for the model in Figure 5.4.

The next step is to treat local dependence induced by unexplained heterogeneity, if possible, by the addition of grouping variables to either the structural or measurement model components. When there are many possible explanatory variables from which to choose, the selection of grouping variables is best guided by a working theory regarding the drivers of the latent constructs as well as the errors of measurement. Age, race, sex, and education are typical correlates of the latent constructs. However, for modeling measurement error, additional variables related to the measurement process should be considered if available. These might include interviewer characteristics, mode of interview, interview setting, and interview task variables, as well as paravariables such as number of attempts to interview and respondent reluctance to be interviewed.

The third step is to determine whether one or more indicator by indicator interaction terms should be added to address any residual local dependence from the first two steps. Fortunately, some excellent diagnostic tools are available to guide this step of the modeling process. Most of these methods compare the observed and model predicted cross-classification frequencies for all $\binom{L}{2}$ possible pairs of indicators. Association in the observed frequency table that is greater than the corresponding association in the predicted frequency table indicates unexplained local dependence in the current model. Various approaches incorporating this idea have been developed by Espeland and Handelman (1989), Garrett and Zeger (2000), Hagenaars (1988b), Qu et al (1996), and Vermunt and Magidson (2001). Sepulveda et al. (2008) provide a discussion of these approaches and propose a new method based on biplots. Their method is useful when it is important to distinguish between simple direct effects and second-order interactions (i.e., between AB and XAB).

One approach that often produces good results is the method proposed by Uebersax (2000) which is a variation of the *log-odds ratio check* (LORC) method of Garrett and Zeger (2000). Suppose that there are four indicators of X denoted by A, B, C, D. Suppose further that some model, say, M, has been fit to the data and we wish to determine whether direct effects should be added to the model to account for local dependence. The Uebersax method is performed on all pairs of indicators AB, AC, AD, BC, BD, and CD. The method will be described for dichotomous indicators, although it can be straightforwardly extended to polytomous indicators.

For each dichotomous indicator pair I_1 and I_2, we proceeds as follows:

1. Construct the observed and model predicted two-way cross-classification frequency table for the pair as in Table 5.6.
2. Calculate the log-odds ratio (LOR) in both the observed and predicted two-way tables where $\text{LOR(obs)} = \log(n_{11}n_{22}/n_{12}n_{21})$ with analogous definition for LOR(pred); that is, $\text{LOR(pred)} = \log(\hat{m}_{11}\hat{m}_{22}/\hat{m}_{12}\hat{m}_{21})$
3. Calculate the standard error of LOR(pred) only using the approximation

$$\text{s.e.}[\text{LOR(pred)}] = \sqrt{\hat{m}_{11}^{-1} + \hat{m}_{12}^{-1} + \hat{m}_{21}^{-1} + \hat{m}_{22}^{-1}} \quad (5.18)$$

4. Compute the standardized normal z statistic given by

$$|z| = \left| \frac{\text{LOR(obs)} - \text{LOR(pred)}}{\text{s.e.}[\text{LOR(pred)}]} \right| \quad (5.19)$$

If $|z|$ exceeds the value 1.96 (i.e., 97.5 percentile of the normal distribution), then I_1 and I_2 are significantly locally dependent and it may be necessary to add the interaction I_1I_2 to the model.

Table 5.6 Observed and Predicted Frequencies for $I_1 \times I_2$

	$I_2 = 1$	$I_2 = 2$
	Observed Frequencies	
$I_1 = 1$	n_{11}	n_{12}
$I_1 = 2$	n_{21}	n_{22}
	Predicted Frequencies	
$I_1 = 1$	$\hat{m}_{11}$	$\hat{m}_{12}$
$I_1 = 2$	$\hat{m}_{21}$	$\hat{m}_{22}$

These steps are then repeated for each pair of indicators in the model: AB, AC, AD, BC, BD, and CD. If none of the pairs are significantly locally dependent (i.e., $|z| < 1.96$ for all pairs), the procedure is completed and no further terms need be added to the model. Otherwise, a direct effect should be added to the model for the indicator pair having the smallest p value in step 4. Call this model M'. The procedure (steps 1–4) is then repeated with this model except the predicted frequencies ($\hat{m}_{ij}, i, j = 1, 2$) from model $M'$are used. If dependencies are found, create the model M'' by adding a direct effect for the pair with the smallest p value in step 4. Continue this process until no pair is locally dependent in step 4.

A useful program for performing the LORC, called CONDEP, is freely available from Uebersax (2000). The LORC procedure is illustrated in the following example.

5.2.5 Example: Locally Dependent Measures of Sexual Assault

This example is based on a real survey, the results of which have yet to be officially released. Consequently, the real data haves been replaced with data expeculated from a model that closely describes the original data (dissimilarity index of less than 1%). In addition, the survey questions have been modified somewhat to avoid confusion with the real survey results. To expeculate the data used in the analysis, a well-specified model was fit to the original data using the techniques described in the previous section. The predicted frequencies for this model were then extracted to play the role of the observed frequencies in the analysis that follows below. The original model was discarded since the expeculated model is by definition the true model for this dataset.

The five questions in Table 5.7 were asked of a general population for the purposes of estimating the prevalence of rape in the population. An examination of the indicators suggested that indicators were bivocal with A, B, and C measuring X defined as a "victim of sexual violence in the past 12 months"

Table 5.7 Five Indicators for the Estimation of Rape Prevalence in a Population

Indicator	Definition
A	Have any of the following occurred in the past 12 months? Touched in a sexual manner with physical force? Touched in a sexual manner when uninvited? Forced you to perform sexual acts such as oral or anal sex? Forced or pressured you to have sexual intercourse (e.g., raped)? =1 if yes =2 if no
B	In the past 12 months, did someone use physical force, pressure you, or make you feel that you had to have any type of sex or sexual contact? =1 if yes =2 if no
C	How long has it been since someone used physical force, pressure you, or make you feel that you had to have any type of sex or sexual contact? =1 if past 12 months =2 if more than 12 months or never
D	In the past 12 months, did someone force or pressure you to have sexual intercourse or rape you? =1 if yes =2 if no
E	How long has it been since someone forced or pressured you to have sexual intercourse (i.e., since you were raped)? =1 if past 12 months =2 if more than 12 months or never

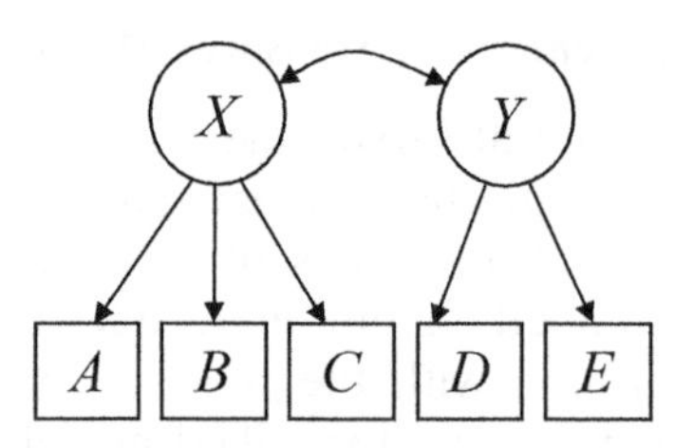

Figure 5.5 Relationship between the five indicators of rape.

and D and E measuring Y defined as "forced or pressured to have sex (e.g., raped) in the past 12 months." The path model of these relationships is shown in Figure 5.5. Note further that the cross-classification of X and Y defines four classes as shown in Table 5.8. However, the cell associated with $Y = 1$ and $X = 2$ is impossible because a person who has been raped has also been sexually assaulted by definition. Thus, there are only three latent classes, and the con-

Table 5.8 Four Classes Defined by Cross-Classifying X and Y

	$X = 1$ (Victim of Sexual Assault)	$X = 2$ (Not a Victim of Sexual Assault)
$Y = 1$ (raped)	Raped	Impossible
$Y = 2$ (not raped)	Victim of assault, but not raped	Never a victim of assault

Table 5.9 Results of the Model-Building Process

Model		df[a]	BIC	L^2	p	d
Model 0	{*XY*} {*XA XB XC YD YE*}	111	−53.5	1054.8	0.00	0.03
Model 1	{*XYFG*}{*XFA XGA*} {*XFB XGB*} {*XFC XGC*}{*YFD YGD*} {*YFE YGE*}	82	−658.4	160.4	0.00	0.005
Model 2	Model 1 + {*BD*}	81	−654.5	154.3	0.00	0.004
Model 3	Model 2 + {*CE*}	80	−735.2	63.6	0.91	0.003

[a]Degrees of freedom.

straint $\Pr(X = 2|Y = 1) = 0$ must be added to the model as a *structural zero* cell in order for the model to be properly specified.

The model-building process proceeded as follows. First, the model {*XY XA XB XC YD YE*} denoted by model 0 in Table 5.9 was fit. This model is shown in Figure 5.5. Next, a number of grouping variables were considered, including the demographic characteristics, prior victimizations, and variables summarizing the respondent's attitude toward the interview, the interview setting (privacy, distractions, etc.). Two variables determined to be the most predictive of both the structural and measurement components of the model were added to the model to account for heterogeneity producing model 1 in the table. This model is depicted in Figure 5.6.

Then the LORC procedure was performed using the observed and model 1 predicted frequencies for the *ABCD* table. The first pass of the LORC found three statistically significant dependencies: *AD*, *BD* and *CE*. Because the pair *BD* has the highest z value, this direct effect was added to the submodel for D to produce model 2. The LORC procedure was repeated for model 2, and the pair *AD* and *CE* were significantly dependent with the latter having the highest z value. Thus, *CE* was added to the submodel for E to produce model 3 shown in Figure 5.7. Applying the LORC test again for model 3 did not identify any significant dependences, and thus LORC procedure was terminated. The LORC results, obtained from the ℓEM input statements in Figure 5.8, are shown in Table 5.10.

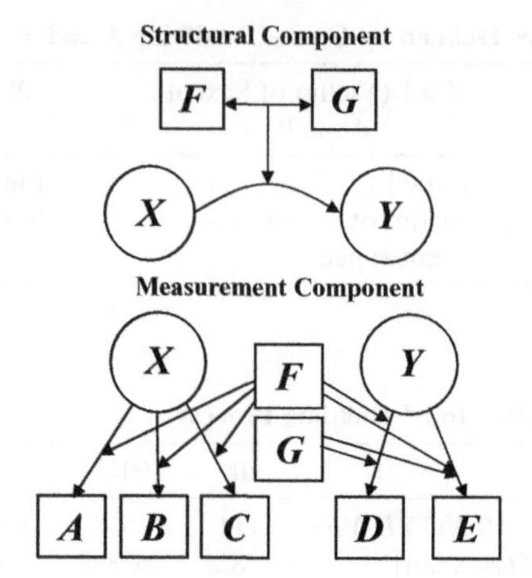

Figure 5.6 Path diagram for model 1 in Table 5.9. Structural component is {*FGXY*}; measurement component is {*XFA XGA*}{*XFB XGB*}{*XFC XGC*}{*YFD YGD*}{*YFE YGE*}.

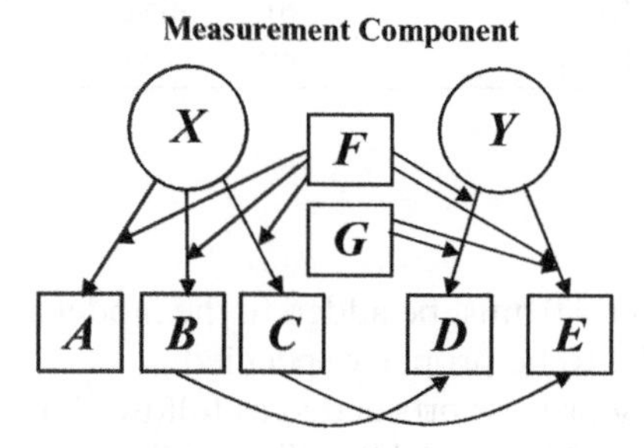

Figure 5.7 Path diagram for the measurement component of model 3 in Table 5.9. Structural component is the same as in Figure 5.7; measurement component is {*XFA XGA*}{*XFB XGB*}{*XFC XGC*}{*YFD YGD BD*}{*YFE YGE CE*}

Finally, the estimates of rape prevalence under model 3 and the corresponding true, simulation parameters are compared in Table 5.11. The model-building process was successful at identifying some of the local dependences in the underlying model, but not all. Still the model parameter estimates are in fairly close agreement with the true parameters. The exceptions are estimates of the terms involving Y that exhibit fairly important biases. The likely reason for this is too few indicators of Y. Since three indicators are needed for the standard latent class model to be identifiable, estimates of Y parameters must draw information from A, B, and C, which are only indirectly related to D and E through X and Y. Thus, the accuracy of the Y parameters depends to a large extent on the strength of the correlation between X and Y. For example, if this correlation were 1, then all five indicators would be univocal and locally independent and, thus, the parameter estimates would be unbiased. As this correlation departs from 1, estimation accuracy deteriorates. This suggests that if bivocality is unavoidable in the construction of multiple indicators, the underlying latent variables should be made as highly correlated as possible.

```
lat 2                              * Two latent variables for bivocal model
man 7
dim 2 2 2 2  2 2 2 2 2
lab X Y F G  A B C D E
mod
   FG
   XY|FG      {X XYFG wei(XY)}   * Weight vector to apply 0 constraint in
                                   Table 5.8
   A|FGX      {XA XFA XGA}
   B|FGX      {XB XFB XGB}
   C|FGXB     {XC XFC XGC}
   D|FGYAB    {YD YFD YGD BD}
   E|FGYCDB   {YE YFE YGE CE}
rec 128
rco
sta X [.02 .98]
sta wei(XY) [1 1 0 1]              * The weight vector assigns a weight of 0 to the
                                     cell corresponding to X=2,Y=1, a structural 0
                                     (see Table 5.8)
sta A|X [.9 .1 .0001 .9999]        * Starting values to control flippage
sta B|X [.9 .1 .0001 .9999]
sta C|X [.9 .1 .0001 .9999]
sta D|Y [.9 .1 .0001 .9999]
sta E|Y [.9 .1 .0001 .9999]
dat model3.dat
```

Figure 5.8 ℓEM input statements for fitting model 3.

Table 5.10 ***z* Values for the LORC of Indicator Pairs for Three Models**

Pair	Model 1	Model 2	Model 3
AB	0.60	0.38	–0.78
AC	–0.08	–0.54	–0.28
AD	2.18[a]	2.25[a]	1.26
AE	–0.12	–0.29	0.42
BC	–0.89	–1.21	–0.38
BD	4.77[a]	0.25	0.82
BE	0.72	0.71	1.74
CD	–1.41	–1.39	–0.95
CE	3.80[a]	3.46[a]	–0.41
DE	0.06	0.38	0.48

[a]The z value is significant at $\alpha = 0.05$.

5.3 MODELING COMPLEX SURVEY DATA

The methods discussed heretofore assume that the sample is selected by simple random sampling without replacement from a large population. With survey data, this assumption is rarely satisfied. Survey samples usually involve some form of clustered sampling, possibly with stratification and unequal probability sampling of clusters as well as units within clusters. A number of articles in the survey literature show unequivocally that data collected under

Table 5.11 Comparison of Estimates for Model 3 and the True Model

	X	Y	$A\|X$	$B\|X$	$C\|X$	$D\|X$	$E\|X$
Model 3: $\{XYFG\}\{XFA\ XGA\}\{XFB\ XGB\}\{XFC\ XGC\}\{YFD\ YGD\ BD\}$ $\{YFE\ YGE\ CE\}$							
Prevalence	0.022	0.014					
False +			0.003	0.001	0.003	0.002	0.003
False −			0.166	0.360	0.181	0.559	0.375
True Model: $\{XYFG\}\{XFA\ XGA\}\{XFB\ XGB\}\{XFC\ XGC\}\{YFD\ YGD\ BD\ AD\}$ $\{YFE\ YGE\ CE\ DE\}$							
Prevalence	0.022	0.022					
False +			0.003	0.001	0.003	0.002	0.002
False −			0.187	0.379	0.192	0.423	0.606

a complex sampling design and analyzed as using simple random sampling methods will often produce biased estimates and standard errors [see, e.g., Korn and Graubard (1999, pp. 159–172), Stapleton (2008), Hansen et al. (1983), and Smith (1990)]. There is also evidence that complex sample designs adversely affect LCA parameter estimates, particularly estimates of latent class probabilities (Patterson et al. 2002). In this section, we consider some options for dealing with complex samples in an LCA. But realize that complex data analysis of LCA is still an underdeveloped area and much more research is needed. Before discussing the methods, the purposes of weighting and cluster sampling are briefly reviewed.

5.3.1 Objectives of Survey Weighting

For the purposes of describing the objectives of weighting, it is useful to consider three populations encountered in survey sampling: the *target population* (or *universe*), the *frame population*, and the *respondent population*. As shown in Figure 5.9, these populations are nested within one another; the target population encompasses the frame population, which, in turn, encompasses the respondent population. The sample is usually a tiny subset of the respondent population. Weighting attempts to take this tiny sample and strategically adjust it to represent the target population (the largest rectangle in Figure 5.9).

The target population is sometimes also referred to as the *inferential population*. It is the population to be studied in the survey and for which the basic inferences from the survey will be made. For example, a study of child health may infer to the population of all children in the country between the ages of 0 and 14 years. A study of biohazard waste material may infer to all manufacturers and industrial establishments that produce such matter as byproducts

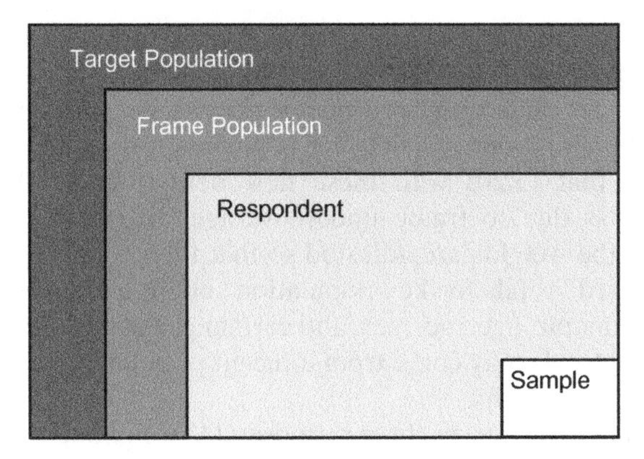

Figure 5.9 Target, frame, and respondent populations and the sample.

of manufacturing operations. The target population is regarded as the ideal population to be studied. In practice, however, this ideal is seldom achieved due to frame undercoverage and nonresponse.

The respondent population is a purely hypothetical concept; it is impossible to identify all the members of this population. Still it is quite useful for conceptualizing the potential bias due to nonresponse. It is defined as that subset of the frame population that is represented by units who *would respond to the survey if selected*. This definition was posited by Cochran (1977) to illustrate in a simple way how nonresponse may bias survey estimates. Cochran considered the frame as consisting of two strata, which we will denote by $R = 1$ for members of the respondent stratum and $R = 2$ members of the nonrespondent stratum. Of course, R is unknown during the sampling process and, thus, the sample will consist of units from both strata. By definition, units with $R = 1$ respond and units with $R = 2$ will not. This deterministic model yields the following formula for the bias due to nonresponse for estimating the prevalence probability, π_1^X [see, e.g., Biemer and Lyberg (2003)]:

$$Bias = \pi_2^R(\pi_{1|1}^{X|R} - \pi_{1|2}^{X|R}) \tag{5.20}$$

In other words, the bias in the usual estimator of π_1^X is equal to the probability of nonresponse (i.e., the expected nonresponse rate for the survey) times the difference in the prevalence of X for respondents and nonrespondents. Nonresponse weighting adjustments attempt to remove this bias.

Weighting can also reduce the bias due to frame error. As discussed in Chapter 1, a common problem in surveys is frame undercoverage, where, population units are missing from the frame and therefore have zero probabilities of selection. If frame undercoverage is substantial (for, e.g., ≥15%), it can cause biases due to the failure to represent the noncovered target population members. A bias that has essentially the same form as (5.20) may occur as a result of frame undercoverage. For example, let $R = 1$ for population members covered by the frame and $R = 2$ for population members that are not covered. Now $\pi_{1|1}^{X|R}$ and $\pi_{1|2}^{X|R}$ denote the prevalence probabilities for the covered and noncovered subpopulations, respectively. It can be shown [see, e.g., Biemer and Lyberg (2003)] that (5.20) with these new definitions of the parameters describes the bias due to frame undercoverage. To compensate for frame undercoverage, the weights are adjusted so that their totals agree with external, "gold standard" totals for key population subgroups that not covered by the frame, for example, age, race, sex, and certain geographic areas. These gold standard control totals may come from a recent population census or a large-scale survey of high quality.

Weighting, therefore, serves three purposes: (1) weighting by the inverse of the selection probabilities (referred to as the *base or weights*) allows valid inference to be extended from the sample to the respondent population, (2) performing a nonresponse adjustment to the base weight is intended to extend valid inference to the next level—the frame population level, and (3) performing a *poststratification adjustment* to the product of the base weight and the nonresponse adjustment factor carries inference to the highest level—the target population. It is important to note that, even if samples are selected with an equal-probability selection method (EPSEM) such as simple random sampling, the observations may still be weighted to account for nonresponse and frame undercoverage or to improve the precision of the estimates.

A fourth type of weighting adjustment is sometimes performed to improve the precision of the estimates (rather than to remove the effects of bias). This weighting adjustment is usually applied in conjunction with the frame coverage weighting adjustment. It incorporates multiplicative factors derived from ratio estimation that can result in substantial variance reduction. A ratio estimator uses *auxiliary information* to form a weight that increases the efficiency of an estimator. This auxiliary information may be another characteristic, say, Y, defined for each member of the target population that is correlated with X and whose values are known for all sample members and whose target population total Y_T is also known.

For a random sample of n respondents, a general estimator of the population total X_T is the well-known Horvitz–Thompson (HT) estimator [see, e.g, Cochran (1977)]

$$\hat{X}_{\text{HT}} = \sum_{i=1}^{n} w_i x_i \tag{5.21}$$

Here, x_i is the observed value of the characteristic X for the ith unit, w_i is a weight given by

$$w_i = w_{Bi} \times w_{\text{NR}i} \times w_{\text{PS}i} \tag{5.22}$$

w_{Bi} is the base weight (i.e., the inverse of the sample inclusion probability for unit i), w_{NRi} is the nonresponse adjustment weight (i.e., the inverse of the estimated probability that unit i responds to the survey), and w_{PSi} is the poststratification adjustment, which includes factors to adjust for frame coverage as well as ratio adjustments for precision enhancement. Further discussion of the concepts of survey weighting can be found in Biemer and Christ (2008).

We can derive simple formula that shows the potential impact of unequal probability sampling on the variance of the estimator of the total. Assume that the observations are mutually independent as well as independent of weights w_i. Note that these assumptions seldom hold in practice. For example, the former assumption is not true for cluster sampling, and the latter assumption is seldom satisfied if units are selected with probabilities proportional to size (PPS) and observations on the units are correlated with these sampling size measures. Nevertheless, the approximate expression for the variance that they yield is still quite useful for understanding the effects of weighting on the variance of estimates.

Under these assumptions, the variance of (5.21) can be written as

$$\begin{aligned} Var\left(\hat{X}_{\mathrm{HT}}\right) &= Var\left(\sum_{i=1}^{n} w_i x_i\right) \\ &= \sigma_X^2 \sum_{i=1}^{n} w_i^2 \\ &= n\sigma_X^2\left(\bar{w}^2 + \frac{\sum_{i=1}^{n} w_i^2 - n\bar{w}^2}{n}\right) \\ &= n^2\bar{w}^2\frac{\sigma_X^2}{n}\left\{1+\frac{s_w^2}{\bar{w}^2}\right\} \end{aligned} \tag{5.23}$$

where $\bar{w}^2$ and s_w^2 are the sample mean and variance of the weights . Note that $n\bar{w} = \sum_i w_i$ is approximately equal to the population total N; therefore $n^2\bar{w}^2\sigma_X^2/n$ is approximately equal to the variance of the estimator of the total of X_T under simple random sampling. Denote this variance by Var_{SRS}. It then follows that

$$Var\left(\hat{X}_{\mathrm{HT}}\right) \doteq Var_{\mathrm{SRS}}[1+(\mathrm{CV})^2] \tag{5.24}$$

where CV denotes the coefficient of variation of the weights, w_i. The term $1 + (\mathrm{CV})^2$ is sometimes referred to as the *unequal weighting effect* (UWE). It is can be very large (say, 3 or more) for samples in which relatively small subpopulations are oversampled. Because of the assumptions made for (5.24), the UWE may be regarded as the maximum increase in variance due to unequal weighting. The increase could be less, for example, if the weights are positively correlated with the x values. Thus we see that weighting can increase the variance of the total estimator (compared to simple random sampling) by the factor UWE. Note that an EPSEM design will have an UWE of 1 before nonresponse and coverage adjustment factors are applied; however, the final postsurvey adjusted weights will have an UWE larger than 1. Analogously, unequal weighting can increase the variances of other types of estimators, including logistic regression modeling and loglinear modeling parameters (Skinner 1989; Korn and Graubard 1999). For this reason, ignoring the weights in calculations of standard errors will often result in underestimating the standard errors. The bias can be substantial for large, national surveys where values of UWE are frequently in the range of 2–4.

In multistage sample selection schemes the variance can be further increased by positive correlations between the observations induced by the clustering of units with similar characteristics within the higher-stage units. For example, a fairly common design is to select counties or provinces as the first or primary-stage sampling units (PSUs) and housing units within counties as second stage units. Many household characteristics are more similar within counties than across counties creating intracluster (within-county) correlations (ICCs) for these characteristics. ICCs may be small (say, 0.1), but their effects on the variance can be substantial.

The well-known expression for the effect of clustering on the variance of the total in (5.21) is the *clustering design effect* or deff given by

$$\text{deff} = 1 + \text{ICC}(m-1) \tag{5.25}$$

where m is the average within PSU sample size [see, e.g., Kish (1965) and Lohr (1999)]. For all practical purposes, the ICC is between 0 and 1. A zero ICC implies no clustering effect and a deff of 1. An ICC equal to 0.1 is considered rather large. As an example, suppose that ICC = 0.1 for a survey with $n = 5000$ selected in a sample of 100 PSUs (i.e., $m = 50$). The deff is 1 + 0.1(50-1) = 1.5. Thus, the sampling variance for this design is 1.5 (i.e., 50%) larger than the variance of a simple random sample of size n. If clustering is ignored in the calculation of the variance, say, by assuming simple random sampling, the variance will be understated by roughly 50%. Added to this bias is the bias of the variance estimator due to ignoring the unequal weighting. Chromy and Meyers (2001) give the following approximate expression for the overall deff (denoted deff_{all}), which accounts for both clustering and unequal weighting:

$$\text{deff}_{\text{all}} \doteq \text{UWE} \times \text{deff} \tag{5.26}$$

As an example, if the UWE is 2 and the deff is 1.5, then $\text{deff}_{\text{all}} = 3$. Clearly the risk of underestimating the variance of model parameters as a result of ignoring the survey design can be substantial. The next section discusses some methods for dealing with the complex survey design in LCA.

5.3.2 LCA with Complex Survey Data

The previous section discussed some of the risks of ignoring the complex sample design in a data analysis. These same risks apply to LCA, although the situation is much less transparent for some LC model parameters. Perhaps the greatest risk in ignoring sampling weights is invalid inference for the latent class probabilities π_x^X. To illustrate, consider a stratified sample with two strata denoted by $H = 1$ and $H = 2$. Suppose that 50% of the population is in each stratum but only 20% of the sample falls in stratum $H = 1$. If the latent class probabilities, $\pi_{x|h}^{X|H}$ differ by H, ignoring H in the LC model will produce an estimate of π_1^X formed by weighting together the stratum estimates by the proportion of the *sample* in each stratum: $\bar{\pi} = 0.2\pi_{x|1}^{X|H} + 0.8\pi_{x|2}^{X|H}$. However, the population parameter is appropriately weighted by the proportion of the *population* in each stratum: $\pi_x^X = 0.5\pi_{x|1}^{X|H} + 0.5\pi_{x|2}^{X|H}$. The bias due to ignoring the stratifier in the LCA is $B(\bar{\pi}) = 0.3(\pi_{x|2}^{X|H} - \pi_{x|1}^{X|H})$ [This can be compared with (5.20).] This can be considerable for latent class x if the difference $\pi_{x|2}^{X|H} - \pi_{x|1}^{X|H}$ is large.

The estimates of the error probabilities may also be biased if they also differ by H (i.e., heterogeneity). It is not uncommon, however, for error probabilities to be homogeneous even though the latent class probabilities differ by H. The recent literature suggests that response probability estimates may be much less affected by the survey design than latent class probability estimates [see, e.g., Vermunt (2007)]. Nevertheless, it is usually good practice to include the stratifying variables in the LC model to address this potential heterogeneity unless the number of additional parameters that this will generate is too large.

Ignoring the sampling weights may also yield standard errors that are biased (usually downward) as noted in the previous section. Ignoring the effects of clustering can further negatively bias the standard errors because clustering tends to induce positive correlations among the observations, which can increase the variances of the model estimators. The risk of violating the assumption of local independence would seem to be higher for cluster sampling but, to date, that issue has not been investigated in the LCA literature. It seems reasonable that the methods for modeling local dependence described in Section 5.2 could also be applied to eliminate local dependence due to cluster sampling. Even so, the standard errors of the estimates could be understated if clustering is not taken into account in the estimation of standard errors.

In general selecting units with unequal probabilities can distort the distribution of the latent variable within the sample and bias the latent class size estimates. For example, if the sample design disproportionately samples unemployed persons, the model estimate of the unemployment latent class will be too large. Weighting corrects for this. On the other hand, the error probabilities, which are conditioned on these latent variables, and may be unaffected by the disproportionate sampling. Vermunt and Magidson (2007) suggest that ignoring the weights may be acceptable if the goal of LCA is solely the evaluation of classification error. They argue that survey weighting does appropriately account for the heterogeneity in the error probabilities induced by unequal probability sampling. Instead, heterogeneity is better treated by multigroup LCA using grouping variables that effectively account for this type of heterogeneity.

For example, suppose that S is a stratification variable and the sample is disproportionately allocated to strata. In this situation, the survey weights will differ by strata. Vermunt and Magidson suggest adding the interactions XSA, XSB, and so on to the model rather than weighting the data to account for any heterogeneity in the error probabilities XA, XB, and so forth across strata. Haertel (1984a, 1984b, 1989) has also argued that accounting for the complex survey design may not be appropriate when the LCA is focused on the measurement component rather than the structural components. However, this advice is in no way definitive, and additional research is this area would be welcome.

With regard to clustering, even if the latent characteristics (true values) are geographically clustered in the population, classification errors may not be. Patterson et al. (2002) provide some evidence of this in their study of the survey design effects on LCA estimates for dietary data. Their findings are considered in some detail in Section 5.3.8. If classification errors are not clustered within PSUs, the standard errors of the response probability estimates will not be affected by cluster sampling. However, the estimates of latent class probabilities are still likely to be.

Until more research can be conducted to evaluate the options for dealing with complex survey data in LCA, we advocate the use of LCA methods that at a minimum account for unequal probability sampling, especially for unbiased estimation of the latent class probabilities. In the remainder of this chapter, the following five methods for incorporating survey weights in the analysis will be described:

- Including design variables in the fitted model
- Analyzing weighted and rescaled frequencies
- Pseudo-maximum-likelihood estimation
- Treating the cell weight as an offset parameter
- Two-step estimation

The strengths and weaknesses of each approach will be considered. We shall show that the choice of which method to use depends on several factors. Some methods are appropriate if the sample design only involves stratification or if there is a small number (say, 10 or fewer) clusters. Some approaches work well for unequal probability sampling but do not account for sample clustering effects. One method (pseudo-maximum-likelihood estimation) will handle both clustering and unequal probability sampling, but it requires special commercial software to implement. In addition, there is still much controversy as to whether it appropriately accounts for sample design effects for the measurement component of an LC model. The section concludes with the lament: that there is still much work to be done in the area of LCA with complex samples.

To simplify the exposition, these methods will be described assuming a simple three-indicator LC model, possibly with grouping variables. However, as we have seen throughout this book, the results for this special case can easily be extended to LC models with any number of indicators and other interaction terms. The next section discusses a method for dealing with simple non-EPSEM designs such as stratified sampling or clustering sampling with few strata.

5.3.3 Including Design Variables in the Fitted Model

In stratified sampling, the sampling frame is partitioned into L groups by some stratifying variable H. Let N_h denote the number of units and let n_h denote the sample size, respectively, in the hth stratum where $N = \Sigma_h N_h$ and $n = \Sigma_h n_h$. If $n_h = nN_h/N$, the sample is said to be *proportionally allocated* to strata and the sample weights are equal across strata (i.e., the design is EPSEM). Otherwise, sampling is disproportionate and the stratum weights $w_h = N_h/n_h$ must be applied to each unit in stratum h for $h = 1,\ldots, L$. If L is relatively small (no more than 10, say) adding the stratifying variable to the model (including the stratum by latent variable interaction) will appropriately account for the weight and an unweighted analysis can be done. This is analogous to treating H as a grouping variable in the analysis.

As an example, instead of fitting $\{XA\ XB\ XC\}$, the model $\{XH\ XA\ XB\ XC\}$ would be fit. This model assumes the latent variable proportions vary by stratum, otherwise the inclusion of H would have no effect on the estimates of π_x^X. If the interaction, XH, is not significant, it is possible to ignore H in the model. Some analysts recommend always including XH as a precaution against model misspecification because its presence is dictated by the sample design.

To estimate the population proportions, π_x^X, on should weight the estimates of $\pi_{x|h}^{X|H}$ together using the population stratum proportions, $W_h = N_h/N$ as follows:

$$\hat{\pi}_x^X = \sum_{h=1}^{L} W_h \hat{\pi}_{x|h}^{X|H} \tag{5.27}$$

The standard error can be computed as the square root of

$$v(\hat{\pi}_x^X) = \sum_{h=1}^{L} W_h^2 v(\hat{\pi}_{x|h}^{X|H}) \tag{5.28}$$

where $v(\hat{\pi}_{x|h}^{X|H})$ is an estimate of the $Var(\hat{\pi}_{x|h}^{X|H})$ provided by the software for the unweighted data. These population-level estimates are normally computed outside the LCA software does not have access to the stratum weights. The estimate of π_x^X the software will compute uses the weights n_h/n in place of W_h in (5.27) which will yield a biased estimate in general.

The specification {*XH XA XB XC*} assumes that the error probabilities are homogeneous with respect to H. The model {*XHA XHB XHC*} allows for the possibility of stratum-heterogeneous errors, and it is good practice to test this model against the homogeneous alternative. In that case, for example, *XHA* is significant, the estimate of $\pi_{a|x}^{A|X}$ must be appropriately weighted as in (5.27). A large number of strata can result in data sparseness issues that create problems for this approach (see Section 5.1.4). It may be possible to collapse strata having very similar values of W_h to reduce the number of strata (i.e., reduce the dimension to, say, $H' < H$). This is usually better than ignoring the strata; however, the approach depends on how well the H' strata represent the original H strata, particularly with respect to X.

A better alternative is discussed in the next section. This method works well for unequal probability sample with no (or very little) clustering in the sample.

5.3.4 Weighted and Rescaled Frequencies

As noted previously, large-scale surveys typically involve stratification with multiple stages (primary, secondary, tertiary, etc.) of sampling within strata. PSUs are typically selected with probabilities proportional to size (PPS) where the measure of size is related to the number of target population units (households, persons, employees, patients, etc.). If the number of secondary-stage units selected within each PSU is the same in all PSUs within a stratum, the design is approximately EPSEM within strata and the method described in the previous section can be used. However, more often, selection probabilities vary across the units within a stratum, especially when postsurvey adjustments are applied. If all the variables that were used to determine the survey weights were known and could be included in the model as grouping variables, inference could proceed using the design variables method described in the previous section. However, this could potentially include a large number of variables. For instance, data sparseness, model instability, and boundary estimates could be problematic, and the method could become inherently infeasible.

A simple method for dealing with survey weights advocated by Clogg and Eliason (1987) for loglinear models is to use weighted cell frequencies that have been rescaled so that their sum is equal to the original sample size, n. LCA then proceeds as if the sample were drawn by simple random sampling. This approach was used in some of the previous examples and illustrations. The advantage of this approach is that it produces model unbiased parameter estimates without the need to add design variables to the model. Its disadvantage is the likely underestimation of standard errors caused by ignoring the sample design. The usual diagnostics for assessing model adequacy (X^2,

L^2, BIC, AIC, etc.) are also no longer valid, strictly speaking, since the cell frequencies no longer refer to the number of independent observations in the cells. Because the unequal probabilities of selection will generally yield increased standard errors (through the UWE), ignoring them will typically underestimate standard errors. Consequently, there is a tendency of chi-square diagnostics to reject models that are correctly specified with this approach. As an example, the model L^2 p value may be only 0.01 (which would reject the model) while the correct p value may be much larger, say, 0.1. This may lead an analyst to add more variables to the model to increase the p value, resulting in model overfitting or compromising model parsimony. An approximate method is to multiply the variance estimates by the UWE [i.e., $1 + (\text{CV})^2$] as shown in (5.24).

To use the weighting and rescaling method, let (a,b,c) denote an arbitrary cell of the ABC table and let n_{abc} denote its unweighted frequency. Let w_{abc} denote the sum of the survey weights (i.e., base weights adjusted for nonresponse and noncoverage) over the observations in the cell

$$w_{abc} = \sum_{i=1}^{n_{abc}} w_i \tag{5.29}$$

where w_i is the weight for observation i. Let $w_{+++} = \Sigma_{abc} w_{abc}$ denote the sum of the cell weights across all the cells in the table. Then the weighted and rescaled frequency for cell (a,b,c) is

$$n'_{abc} = n\frac{w_{abc}}{w_{+++}} \tag{5.30}$$

Note that $\Sigma_{abc} n'_{abc} = n$ as desired. Let ABC′ denote the ABC table after replacing n_{abc} by n'_{abc}. The LCA now proceeds using ABC′ under the assumption of simple random sampling.

Rescaling the weights ensures that the LC model software calculates the model diagnostics using the original sample size n. Otherwise, the software would use the sum of the unrescaled frequencies, $\Sigma_{abc} w_{abc} \doteq N$, that is, the population size, as the sample size. In that case, the model goodness-of-fit statistics would be useless. Rescaling increases their usefulness; however, they should still not be trusted since, as noted previously, chi-square distribution assumptions are no longer valid for the weighted analysis. In many applications, the diagnostics are still used albeit with caution, recognizing their potential bias.

A frequently used analysis strategy is to fit the same LC model to both unweighted and weighted tables. If the estimates are fairly close, then one may conclude that the survey design has no apparent effect on the estimates. In that case, the design is *ignorable* with respect to the parameters of interest.

Patterson et al. (2002) used a modified Wald statistic (see Section 4.4.4) to test whether the weighted and unweighted estimates are statistically significantly different. If the design is ignorable, the unweighted analysis is preferred since the estimates often have smaller standard errors and the models exhibit greater stability. In addition, the model fit diagnostics will be valid. If the weighted and unweighted estimates are significantly and substantively different, the analyst should try to discover the reasons for the differences. One possibility is the unweighted model may have been misspecified. Augmenting the model with additional grouping variables, direct effects, and higher-order interactions could bring the weighted and unweighted estimates into much closer agreement. If these actions are unsuccessful and the estimates still differ, then the weighted model estimates should be used because it is likely that the weights correct for unobserved sample selection bias.

If the respecification is successful and the unweighted model is used, the standard errors may still be incorrect if the sample is clustered unless additional steps are taken to correct for this. The options for providing design-corrected standard errors are quite limited. The preferred approach is to employ pseudoreplication variance estimation methods, such as the jackknife or the bootstrap. These methods draw repeated subsamples from the original sample that are intended to replicate original sample's design. The estimation process is applied to each subsample to create multiple estimates have the same functional form as the original estimator. Then, the variance of these replicated estimates can be regarded as a robust estimator of the variance of the original estimator. This method is considered further in Section 5.3.8. It can also be used to provide design-corrected goodness-of-fit measures [see, e.g., Langeheine et al. (1996)]. A number of packages including Latent GOLD, MPlus, and PANMARK have implemented this approach.

A rather crude correction to the standard errors from an LCA is to multiply the model standard errors by $\sqrt{\text{deff}_{\text{all}}}$ [see (5.26)] if it is known. This method was employed by Biemer (2001) as well as Haertel (1984a, 1984b, 1989). If the UWE is small and deff_{all} is due mostly to clustering, this correction will likely result in a conservative estimate of the standard error for error probabilities that have been appropriately modeled for heterogeneity. In general, the correction tends to be conservative (Skinner 1989). If no other options are available and the sample design is nonignorable, this method is preferred to using standard errors that are unadjusted for the sample design.

5.3.5 Pseudo-Maximum-Likelihood Estimation

An important, recent advance in the analysis of complex survey data is the development of the *pseudo-maximum-likelihood* (PML) method (Binder 1983; Skinner 1989; Pfeffermann 1993). The PML method has been implemented in at least three latent class software packages: Latent GOLD (Vermunt and Magidson 2005a, 2005b), MPlus (Muthén and Muthén 1998–2005) and GLLAMM (Rabe-Hesketh et al. 2004). These packages also take clustering into account when estimating standard errors using either replication methods such as the jackknife or the sandwich (linearization) variance estimator, to be described subsequently.

For the PML method, consider an LCA with three dichotomous indicators (generalizations to four or more polytomous indicators are straightforward). Under simple random sampling, the loglikelihood kernel for the ABC table is

$$\ell = \sum_a \sum_b \sum_c n_{abc} \log(\pi_{abc}^{ABC}) \tag{5.31}$$

which is written in terms of cell frequencies n_{abc}. To write this likelihood in terms of the individual observations, we let i denote an individual in cell (a,b,c) and rewrite (5.31) as

$$\ell = \sum_a \sum_b \sum_c \sum_{i=1}^{n_{abc}} \log(\dot{\pi}_i \pi_{abc}^{ABC}) \tag{5.32}$$

where $\dot{\pi}_i$ is the probability of selection for the ith unit. For example, for simple random sampling and other EPSEM designs, $\dot{\pi}_i = n/N$. In that case, $\log(\dot{\pi}_i)$ is constant and can be absorbed in the model intercept term. In general, the $\dot{\pi}_i$ values vary across the units. The units may have also been selected from the same PSUs, which can introduce intercorrelations that can violate local independence as well as serious inflate the standard errors of the estimates. For many complex sample designs, specifying the likelihood taking into account these sampling complexities is intractable except for the simplest designs. The PML method addresses this complexity by first considering the likelihood under the assumption that every individual in the population were selected—that is, as if a complete census were taken. Then this likelihood will be estimated using the usual Horvitz–Thompson estimator in (5.21).

The likelihood for a complete census follows directly from (5.32) by replacing n_{abc} by N_{abc}, the number of individuals in the population that would be classified in cell (a,b,c) by the three classifiers if the classification process were to be applied to all N individuals in the population. In a census, $\dot{\pi}_i = 1$ for all i. This gives rise to the *census loglikelihood* given by

$$\ell_{\text{census}} = \sum_a \sum_b \sum_c \sum_{i=1}^{N_{abc}} \log(\pi_{abc}^{ABC}) \tag{5.33}$$

Maximizing this loglikelihood with respect to the parameters would provide model-consistent estimators; however, the population totals, $\sum_{i=1}^{N_{abc}} \pi_{abc}^{ABC} = N_{abc}\pi_{abc}^{ABC}$, are unknown for all (a,b,c). Nevertheless, by applying the Horvitz–Thompson estimator in (5.21), one can estimated them from the sample by $\sum_{i=1}^{n_{abc}} w_i \pi_{abc}^{ABC}$ when $w_i = \dot{\pi}_i^{-1}$. This yields the *pseudolikelihood* given by

$$\begin{aligned}\ell_{\text{pseudo}} &= \sum_a \sum_b \sum_c \sum_{i=1}^{n_{abc}} w_i \log(\pi_{abc}^{ABC}) \\ &= \sum_a \sum_b \sum_c w_{abc} \log(\pi_{abc}^{ABC})\end{aligned} \tag{5.34}$$

where $w_{abc} = \sum_{i=1}^{n_{abc}} w_i$ is the sum of the weights for the observations in cell (a,b,c). Apart from a normalizing constant, (5.34) can be rewritten as

$$\ell_{\text{pseudo}} = \sum_a \sum_b \sum_c n'_{abc} \log(\pi_{abc}^{ABC}) \tag{5.35}$$

where n'_{abc} is as defined in (5.30). The PML loglikelihood in (5.35) is identical to the simple random sampling loglikelihood for the ABC′ table defined for the weighting and rescaling method of the previous section. This shows that the PML method yields exactly the same estimates as the weighting and rescaling method.

The PML estimates are not MLEs, and although they are consistent (i.e., asymptotically unbiased) estimators, they do not share other properties of MLEs such as efficiency and uniqueness. Nevertheless, they and their standard error estimators tend to be quite robust to model misspecification. To obtain the standard error under the PML method, the "sandwich" (also referred to as the *robust* or *linearization*) *estimator* can be used, denoted by $v(\hat{\boldsymbol{\pi}}_{\text{PML}})$. Its matrix form is

$$v(\hat{\boldsymbol{\pi}}_{\text{PML}}) = \mathbf{I}^{-1}\mathbf{V}\mathbf{I}^{-1} \tag{5.36}$$

Note that $\mathbf{V}$ is "sandwiched" between two inverse matrices, where $\mathbf{I}$ is the information matrix, slightly different from the usual $\mathbf{I}$ since it is obtained under the pseudolikelihood ℓ_{pseudo} and $\mathbf{V}$ is a npar × npar sample covariance matrix that reflects the PSU-to-PSU variation (within strata) of pseudolikelihood first derivatives with respect to the model parameters (Skinner 1989).

For example, consider a stratified two-stage design with H strata. Let N_h denote the number of PSUs in stratum h of which n_h PSUs are sampled. A typical element (k,k') of $\mathbf{V}$ has the form

$$v_{kk'} = \sum_{h=1}^{H} \left(1 - \frac{n_h}{N_h}\right) n_h s_{hkk'} \tag{5.37}$$

where $s_{hkk'}$ is the within-stratum h sample covariance of the PSU quantities d_{hik} given by

$$d_{hik} = \sum_{j=1}^{m_{hi}} w_{hij} \frac{\partial \log \pi_{abc}}{\partial \pi_k} \tag{5.38}$$

for an arbitrary parameter π_k, describing the (a,b,c) cell frequency, π_{abc}, where m_{hi} is the number of observations in PSU hi. As seen from these expressions, computing **V** requires knowledge of the stratum, PSU, and the sampling weight for each observation in the sample as well as the stratum sampling fractions n_h/N_h. If any of these quantities are not known by the user, (5.36) cannot be computed.

The sandwich variance estimator takes into account all the relevant survey design features, including stratification, clustering within PSUs, unequal weighting at the level of the observations, and without replacement sampling with nonnegligible sampling fractions. In addition, it is robust to model misspecification. See Skinner (1989), Patterson et al. (2002), and Wedel et al. (1998) for additional details regarding the PML method.

One disadvantage of the PML method is that, unlike the other methods considered here, it cannot be applied to a simple table of cell frequencies or a standard cross-classification table. As noted previously, the computations for (5.36) requires considerably more information about the survey design than may be available to the data analyst. Another disadvantage is, as noted for the weighting-rescaling method, standard goodness-of-fit tests and related measures such as BIC and AIC are no longer valid for the weighted tables. The next method discussed provides valid goodness-of-fit tests while accounting for unequal weighting. One of its disadvantages is that it does not address the problem of cluster sampling.

5.3.6 Treating the Average Cell Weight as an Offset Parameter

Data analysts should have little interest in the model that describes the sample unless that model also describes the population. For simple random sampling and other EPSEM sample designs, the sample and the population models are the same. For unequal probability sampling schemes, they may be quite different. Appropriately incorporating the survey weights into an LCA eliminates this difference. However, as noted previously,, these methods do not provide valid fit diagnostics. Clogg and Eliason (1987) proposed an approach for accommodating survey weights in loglinear modeling by a simple alteration of the standard loglinear model. Under certain conditions, their approach provides unbiased parameter estimates as well as valid chi-square, BIC and AIC measures.

Clogg and Eliason expressed the weighted cell frequencies as the product of the unweighted frequency and the average cell weight, that is, $w_{abc} = z_{abc}n_{abc}$, where $z_{abc} = w_{abc}/n_{abc}$. Loglinear analysis can now proceed using the unweighted data table after augmenting the model by an offset term, namely, $\log(z_{abc})$. This method was extended to LC models in Vermunt (2002, 2007).

To see how this method works for the standard three-indicator LC model, let m_{abc} denote the expected frequency for cell (a,b,c) in the unweighted ABC table. Let m'_{abc} denote the expected frequency for the census version of the model defined in the previous section; specifically, m'_{abc} denotes the number of observations that would fall in to cell (a,b,c) if the entire population were classified by the three classifiers, A, B, and C. We can then write

$$m_{abc} = \frac{m'_{abc}}{z_{abc}} \tag{5.39}$$

that is, the expected sample frequency is equal to the expected census (or population) frequency divided by the average cell weight. It follows that

$$\sum_x m_{xabc} = \frac{\sum_x m'_{xabc}}{z_{abc}} \tag{5.40}$$

where m_{xabc} and m'_{xabc} are the corresponding expected frequencies for the complete table that has been augmented by the latent variable X. Thus, the relationship between the expected weighted and unweighted frequencies still holds for the complete XABC table as the X variable is integrated out in the incomplete data likelihood.

Under a Poisson[2] sampling (loglinear) model (see Appendix B), the LC model for the expected population frequency for cell (x,a,b,c) is

$$\log(m'_{xabc}) = u + u_a^A + u_b^B + u_c^C + u_x^X + u_{ax}^{AX} + u_{bx}^{BX} + u_{cx}^{CX} \tag{5.41}$$

Using (5.40), we can replace m'_{xabc} by $z_{abc}m_{xabc}$ and therefore

$$\log(m'_{xabc}) = \log(z_{abc}m_{xabc}) = \log(z_{abc}) + \log(m_{xabc}) \tag{5.42}$$

Thus, substituting (5.42) for the left hand side of (5.41), we can rewrite $\log(m_{xabc})$ as

$$\log(m_{xabc}) = -\log z_{abc} + u + u_a^A + u_b^B + u_c^C + u_x^X + u_{ax}^{AX} + u_{bx}^{BX} + u_{cx}^{CX} \tag{5.43}$$

The $\log(z_{abc})$ term is called an *offset* [see, e.g., Agresti (2002, p. 385)]. It is fit as a constant in the model (i.e., with coefficient fixed to 1). Since an average

[2]Poisson sampling model should not be confused with the Poisson sampling design (Hájek 1981). In the latter, units are selected independently with unequal probabilities (for example, PPS with no clustering of units).

cell weight cannot be computed for zero frequency cells, z_{abc} is not defined for zero cells. Vermunt (2002) suggests setting z_{abc} to 1 for such cells. Thus, the LCA proceeds by fitting {*XA XB XC*} to the unweighted table using the average cell weight as an offset term. This will produce estimates of the census model *u* parameters as desired. This model can be fit in ℓEM using a weight vector [i.e., `wei()` command[3] (Vermunt, 1997b, p. 18). Latent GOLD, MPlus, GLLAMM, and several other packages mentioned in Section 4.1.3 also provides an option for fitting LC models with offset terms.

Vermunt (2002) and Vermunt and Magidson (2007) claim that, under a correctly specified population model, this method will produce consistent estimates of the model parameters. In addition, the usual chi-square statistics and other fit diagnostics will be valid. However, if the population model is misspecified, the estimates will not be consistent. For this reason, the PML method is often preferred because its estimates will be approximately unbiased for values of the population model parameters regardless of whether the model was correctly specified.

Skinner and Vallet (in press) caution against the use of this method since in most practical situations, it produces invalid standard errors. This is because the method does not take into account the within-cell weight variation. Unless all the observations in cell (*a*,*b*,*c*) have the same weight for all *a*, *b*, and *c*, the method will underestimate the true standard errors. However, in that case, including survey design variables in the model (see Section 5.3.3) is usually preferred. When there is substantial weight variation within cells, the offset parameter approach can substantially underestimate the standard errors. In their empirical studies, Skinner and Vallet found that the offset method produced standard errors that were virtually identical to those produced by the unweighted analysis. They also provide evidence disputing the claims that the usual LCA fit diagnostics are valid under the offset model. Further, the offset parameter method takes no account of the potential clustering of the observations in complex designs. In light of these findings, there seems to be no advantage of this method over the PML methods described in the previous two sections. Skinner and Vallet discourage its use with complex samples for loglinear modeling with or without latent variables.

5.3.7 Two-Step Estimation

This method uses an unweighted LCA to first estimate the error probabilities and then, fixing these probabilities to the estimated values, performs a second LCA to estimate the latent class probabilities. The method addresses the potential bias in the estimates of π_x^X when sampling weights are correlated

[3]The ℓEM `wei()` command can be used for a number of purposes. Previous examples used this command to fix structural zeroes in the data tables and to zero out unwanted probabilities and model effects.

with X. However, it also addresses the concern expressed above that weighting the observations by the selection probabilities does not appropriately address the problem of heterogeneity in the error probabilities caused by disproportionate sampling.

To examine this last point further, consider the following simple illustration. Suppose that the population consists of two strata where, using the notation developed in Section 5.3.3, $N_1 = 200{,}000$ and $N_2 = 80{,}000$. A stratified random sample $n = 5000$ is selected with equal allocation to the two strata: $n_1 = n_2 = 2500$. This design yields two sampling weights: $w_1 = 20{,}000/2500 = 8$ and for $w_2 = 80{,}000/2500 = 32$. Suppose that three parallel indicators (A, B, and C) are applied to this sample with error probabilities that differ between strata (creating heterogeneity across strata). Would the correct error probabilities be estimated by sampling weighting the observations by these sampling weights?

To answer this question, a contrived dataset was expeculated using the model $\{XHA\ XHB\ XHC\}$ with parameters in the first two rows of Table 5.12. The population values in the third row were obtained as the weighted average of the parameters of the two strata; for example, $\pi_1^X = (8\pi_{1|1}^{X|H} + 32\pi_{1|2}^{X|H})/40 = 0.34$. The fourth row of the Table 5.12 shows the estimates for the correct stratified, unweighted model, the next row shows the unweighted estimates from the model ignoring the strata (i.e., $\{XA\ XB\ XC\}$), and the last row shows the estimates from this model using weighted and rescaled frequencies.

The only model that appropriately accounts for the stratified design is the stratified unweighted model. The model that ignores the strata is quite biased when applied to unweighted frequencies. This same model applied to weighted frequencies is less biased for π_1^X but the error probability estimates still exhibit appreciable biases. Using survey weights does not adequately account for heterogeneity in this example. In fact, survey weighting was never intended to be used for the purpose of modeling heterogeneity in the error probabilities caused by unequal probability sampling.

Vermunt (2002) advocates a two-step method for dealing with unequal weighting that has also been implemented in Latent GOLD as follows:

Table 5.12 Expeculation Parameters and Their Estimates by Three Methods

Parameter	$\pi_{1\|h}^{X\|H}$	$\pi_{1\|h2}^{A\|HX} = \pi_{1\|h2}^{B\|HX} = \pi_{1\|h2}^{C\|HX}$	$\pi_{1\|h2}^{A\|HX} = \pi_{1\|h2}^{B\|HX} = \pi_{1\|h2}^{C\|HX}$
Stratum 1 ($H = 1$)	0.10	0.30	0.10
Stratum 2 ($H = 2$)	0.40	0.20	0.15
Population	0.34	0.16	0.26
Stratified unweighted	0.34	0.16	0.26
Unweighted	0.28	0.16	0.16
Weighted	0.37	0.13	0.22

1. Let M denote an LC model that has been specified to account for any possible heterogeneity in the population. Fit M to the unweighted table T, and obtain estimates of the response probability vector $\boldsymbol{\pi}_e$ for all indicators in the model. Denote this estimate by $\hat{\boldsymbol{\pi}}_e^{(1)}$ where the superscript (1) indicates step 1.
2. Let T′ denote the table formed by weighting and rescaling the frequencies for the cells in T by the survey weights. Fit the model M to T′ with the constraints $\boldsymbol{\pi}_e = \hat{\boldsymbol{\pi}}_e^{(1)}$. This will yield estimates of the latent class probabilities denoted by $\hat{\boldsymbol{\pi}}_x^{(2)}$ where the superscript 2 indicates step 2.

The set of parameter estimates $\{\hat{\boldsymbol{\pi}}_e^{(1)}, \hat{\boldsymbol{\pi}}_x^{(2)}\}$ represents the estimates for the model M. According to Vermunt (2002), the unweighted estimates of $\boldsymbol{\pi}_e$ will usually have smaller variance than will their weighted counterparts. In addition, if heterogeneity has been effectively accounted for in the model using grouping variables, there is no reason to use weighted frequencies to estimate $\boldsymbol{\pi}_e$. If some heterogeneity remains, weighting is not an effective means for removing it. In either case, the two-step method should be an improvement over the one-step approach because of the variance reduction for the estimates of $\boldsymbol{\pi}_e$. In the study by Patterson et al. (2002) (discussed in the next section), the standard error was doubled for some estimates of $\boldsymbol{\pi}_e$ by weighting.

For the standard errors of the error probabilities, the estimates reported in step 1 for these parameters can be used since, under the assumption of simple random sampling for these parameters, they are still valid. This will often be sufficient if the focus of the study is on $\boldsymbol{\pi}_e$. If $\boldsymbol{\pi}_x$ is also of interest, the standard errors obtained in step 2 are not valid because of (1) the model restrictions imposed to fix $\boldsymbol{\pi}_e$ to their step 1 values and (2) ignoring any clustering by the survey design in the modeling. In that case, standard errors for $\{\hat{\boldsymbol{\pi}}_e^{(1)}, \hat{\boldsymbol{\pi}}_x^{(2)}\}$ should be estimated using variance replication methods. Note that, for each replicate, both steps of the two-step approach must be implemented.

5.3.8 Illustration of Weighted and Unweighted Analyses

Patterson et al. (2002) compared the PML estimates with those from an unweighted analysis for a two-class LC model with four indicators. The data were taken from the Continuing Survey of Food Intake by Individuals (CSFII)—a multistage stratified household sample of women aged 19–50. Data were collected from 1028 women in 120 PSUs and 60 strata in a design of two PSUs per stratum. The sample was weighted for selection probabilities (although it was designed to be approximately EPSEM) as well as for nonresponse and noncoverage. All four indicators were dichotomous variables over six interviews spaced approximately 2 months apart over a 12-month data collection period. As the focus of the study was on irregular or infrequent vegetable eaters, a "1" for an indicator denoted no consumption of vegetables during the preceding 24-hour period and a "2" denoted consumption of at least one vegetable.[4] The standard LC model ($\{XA\ XB\ XC\ XD\}$) was fit to the data where A, B, C, and D denote the indicators and X denotes the true status of vegetable consumption.

Rather than use the PML sandwich variance estimator for the PML estimates, Patterson et al. opted to use a jackknife replication method [see, e.g., Wolter (2007), Chapter 4)]. This involves constructing replicate sets of survey weights $\{w\}_{hi}$ corresponding to each PSU i in stratum h, for $i = 1,\ldots, n_h, h = 1,\ldots, L$. This set of weights is constructed by removing from the sample all units from the ith PSU in stratum h and inflating the weights of the other $(n_h - 1)$ units in this stratum by the factor $n_h/(n_h - 1)$. The process is repeated for all $n = \Sigma_h n_h$ PSUs in the sample. For each set of replicate weights, $\{w\}_{hi}$, the LC model is recomputed and new estimates of the parameters are obtained. Let $\hat{\boldsymbol{\pi}}_{\{hi\}}$ denote the vector of parameter estimates for the replicate weight set $\{w\}_{hi}$. The jackknife estimator of the variance-covariance matrix of the PML estimate, $\hat{\boldsymbol{\pi}}$, is $\hat{\mathbf{V}}_{JK}$ with component $v_{kk'}$ corresponding to parameters k and k' given by

$$v_{kk'} = \sum_{h=1}^{L}\left(\frac{n_h}{n_h - 1}\right)\sum_{i=1}^{n_h}(\hat{\pi}_{\{hi\}k} - \hat{\pi}_k)(\hat{\pi}_{\{hi\}k'} - \hat{\pi}_{k'}) \tag{5.44}$$

The use of the jackknife for estimating the variance of LC model parameters estimates was studied empirically by Bolesta (1998) under simple random sampling. A simulation study by Patterson et al. (2002) found that the jackknife variances tended to slightly (less than 10%) overestimate the true variances.

To test the fit of the model, Patterson et al. (2002) used a Wald statistic such as (4.52), except it was modified for cluster-correlated observations given by

$$W_C = (\hat{\boldsymbol{\pi}} - \boldsymbol{\pi}_0)'\mathbf{V}_{JK}^{-1}(\hat{\boldsymbol{\pi}} - \boldsymbol{\pi}_0) \tag{5.45}$$

where $\boldsymbol{\pi}_0$ is the value of $\boldsymbol{\pi}$ under the null hypothesis. Apart from a constant factor, W_C is distributed approximately as an F-distribution when the null hypthesis is true (Korn and Graubard 1999, pp. 91–93). Using this approach, the model $\{XA\ XB\ XC\ XD\}$ fit the weighted data quite well ($p = 0.75$). For comparison purposes, the same model was fit to the unweighted frequencies. The two sets of estimates are shown in Table 5.13.

The following observations can be made from the results of Table 5.13:

[4]As noted by Vermunt (2002) in his commentary for this paper, these data could more appropriately fit a Markov latent class model (see Chapter 6) to allow for real changes in dietary intake over time.

Table 5.13 Comparison of Weighted and Unweighted Latent Class Estimates and Their Jackknife Standard Errors

	Weighted		Unweighted		
Parameter	Estimate	Jackknife Standard Error	Estimate	Jackknife Standard Error	Ratio of Standard Errors (Weighted over Unweighted)
π_1^X	0.1784	0.1275	0.3309	0.1374	0.93
$\pi_{1\|2}^{A\|X}$	0.4563	0.2000	0.6039	0.0778	2.57
$\pi_{1\|2}^{B\|X}$	0.3913	0.2273	0.5099	0.0943	2.41
$\pi_{1\|2}^{C\|X}$	0.2755	0.1134	0.3960	0.0817	1.39
$\pi_{1\|2}^{D\|X}$	0.3921	0.1479	0.4636	0.0744	1.99
$\pi_{1\|2}^{A\|X}$	0.2187	0.0214	0.1995	0.0191	1.12
$\pi_{1\|2}^{B\|X}$	0.2364	0.0300	0.1821	0.0336	0.89
$\pi_{1\|2}^{C\|X}$	0.2341	0.0693	0.1898	0.0648	1.07
$\pi_{1\|2}^{D\|X}$	0.2705	0.0404	0.2130	0.0457	0.88

Source: Patterson et al. (2002), Table 2.

1. The standard errors of the weighted estimates are quite large. On average, they are about 1.5 times larger than those of the unweighted estimates (see the last column of the table).
2. The weighted and unweighted estimates of π_1^X (the true proportion of persons who do not consume vegetables) differ markedly. However, the Wald significance test performed by Patterson et al. was not significant ($p = 0.16$).
3. Weighted and unweighted estimates of the error probabilities were more similar, especially the false-positive probabilities ($\pi_{1|2}^{A|X}$, etc.). Again, they were not significant ($p = 0.49$).

Patterson et al. also calculated estimates of standard errors for the PML estimates assuming simple random sampling. These were about one-half the size of the jackknife standard errors. As an example, the simple random sampling standard error for π_1^X was 0.07 compared to 0.13 from the jackknife. This translates into a deff_{all} of $(0.13/0.07)^2 = 3.45$. They further note that the average deff_{all} for the survey was about 1.43 (mostly as the result of unequal weighting). Thus, multiplying the simple random sampling standard errors by $\sqrt{\text{deff}_{\text{all}}}$ as suggested in Section 5.3.4 would have *undercorrected* the simple random sampling standard errors for clustering and unequal weighting in this case.

5.3.9 Conclusions and Recommendations

Dealing appropriately with the complex survey design in data analysis is extremely important since complex sampling often provides a distorted view of the population. Weighting the sample data is analogous to viewing the sample through a corrective lens that removes these distortions. Weighting also tends to increase the variance of the estimates, so any unnecessary use of weighting should be avoided. In addition, particularly for face-to-face surveys, the samples may be highly clustered to reduce data collection costs. Variance estimators that assume simple random sampling are likely to underestimate the true sampling variance since they ignore the ICC induced by cluster sampling. Currently the PML method is the preferred method for dealing with complex sample designs since it addresses both the weighting and clustering features of a survey design. An increasing number of statistical packages are incorporating methods for accounting for the sample design, which is good news for data analysts.

In the field of LCA, these developments have lagged somewhat. Two packages, Latent GOLD and MPlus, implement the PML method for LCA. However, more options are needed for the casual user of LCA, or in situations where detailed information on the survey design required by PML is not available. In this chapter, we considered four methods that can be used with LCA software packages that have not implemented PML. Three of these (design variable, weighting and rescaling, and two-step methods) seem to be viable alternatives to PML for these users. However, more work is needed on these methods as well as PML to better understand how and when they should be used.

For simple designs, the design variable approach seems to be the best choice since it accommodates the survey design while preserving the goodness-of-fit diagnostics. For more complex samples, the PML approach seems to be the best alternative; however, weighting and scaling combined with replicate variance estimation and bootstrap fit diagnostics can provide results that are essentially equivalent to those of PML. The weighting–rescaling method with the $\sqrt{\text{deff}_{\text{all}}}$ correction to the standard error is a good choice for the novice user. This method is closely related to the two-step approach. The key difference is that an unweighted analysis is not conducted to produce the error probabilities as in the case of the two-step method. Although the arguments in favor of the two-step method are sound, more research is needed before it can be recommended. In general, there is still much research to be done for applying LCA to complex survey data, particularly for goodness-of-fit diagnostics such as BIC, AIC, and the chi-square tests for model adequacy.

CHAPTER 6

Latent Class Models for Special Applications

This chapter considers several special LC models that have some utility in the evaluation of survey error. One motivation for including these models is to show the versatility of the LCA approach. Another is that the methods and examples illustrating their use are helpful to understand the how LCA can be applied to extract valid information on classification error. In addition, the applications described here arise frequently in survey error evaluation studies.

Section 6.1 discusses some of the issues in applying the standard LC model, which assumes nominal data, to ordinal variables such as Likert scales. Section 6.2 describes how LCA can be used to estimate reliability as an alternative to the methods discussed in Chapters 2 and 3. Finally, Section 6.3 describes a model that can be used for estimating population size that has been applied in census undercount evaluations.

6.1 MODELS FOR ORDINAL DATA

Thus far, X has been a nominal latent variable and likewise, its indicators were all inherently nominal. What is a proper LCA of classification error when X is an ordinal variable? In this section, we consider this question. For example, X may correspond to a Likert scale with $K = 5$ categories: "strongly disagree" ($X = 1$), "disagree" ($X = 2$), "neither agree nor disagree" ($X = 3$), "agree" ($X = 4$), and "strongly agree" ($X = 5$). For purposes of classification error modeling, the five categories define five subpopulations, and we wish to determine the error probabilities associated with indicators of X arising from alternate question versions or data collection methods.

If X is treated as nominal, there is no a priori reason to expect persons in category 5 to have a higher or lower probability of misclassification in any of the other four categories. For example, with nominal data, there is no theoretical reason to expect $\pi_{4|5}^{A|X}$ to be higher than $\pi_{1|5}^{A|X}$. However, this is not the case for ordinal data where misclassification might be higher in categories adjacent to the true category than nonadjacent ones. One might hypothesize that the probability an individual in category x is classified in category x' is zero or quite small if x and x' differ by more than say, 1 or 2 scale points. For example, for an individual whose true opinion is "strongly disagree," an ordinal model could specify that the probability of being classified as "strongly agree" is much less than the chance of being classified as "disagree" since "disagree" is closer on the ordinal scale to "strongly disagree." For ordinal categories, Clogg (1979) suggested that the probabilities of being misclassified to nonadjacent categories be set to 0. However, for purposes of estimating error probabilities associated with specific Likert scales, imposing such constraints on the probabilities is not justified. Doing so produces biased estimates of the error probabilities.

As an example, it is not uncommon for respondents to misunderstand attitudinal questions due to faulty question wording. A person who misunderstands a question could respond "strongly disagree" to a question to which they might otherwise respond "agree" or "strongly agree." An unconstrained LC model would capture these errors. An ordinal model that only allows misclassification to adjacent categories would underestimate these errors. Forcing misclassification probabilities to be zero is not advisable unless there is good reason to believe that they are zero.

Another approach for constraining the error probabilities for ordinal latent variables was developed by Croon (1990, 1991, 2002). Consider a three-category classification scheme for past-month marijuana use where the categories are nonuser, infrequent user, and frequent user. A *nonuser* is defined as a person who has not smoked marijuana in the past month, an *infrequent user* is a person who smoked marijuana a few times in the past month, and a *frequent user* is a person who smoked marijuana more than a few times in the past month. Let X denote an individual's true classification as nonuser ($X = 1$), infrequent user ($X = 2$), and frequent user ($X = 3$). Suppose further than three trichotomous indicators of X, denoted by A, B, and C, are to be compared to determine which is best in terms of misclassification probability. The standard LC model $\{XA\ XB\ XC\}$ can be fit without constraints and will provide valid estimates of the error probabilities, even when X is ordinal. However, it may be of interest to consider the ordinal nature of X by imposing certain constraints which limit the magnitudes of the error probabilities.

For trichotomous latent class and indicator variables, Croon's constraints specify the following:

$$\begin{aligned} \pi_{3|1}^{A|X} &\leq \pi_{3|2}^{A|X} \leq \pi_{3|3}^{A|X} \\ \pi_{1|3}^{A|X} &\leq \pi_{1|2}^{A|X} \leq \pi_{1|1}^{A|X} \end{aligned} \tag{6.1}$$

```
* LEM Code for Fitting Ordinal Model for Trichotomous Data in Table 5.2
lat 1
man 3
dim 3 3 3 3
lab X A B C
mod X
    A|X or1
    B|X or1
    C|X or1
rec 27
rco
sta X [.5 .3 .2]
sta A|X [.7 .2 .1 .1 .8 .1 .25 .1 .65]
dat table_5-2.dat
```

Figure 6.1 ℓEM input statements for Croon's ordinal constraints.

The first set of inequality constraints states that the proportion of nonusers who are classified as frequent users is smaller than the corresponding proportion of infrequent users. Next, these two proportions are both smaller than the proportion correctly classified as frequent users. Likewise, the second set of constraints specifies that the proportion of frequent users who are classified as nonusers is smaller than the corresponding proportion of infrequent users. These proportions are both smaller than the proportion correctly classified as nonusers. Note that there are no constraints concerning persons classified as infrequent users. Croon's constraints are much less restrictive than Clogg's since all types of misclassifications are possible, not just misclassification to adjacent categories.

Still, Croon's approach is more appropriately applied in situations where the number of latent classes is not known a priori, as in exploratory typological analysis (see, e.g., Section 4.1.2). Forcing an ordering on the latent classes eliminates many parameter combinations from being considered that do not satisfy constraints such as (6.1). This could be desirable in some situations where the focus is on the structural part of the model. In other situations, for example, when the number of latent classes is known a priori (as in measurement error evaluation), such solutions have limited applicability. Indeed, Croon's inequalities, like Clogg's, may impart a bias in the error parameter estimates in cases where the constraints are violated by the observed data.

To see this, suppose that the true population model is the standard three-indicator LCA ({*XA XB XC*}) with error probabilities for *XA* given by the parameters in Table 5.1 (Chapter 5) where $XA = XB = XC$ (parallel measurements). Note that the conditional probabilities in this table violate the first constraint in (6.1) because $\pi_{1|3}^{A|X} > \pi_{1|2}^{A|X}$. This means, for example, that a frequent marijuana user has a higher probability of being classified as a nonuser than does an infrequent marijuana user. The ℓEM software (using the input code in Figure 6.1) was used to fit this model with Croon's ordinal constraints to illustrate the bias. The resulting parameter estimates are shown in Table 6.1. As noted from Figure 6.1, the parameter values from Table 5.1 were entered into ℓEM as starting values to avoid flippage and possible local maxima.

Table 6.1 ℓEM Estimates of the Parameters in Table 5.1 Using Croon's Ordinal Constraints

```
*** (CONDITIONAL) PROBABILITIES ***

          * P(X) *
1              0.4979
2              0.3353
3              0.1668
        * P(A|X) *
1 | 1          0.7054
2 | 1          0.1915
3 | 1          0.1030
1 | 2          0.1569
2 | 2          0.5464
3 | 2          0.2967
1 | 3          0.1569
2 | 3          0.0993
3 | 3          0.7438
        * P(B|X) *
1 | 1          0.9799
2 | 1          0.0000 *
3 | 1          0.0201
1 | 2          0.1436
2 | 2          0.7530
3 | 2          0.1034
1 | 3          0.1436
2 | 3          0.0000
3 | 3          0.8564
        * P(C|X) *
1 | 1          0.8173
2 | 1          0.0134
3 | 1          0.1693
1 | 2          0.2922
2 | 2          0.5386
3 | 2          0.1693
1 | 3          0.1256
2 | 3          0.0417
3 | 3          0.8327
```

The estimates in Table 6.1 are considerably different from the parameters in Table 5.1. For example, $\hat{\pi}_x^X$, $x = 1,2,3$ are 0.498, 0.335, and 0.167, respectively, while the parameter values are 0.5, 0.3, and 0.2, respectively. Likewise, there are substantial differences among the error probabilities, particularly for indicator B. These results suggest that ordinal constraints may not be appropriate for survey error analysis since they tend to provide biased estimates when the ordinal constraints do not hold. When they do hold, the traditional nominal LC model will still provide correct parameter estimates.

A third approach for LC modeling of ordinal data is the so-called probit latent class model due to Uebersax (1999). This model assumes that the indicators, $A, B, \ldots$, are discretized versions of a latent continuous variable X. The categories of the indicators are defined by fixed thresholds that divide the latent continuous variable into mutually exclusive regions corresponding to the observed response levels. The model further assumes that within each of these latent classes, the latent continuous variables have a multivariate–normal distribution. Further consideration of this model is beyond the scope of this book. However, like Croon's approach, it has limited applicability for survey error evaluations.

6.2 A LATENT CLASS MODEL FOR RELIABILITY

As we have seen in the previous chapters, at least two indicators are required for most survey error analysis. When only two indicators are available, the options for error evaluation are quite limited. Three general approaches were considered in the previous chapters: (1) reliability analysis, which requires parallel indicators, (2) gold standard analysis, which requires known true values, and (3) the Hui–Walter approach, which requires a special type of grouping variable. In this section, we discuss a fourth approach, called the *agreement model* (Guggenmoos-Holzmann 1996; Guggenmoos-Holzmann and Vonk 1998).

The agreement modeling approach is related to both reliability analysis and LCA. It shares two attributes with reliability analysis: (1) it does not require the assumption that a true value exists for the phenomenon under study, and (2) it results in a kappa-like measure of reliability under much less restrictive assumptions than reliability analysis. Like LCA, the agreement model is a LC model. But, unlike LCA, the latent classes are not related to true values or the characteristic being measured. The agreement model is important because it is one of only a few methods that are available to the analyst when only two measurements are available.

In Chapter 2, several estimators of the reliability ratio R were considered when two parallel measurements are available: $\hat{R}$ given by (2.37) and Cohen's κ defined in (2.42). We noted there the remarkable result that κ and $\hat{R}$ have identical computational forms even though they were developed independently and from very different theoretical underpinnings. A third estimator of R was given in (2.40) that is the product moment correlation for two measures. A fourth estimator of R based on the Hui–Walter model was considered in Chapter 3. The estimator was obtained by replacing the parameters π, θ, and ϕ (in Bross' notation) in the expression for I in (3.52) by their LC model estimators. Below we show that under certain restrictions, the agreement model provides a fifth estimator of R that has some important advantages over the estimators discussed thus far.

The estimator of R that we will derive has an interpretation that is very different from that of either κ and $\hat{R}$. Recall from Chapter 2 that former is the expected agreement rate beyond chance agreement and the latter is $1 - I$, where I is the proportion of the total variance that is simple response variance. Guggenmoos-Holzmann (1996) and Guggenmoos-Holzmann and Vonk (1998) develop a third interpretation of R: the proportion of the population that is "conclusive," that is, who will respond perfectly consistently to the question. They propose a model to estimate this proportion and show that their estimator is equivalent to κ when certain parameters of the model are constrained. They also show that these constraints are rather restrictive in practical applications. As a result, κ is not a useful measure of data quality. But by relaxing these assumptions, a new class of reliability measures can be derived, which they refer to as *generalized kappa statistics*. They argue that these new measures are more appropriate for error evaluations. For a number of important cases, only two measurements are required to estimate these kappa statistics.

Guggenmoos-Holzmann's application of LCA is quite different from any other considered in this book. One important difference is that there is no latent variable to represent the true value of the characteristic. In fact, as was the case in the discussion of reliability analysis in Chapter 2, the concept of a true value is not needed for the agreement model. Likewise, the error probabilities for the indicators are not the focus here. Rather, the role that the indicators play in the LCA is to determine the size of the group of *conclusive* persons in the population. The proportion of the population belonging to this group is Guggenmoos-Holzmann and Vonk's definition of reliability. Subtracting this proportion from 1 produces the inconsistency index, or the proportion of *inconclusive* persons.

6.2.1 Generalized Kappa Statistics

Recall from Chapter 2 that the index of inconsistency for two measures, denoted by $\hat{I}$, is based on the statistic g, referred to as the *gross difference rate*, where g is the proportion of the sample whose interview and reinterview responses for the survey variable (y) disagree. The basic idea of the agreement model is that the inconsistent respondents that make up g belong to a latent class of inconsistent or "inconclusive" persons with respect to y. However, the size of this group may be bigger than g since inconclusive persons can respond consistently by chance. In that case, membership in the inconclusive group cannot be observed directly and, thus, constitutes a latent class.

Denote membership in the conclusive latent class by $H = 1$ and membership in the inconclusive latent class by $H = 2$. Suppose that A and B are two dichotomous, parallel measurements. (These assumptions will be relaxed later.) By definition, individuals belonging to the conclusive class respond consistently to A and B with probability 1:

$$\Pr(B=1 \mid A=1, H=1) = \Pr(B=2 \mid A=2, H=1) = 1 \tag{6.2}$$

Or, in words, given that an individual is a conclusive ($H = 1$) and responds positively (negatively) to A, the probability that he/she responds positively (negatively) to B is 1.

For inconclusive persons ($H = 2$), the probability of consistent response is strictly less than 1:

$$\Pr(B=1 \mid A=1, H=2) < 1 \quad \text{and} \quad \Pr(B=2 \mid A=2, H=2) < 1 \tag{6.3}$$

In other words, an individual is inconclusive if and only if there is a positive probability that their responses to A and B will disagree.

Let π_1^H denote the proportion of the population in the conclusive group, and let $\pi_2^H = 1 - \pi_1^H$ denote the proportion in the inconclusive group. According to Guggenmoos–Holzmann (1996), π_1^H may be interpreted as the *reliability* of the classification process, and thus π_2^H is a measure of response *inconsistency*. Let $\pi_{1|1}^{A|H}$ and $\pi_{2|1}^{A|H}$ denote the probabilities of being classified as a positive and a negative, respectively, by classifier A for conclusive population members and $\pi_{1|2}^{A|H}$ and $\pi_{2|2}^{A|H}$ analogously for inconclusive persons. Further assume that A and B are parallel, locally independent measures; that is, assume

$$\pi_{a|h}^{A|H} = \pi_{b|h}^{B|H} \tag{6.4}$$

for all a, b, and h and

$$\pi_{ab|h}^{AB|H} = \pi_{a|h}^{A|H} \pi_{b|h}^{B|H} \tag{6.5}$$

where the $\pi_{b|h}^{B|H}$ are defined in analogy to $\pi_{a|h}^{A|H}$. As in the discussion of reliability in Chapter 2, we see the concept of a true value is not required under the agreement model.

Under these assumption, we derive expressions for the cell probabilities, π_{ab}^{AB}, of the 2×2 AB table. Note that, because $\pi_{1|11}^{B|AH} = 1$ and $\pi_{1|12}^{B|AH} = \pi_{1|2}^{B|H} = \pi_{1|2}^{A|H}$, we can write

$$\begin{aligned} \pi_{11}^{AB} &= \pi_1^H \pi_{11|1}^{AB|H} + \pi_2^H \pi_{11|2}^{AB|H} \pi_1^H \pi_{1|1}^{A|H} \pi_{1|11}^{B|AH} + \pi_2^H \pi_{1|2}^{A|H} \pi_{1|12}^{B|AH} \\ &= \pi_1^H \pi_{1|1}^{A|H} + (1 - \pi_1^H)(\pi_{1|2}^{A|H})^2 \end{aligned} \tag{6.6}$$

Likewise, because $\pi_{2|11}^{B|AH} = 0$ and $\pi_{2|12}^{B|AH} = \pi_{2|2}^{B|H} = \pi_{2|2}^{A|H}$, we can write

$$\begin{aligned} \pi_{12}^{AB} &= \pi_1^H \pi_{1|1}^{A|H} \pi_{2|11}^{B|AH} + \pi_2^H \pi_{1|2}^{A|H} \pi_{2|12}^{B|AH} \\ &= (1 - \pi_1^H) \pi_{1|2}^{A|H} (1 - \pi_{1|2}^{A|H}) \end{aligned} \tag{6.7}$$

Because of the symmetry of the table, (6.7) is also the expression for π_{21}^{AB}. Using the same type of arguments, it then follows that

$$\begin{aligned}\pi_{22}^{AB} &= \pi_1^H \pi_{2|1}^{A|H} \pi_{2|21}^{B|AH} + \pi_2^H \pi_{2|2}^{A|H} \pi_{2|22}^{B|AH} \\ &= \pi_1^H (1-\pi_{1|1}^{A|H}) + (1-\pi_1^H)(1-\pi_{1|2}^{A|H})^2\end{aligned} \tag{6.8}$$

From these equations we see that three parameters need to be estimated: π_1^H, $\pi_{1|1}^{A|H}$, and $\pi_{1|2}^{A|H}$. However, the model is nonidentifiable because there are only 2 degrees of freedom available for estimation. To see this, suppose that π_{11}^{AB} and π_{22}^{AB} are known. Then the off-diagonal cells can be deduced after noting that $\pi_{12}^{AB} = \pi_{21}^{AB} = (1-\pi_{11}^{AB} - \pi_{22}^{AB})/2$. They further show that a constraint that achieves identifiability is

$$\pi_{1|1}^{A|H} = \pi_{1|2}^{A|H} \tag{6.9}$$

which states that the probability of being classified as a positive is the same for both conclusive and inconclusive groups. The authors acknowledge, however, that this constraint is impractical and unlikely to hold in actual practice for reasons that will be discussed subsequently. Guggenmoos-Holzmann and Vonk show that the constraint (6.9) can be lifted when the two indicators are at least trichotomous or when three indicators are available.

Guggenmoos-Holzmann and Vonk compare the agreement model concept of reliability with Cohen's reliability concept. By writing the formula for κ in terms of the agreement model parameters, they consider the conditions that must hold in order that $\pi_1^H = \kappa$, that is, that the proportion conclusives in the population is equal to κ. Recall from (2.42) that the formula for Cohen's kappa is $\kappa = (P_0 - P_e)/(1 - P_e)$, where P_0 is the agreement rate and P_e is the probability of chance agreement assuming A and B are independent. From (6.6) and (6.7), it follows that

$$\begin{aligned}P_0 &= \pi_{11}^{AB} + \pi_{22}^{AB} \\ &= \pi_1^H \pi_{1|1}^{A|H} + \pi_2^H (\pi_{1|2}^{A|H})^2 + \pi_1^H \pi_{2|1}^{A|H} + \pi_2^H (\pi_{2|2}^{A|H})^2 \\ &= \pi_1^H + \pi_2^H [(\pi_{1|2}^{A|H})^2 + (\pi_{2|2}^{A|H})^2]\end{aligned} \tag{6.10}$$

and

$$\begin{aligned}P_e &= \pi_1^A \pi_1^B + \pi_2^A \pi_2^B \\ &= (\pi_1^A)^2 + (\pi_2^A)^2\end{aligned} \tag{6.11}$$

since A and B are parallel and where, applying (6.6) and (6.7) again, we obtain

$$\begin{aligned}\pi_1^A &= \pi_{11}^{AB} + \pi_{12}^{AB} \\ &= \pi_1^H \pi_{1|1}^{A|H} + \pi_2^H (\pi_{1|2}^{A|H})^2 + \pi_2^H \pi_{1|2}^{A|H} \pi_{2|2}^{A|H}\end{aligned} \tag{6.12}$$

and $\pi_2^A = 1 - \pi_1^A$. Guggenmoos-Holzmann and Vonk show that κ simplifies to π_1^H if and only if (6.9) holds and, therefore, κ is just a special case of the agreement model reliability. It is easy to show that if (6.9) holds, then $\kappa = \pi_1^H$, In that case, $P_0 = \pi_1^H + \pi_2^H[(\pi_{1|1}^{A|H})^2 + (\pi_{2|1}^{A|H})^2]$ and $P_e = [(\pi_{1|1}^{A|H})^2 + (\pi_{2|1}^{A|H})^2]$. κ can then be written as

$$\begin{aligned} \kappa &= \frac{\pi_1^H + \pi_2^H[(\pi_{1|1}^{A|H})^2 + (\pi_{2|1}^{A|H})^2] - [(\pi_{1|1}^{A|H})^2 + (\pi_{2|1}^{A|H})^2]}{1 - [(\pi_{1|1}^{A|H})^2 + (\pi_{2|1}^{A|H})^2]} \\ &= \frac{\pi_1^H\{1 - [(\pi_{1|1}^{A|H})^2 + (\pi_{2|1}^{A|H})^2]\}}{1 - [(\pi_{1|1}^{A|H})^2 + (\pi_{2|1}^{A|H})^2]} \\ &= \pi_1^H \end{aligned} \tag{6.13}$$

Showing that $\kappa = \pi_1^H$ implies (6.4) is left as an exercise for the reader. Details can be found in Guggenmoos-Holzmann (1996).

As the authors note, (6.9) is quite implausible for most survey applications and, for that reason, Cohen's κ (and, by association, $\hat{I}$) is an inappropriate measure of reliability. Gugenmoos-Holzmann and Vonk argue that, in general, (6.9) will not be satisfied by most survey processes and, further, that there is no compelling reason why it should be. For this constraint to hold, A and H must be independent random variables, an assumption that seems untenable because individuals having ambiguous classifications ($H = 2$) may have characteristics quite different from those whose classifications are more easily determined ($H = 1$). As an example, suppose that $A = 1$ denotes "employed" and $A = 2$ denotes "not employed." It seems reasonable that employed persons should be more easily classified than the unemployed and therefore, the probability of a positive classification should be higher in the conclusive group than in the inconclusive group; that is, $\pi_{1|1}^{A|H} > \pi_{1|2}^{A|H}$ in violation of (6.9).

One scenario that would yield approximately equal probabilities of positive classifications is if the true status of unemployed persons were so ambiguous that interviewers would have to resort to guesswork. For example, if the classification process has a type of learning mechanism that classifies inconclusive individuals as positives at approximately the same rate as previously encountered conclusive individuals, then it is possible that $\pi_{1|1}^{A|H} = \pi_{1|2}^{A|H}$. This situation might be plausible if a single interviewer conducted all the interviews. For example, as the interviewer encounters conclusive persons, he or she learns that roughly, 80% of the population is positive. Thus, when that interviewer encounters an inconclusive person, the interviewer guesses positive about 80% of the time because experience suggests that roughly 80% of the population is positive.

Guggenmoos-Holzmann and Vonk argue that κ is not a useful measure of reliability for categorical variables and suggest using the more general estimator of $\pi_H = 1$ instead, which removes the constraint in (6.9). They refer to these reliability measures as *kappa-like* agreement indices. Another advantage of the agreement model formulation for R is that, unlike κ, the agreement model estimate of R is always between 0 and 1, the parameter space of reliability and inconsistency measures, because it is obtained using maximum-likelihood estimation, which constrains the parameter space to the unit interval.

6.2.2 Comparison of Error Model and Agreement Model Concepts of Reliability

Standard latent class analysis software can be used to fit the agreement model under various constraints, as we will now show. By comparison, we will also show how the traditional latent model, referred to by Guggenmoos-Holzmann and Vonk as the error model, can also yield parameter estimates that can be used to produce an estimate of R indirectly. To illustrate the use of the ℓEM software for the agreement model, the data in Table I of Guggenmoos-Holzmann and Vonk (1998) will be analyzed under both models as it was in the referenced article.

To specify the agreement model for ℓEM, start by defining two latent variables, H and Y, where $H = 1$ and $H = 2$ denote the conclusive and inconclusive groups, respectively; $Y = 1$ denotes the group consistently classified as positive; and $Y = 2$, the group consistently classified as negative. The cross-classification of the latent variables defines four latent classes as shown in Table 6.2. In addition, define three dichotomous parallel measures, A, B, and C, whose reliability will be evaluated in the illustration to follow. The following constraints are required for Cohen's kappa:

1. $\pi_{1|11}^{A|HY} = 1$: conclusively "positive" persons are classified as "1" by A with certainty.
2. $\pi_{2|12}^{A|HY} = 1$: conclusively "negative" persons are classified as "2" by A with certainty.
3. $\pi_{1|21}^{A|HY} = \pi_{1|22}^{A|HY} = \pi_{1|1}^{Y|H}$: inconclusive persons are classified as "1" with probability equal to the proportion of conclusive persons who are classified as "1." Moreover, this probability is independent of Y within the inconclusive group.
4. $\pi_{a|hy}^{A|HY} = \pi_{b|hy}^{B|HY} = \pi_{c|hy}^{C|HY}$, for all $a = b = c$: A, B, and C are parallel.

Table 6.2 Agreement Model Latent Classes for the ℓEM Constraints

	$Y = 1$	$Y = 2$
$H = 1$	Conclusively positive	Conclusively negative
$H = 2$	Inconclusively positive	Inconclusively negative

As an illustration of constraints 4, suppose that 25% of the conclusive group is positive and therefore classified as "1" by A. Then every inconclusive person has probability 0.25 of being classified as positive and probability 0.75 of being classified as "2" by A.

Constraints 1–3 can be implemented in ℓEM using the `eq2` command in conjunction with the design (`des`) and starting value (`sta`) statements. A positive integer in the design vector assigns the integer to the corresponding parameter as a label for later reference. So, for example, the first "1" encountered in the `des` vector assigns the label "1" to the parameter $\pi_{111}^{Y|H}$. The following three 0s designate the next three parameters, namely, $\pi_{211}^{Y|H}$, $\pi_{112}^{Y|H}$, and $\pi_{212}^{Y|H}$, as free, unconstrained, and unreferenced parameters. Finally, the last eight integers in the design vector refer to the probabilities for the term $A|XY$ in the model.

With ℓEM, a probability can be constrained to a specified value (such as 1 or 0) by placing a "–1" in the parameter's position in the design vector. Then, the value for the constraint must be placed in the parameter's position in the starting values vector. To apply constraint 1 above, "–1" is specified in the position for $\pi_{1111}^{A|HY}$ in the `des` vector with "1" is the first position in the vector following `sta A|XY`. Once $\pi_{1111}^{A|HY}$ is constrained to 1, $\pi_{2111}^{A|HY}$ is necessarily constrained to 0; so this parameter remains unconstrained in the `des` vector (specified by "0"). Likewise, for constraint 2, $\pi_{1112}^{A|HY}$ and $\pi_{2112}^{A|HY}$ are fixed at 0 and 1, accomplished by placing "0" and "1" in their respective positions in the `des` vector (to denote "free" and "constrained," respectively) and again in the `sta A|XY` vector (to denote their constrained values, 0 and 1, respectively). Finally, to specify constraints 3, we place the label assigned to $\pi_{111}^{Y|H}$ (viz., "1") in the positions for $\pi_{1121}^{A|HY}$ and $\pi_{1122}^{A|HY}$ in the `des` vector. To complete constraint 3, any reasonable starting values can be used for $\pi_{111}^{Y|H}$, $\pi_{1121}^{A|HY}$, and $\pi_{1122}^{A|HY}$ as long as they satisfy the constraints; for the current situation, that implies they must be the same starting value. Using starting values of "1" has the same effect as not specifying starting values (i.e., random). Finally, constraint 4 is specified using the `eq1` command.

The data in Table I of Guggenmoos-Holzmann and Vonk (1998) will be used to illustrate the estimation process. In that table, four frequencies are provided (30, 9, 2, 29) corresponding to the number of times (0, 1, 2, and 3, respectively) that positive ratings were recorded for three pathologists (A, B, and C) who judged the presence or absence of cancer in 70 prostate biopsies. These data were converted to the input expected by ℓEM by dividing the counts into eight frequencies corresponding to the eight cells of the ABC cross-classification table. In doing this, the patterns "111" and "222" were given the frequencies 30 and 29, respectively, and the mixed patterns divided up the remaining counts consistent with the number of positive ratings in their Table I.

Note that, with Cohen's constraints in place (Eq. 6.9), there are only two identifiable parameters to estimate: π_1^H and $\pi_{111}^{Y|H}$. Although ℓEM provides an estimate for $\pi_{112}^{Y|H}$, it is meaningless and should be discarded since it is

Table 6.3 Conditional Probabilities for the Agreement Model Using the ℓEM Input Statements in Figure 6.2

```
*** (CONDITIONAL) PROBABILITIES ***
              * P(H) *
      1                  0.7898
      2                  0.2102
             * P(Y|H) *
      1 | 1              0.4526
      2 | 1              0.5474
      1 | 2              0.5506
      2 | 2              0.4494
             * P(A|HY) *
      1 | 1 1            1.0000 *
      2 | 1 1            0.0000 *
      1 | 1 2            0.0000 *
      2 | 1 2            1.0000 *
      1 | 2 1            0.4526
      2 | 2 1            0.5474
      1 | 2 2            0.4526
      2 | 2 2            0.5474
```

nonidentifiable. This can be verified by running the software multiple times to obtain different estimates of $\pi_{1|2}^{Y|H}$ with the same maximum value of the likelihood and the identifiable parameter estimates unaltered. Although ℓEM provides fit statistics associated with model degrees of freedom equal to 4, these too should be ignored as the model degrees of freedom are actually 1. This is because, as shown in Guggenmoos-Holzmann and Vonk's Table I, the minimally sufficient statistics for this analysis consists of four frequencies associated with 0, 1, 2, and 3 positive classifications across the three raters. Thus, after estimating two parameters, only 1 degree of freedom remains.

Table 6.3 shows the parameter estimates from the ℓEM output. The value of κ is $\hat{\pi}_1^H = 0.79$ and $\hat{\pi}_{1|1}^{Y|X} = 0.45$. Although standard errors are not reported by ℓEM, Guggenmoos-Holzmann and Vonk report these as 0.06 and 0.05, respectively. With constraint 3 (above) removed, a third parameter can be estimated: $\pi_{1|2}^{A|H}$. This is easily accomplished in ℓEM by replacing `1 0 1 0` in the third line in the `des` statement in Figure 6.2 with `2 0 2 0`, which breaks the equality between $\pi_{1|1}^{Y|H}$ and $\pi_{1|2}^{A|H}$. Now the estimate of π_1^H is lowered to 0.65. In addition, the estimate of $\pi_{1|1}^{Y|X}$ is 0.64 and $\pi_{1|2}^{A|H}$ is 0.18. If the generalized kappa estimate (with constraint (c) removed) is taken as the gold standard, kappa is biased upward. Guggenmoos-Holzmann and Vonk show that this is indeed a chronic problem with Cohen's kappa.

To complete this illustration, the standard LC model $\{XA\ XB\ XC\}$ was fit to these data, where X is defined as the true cancer status of a patient. The parallel measures constraints were imposed using the `eq1` command (i.e., `B|X eq1 A|X` and `C|X eq1 A|X`) as shown in Figure 6.2. No other constraints were specified in the model. Estimates of the three model parameters

```
* LEM Code for the Agreement Model with Three Parallel Dichotomous Measures
* Applying the Constraints that Produces Cohen's Kappa
man 3
lat 2
dim 2 2 2 2 2
lab H Y A B C
mod H
    Y|H   eq2
    A|HY  eq2
    B|HY  eq1  A|HY
    C|HY  eq1  A|HY
des [1  0    0  0         * Assigns label '1'to Pr(Y=1|H=1)
    -1  0   -1  0         * Fixes Pr(A=1|H=1,Y=1) and Pr(A=2|H=1,Y=2) to 1
     1  0    1  0]        * Sets Pr(A=1|H=2,Y=1)= Pr(A=1|H=2,Y=1)= Pr(Y=1|H=1)
sta A|HY [1 0 0 1         * Constrained values corresponding to line 2 of
                            des [ ]
          1 1 1 1]        * No starting values specified when H=2.
dat [29 0 1 3 1 3 3 30]   * Data translated from Table I in the reference
```

Figure 6.2 ℓEM input statements for the agreement model.

(viz., π_1^X, $\pi_{2|1}^{A|X}$, and $\pi_{1|2}^{A|X}$) were obtained. An estimate of the inconsistency index $\hat{I}$ was computed using (3.56) and reliability was estimated by $1-\hat{I}$. Thus, the estimate of R is

$$\hat{R} = 1 - \frac{\hat{\pi}_1^X \hat{\pi}_{1|1}^{A|X} \hat{\pi}_{2|1}^{A|X} + \hat{\pi}_2^X \hat{\pi}_{1|2}^{A|X} \hat{\pi}_{2|2}^{A|X}}{\hat{\pi}_1^A \hat{\pi}_2^A} \tag{6.14}$$

The estimates were as follows: $\pi_1^X = 0.57$, $\hat{\pi}_{2|1}^{A|X} = 0.091$, $\hat{\pi}_{1|2}^{A|X} = 0.013$, and $\hat{\pi}_1^A = 0.47$. Plugging these estimates into (6.14) yields $\hat{R} = 0.79$, which is the same value (within roundoff) as κ. This is not surprising since the assumptions of the LC model (in particular, parallel measures and local independence) are quite consistent with those for κ. However, within the LC modeling framework, it is possible to remove the parallel measures constraints and compute separate reliability estimates for each measure. In addition, it may be possible to introduce local dependence terms for the measures, which may also provide a better-fitting model, more accurate estimates of the model parameters, and, as a result, a better estimate of R. The next section describes an application of these concepts to assess the reliability of self-reports of race.

6.2.3 Reliability of Self-Reports of Race

In 1997, the US Office of Management and Budget released new standards for questions about race to better reflect the increasing racial and ethnic diversity of the US population. Under the new standards, federal agencies are required to offer individuals the opportunity to select more than one of five possible race categories: (1) American Indian/Alaska native, (2) Asian, (3) black/African American, (4) native Hawaiian/other Pacific islander, and (5) white. The first nationwide implementation of these standards was the 2000 decennial census.

A pretest of the census operations was conducted in three sites: Columbia, South Carolina; rural South Carolina; and Sacramento, California. A few weeks following the pretest, a reinterview survey [also referred to as the *postenumeration survey* (PES)] was conducted in these areas to evaluate the quality of the census data. PES reinterviews were conducted for $n = 40{,}519$ census pretest respondents. These data will be used to estimate R using the three methods described above: the agreement model, the LC model, and Cohen's κ. The estimates will be based on unweighted data to illustrate the methods, although weighted data yielded very similar results.

Biemer and Woltman (2002) reported the reliability of the various race categories using the method LCA method in (6.14). Biemer (2004b) compared three methods for estimating reliability: κ, the LCA error model (6.14) and the agreement model. The issue considered in some detail is whether persons, when given a choice to check more than one race category, will consistently report multiple races or a single race. Biemer did not consider which particular race or races were selected, just whether one or more races were selected consistently. This issue is of interest because of prior evidence that persons will sometimes indicate they belong to several race groups and sometimes only one, depending on the circumstance.

To estimate the reliability with which an individual will report multiple races versus a single-race category when given the option to report multiple races, Biemer defined an indicator variable for the census, denoted by A, where $A = 1$ if an individual selects at least two race categories in response to the race question and $A = 2$ if only one race category is selected. A second indicator, B, was defined analogously for the PES. Then R denotes the reliability with which individuals report multiple races.

Under the assumption of parallel measurements, A and B have the same reliability and the estimate of R is the reliability for both A and B. However, using the LCA error model with Hui–Walter constraints, separate estimates of the error parameters can be obtained for A and B and, thus, separate reliability ratios, R_A and R_B corresponding to A and B respectively, can be estimated (see Section 3.4.4). Table 6.4, taken from Biemer (2004b), provides the

Table 6.4 Census by PES Multirace Response

	PES Classification (B)		
Census Classification (A)	Multiple	Single	Totals
Multiple	429	2,328	2,757
Single	956	36,806	37,762
Totals	1,385	39,134	40,519

Source: Biemer (2004b).

Census Version Questionnaire	PES Version
What is this person's race? Mark ☒ one or more races to indicate what this person considers himself/herself to be. ☐ White ☐ Black, African Am., or Negro ☐ American Indian or Alaska Native–Print name of enrolled or principal tribe. ↓ ________ ________ ☐ Asian Indian ☐ Native Hawaiian ☐ Chinese ☐ Guamanian or Chamorro ☐ Filipino ☐ Japanese ☐ Samoan ☐ Korean ☐ Other Pacific Islander ☐ Vietnamese Asian – *Print race.* ↓ ☐ Other Asian–*Print race.* ↓ ________ ________ ☐ Some other race–*Print race.* ________	(Show card B.) **Which of these categories best describes (your/...'s) race? You may choose one or more races.** ☐ White ☐ Filipino ☐ Native Hawaiian ☐ Black, African Am., or Negro ☐ Japanese ☐ Guamanian/Chamorro ☐ Asian Indian ☐ Korean ☐ Samoan ☐ Chinese ☐ Vietnamese ☐ Other Pacific Islander ☐ American Indian or Alaska Native–**What is the name of (Your/...'s) enrolled or principal tribe.** ↓ ☐ Other Asian↓ ☐ Some other race **What is this race?** ↓ ________

Figure 6.3 Multiple-race questions from the census and the PES.

AB (interview–reinterview) cross-classification table that provides the sufficient statistics for all the estimation methods that will be considered.

The inconsistencies between the census and PES classifications are apparent from the data in this table. Among the 2757 persons who chose multiple races in the census, an incredible 2328 (or 84%) changed to a single race in the PES! This shift may be due, in part, to the change in mode of data collection for the PES. The census responses were obtained through a self-administered, paper questionnaire while the PES was conducted by face-to-face and telephone interviewing. Figure 6.3 shows the questions for the census and the PES. Although the questions about race are somewhat different in the two questionnaires, these differences are unlikely to account to the large shift in responses observed. Perhaps a better explanation would be interviewer effects, since interviewers can, by the way they ask the question, discourage respondents from selecting more than one race. A convenient way for the interviewer to do this is to simply not ask the second part of the question" "You may choose one or more races." It is not known whether this is the cause, however.

Because 84% of the census multiracial group was reclassified into a single-race category in the PES, the parallel measures assumption clearly does not hold for *A* and *B*. As shown in Chapter 2, the assumption of equal error distributions can be formally tested by the hypothesis H$_0$: $\pi_1^A = \pi_1^B$ versus H$_0$: $\pi_1^A \neq \pi_1^B$, which uses the net difference rate (NDR) as the test statistic. If the

NDR is significantly different from 0, then hypothesis H_0 is rejected, and the parallel measurements assumption must also be rejected. For this table, $\hat{\pi}_1^A = 6.8$ and $\hat{\pi}_1^B = 3.4$ for an NDR of 3.4%, which is highly significantly and, thus, H_0 is rejected. This result implies that estimates of R based on the parallel assumption will be biased and considerably so, judging from the magnitude of the difference. Under these conditions, the Hui–Walter method, which separately estimates R for each measure, is likely to be preferred since it does not require that the two measurements be parallel.

Another motivation for the using the LC error model approach in this study is to test the hypothesis that the postenumeration survey can be regarded as a gold standard for evaluating quality of the race data in the census. This is a common assumption in census evaluation because the postenumeration survey uses the most experienced and competent interviewers, the most preferred mode of interview (face to face), and questions which designed to elicit highly accurate responses. In addition, interviewers are trained to ask probing questions to clarify responses if necessary. With the LC error model approach, separate reliability estimates, denoted by R_A and R_B, as well as separate error probabilities, can be compared to test this assumption.

One problem with the LC model is that it assumes the existence an individual's "true race." However, because race can be subjective and based on personal preference or an individual's racial identity, an individual's true race may not exist. Biemer and Woltman (2002) propose an operational definition by conceptualizing a preferred method of obtaining race data, specifically, a method that is devoid of all influences that would cause instability in responses to the race question. They then interpret the latent variable in the LC error model as the classification obtained under this scenario. Hence, deviations of the indicators A and B from the latent variable can be interpreted as classification error. They note that this concept allows for an examination of the systematic errors in classifications of race that may be related to the interviewing process.

The alternative estimators of R that can be computed from the data in Table 6.4 are shown in Table 6.5. Note that, with only two measurements and dichotomous variables, Cohen's kappa constraints were imposed on the agreement model to obtain identifiability. Thus, the estimates of $\hat{\pi}_1^H$ should be very close to Cohen's κ, and they are. Standard errors of the estimates are not shown but are quite small—less than 0.2 percentage points. To obtain the Hui–Walter model estimates, $\hat{R}_A$ and $\hat{R}_B$, we used a dichotomous grouping variable denoted

Table 6.5 Three Estimates of the Reliability of Multiple-Race Reports in the Census

	κ	$\hat{\pi}_{H=1}$	$\hat{R}_A$	$\hat{R}_B$
Estimate	16.9	16.4	56.4	6.0

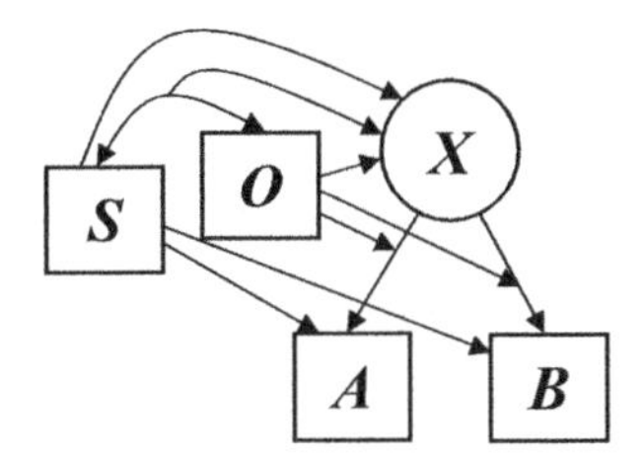

Figure 6.4 Path model for the multiple-race/single-race data analysis.

by O where $O = 1$ if the individual is of Hispanic origin and $O = 2$ if not. Because the prevalence of multiple-race responses differ markedly between Hispanics and non-Hispanics, this choice of grouping variable satisfies one of the Hui–Walter assumptions, namely, that the prevalence rates differ among the groups of the grouping variable. However, the assumption of equal error probabilities across groups may not be tenable. An alternative grouping variable that may better satisfy this assumption is the site variable S, with the three levels mentioned above: Columbia, South Carolina; rural South Carolina; and Sacramento, California. However, this variable is also not ideal because the proportion of respondents classified in multiple-race categories does not differ appreciably across the sites. The best choice of grouping variable among those considered was the joint variable OS, with six levels and this grouping variable was used in the analysis. The final model used was $\{XSO\}\{XOA\ SA\}\{XOB\ SB\}$, which is technically not a Hui–Walter model, due to the presence of the effects SA and SB in the error component. This model is fully saturated; thus, no test of model fit is available. The path diagram for the model is shown in Figure 6.4, and the ℓEM input statements for the model are shown in Figure 6.5. This example clearly illustrates that grouping variables have different purpose in an LC model. As discussed in Section 4.4.3, one needs to consider the structural and measurement components separately when selecting the explanatory variables for the model.

Note from Table 6.5 that κ and $\hat{\pi}_{H=1}$, both of which assume parallel measures, are quite small compared to $\hat{R}_A$, which specifies separate error probabilities for the census and the PES. As previously noted, the parallel assumption does not seem to hold for these data. The PES has a small reliability estimate relative to the census. Note that the average reliability for A and B, specifically, $(\hat{R}_A + \hat{R}_B)/2$, is 31.2, which is still twice the magnitude of the estimates of R that assume parallel measurements. For these data, $\hat{R}_A$ appears to be a much better estimate of the reliability of the census than either κ or $\hat{\pi}_{H=1}$.

For an extensive analysis of these data, including estimates of reliability and measurement bias for each race category and for both the census and the PES, see Biemer and Woltman (2002). Their study concluded that (1) the postenumeration survey should not be considered as a gold standard for the purposes of evaluating the quality of the census race responses, (2) there is evidence that race information collected by different modes yields

```
Hui-Walter Model for Analysis of Multiple vs. Single Race.
* X = True status where 1 is multiple race and 2 is single race
* A = Census response defined analogously to X
* B = PES response defined analogously to X
* S = Site variable with three levels
* O = Hispanic origin variable with two levels
lat 1
man 4
dim 2 3 2 2 2
lab X S O A B
mod SO
    X|SO   {XSO}
    A|SOX  {XA XOA SA} * No XSA interaction for the HW assumption of equal
                         error probs
    B|SOX  {XB XOB SB} * No XSB interaction for the HW assumption of equal
                         error probs
rec 24
rco
sta XA [ .9 .1 .1 .9]
sta XB [ .9 .1 .1 .9]
dat [
1     1     1     1     56
1     1     1     2     218
1     1     2     1     237
1     1     2     2     2386
1     2     1     1     199
1     2     1     2     267
1     2     2     1     425
1     2     2     2     9747
2     1     1     1     9
2     1     1     2     258
2     1     2     1     12
2     1     2     2     160
2     2     1     1     27
2     2     1     2     50
2     2     2     1     88
2     2     2     2     12107
3     1     1     1     103
3     1     1     2     1486
3     1     2     1     14
3     1     2     2     222
3     2     1     1     35
3     2     1     2     49
3     2     2     1     180
3     2     2     2     12184]
```

Figure 6.5 ℓEM input statements for the path model in Figure 6.4.

difference reliabilities, (3) the assumption of parallel measurements for interview and PES is violated from these data, and (4) the LCA error model is a more appropriate estimation method for these data since it does not rely on the assumption of parallel measurements.

6.3 CAPTURE–RECAPTURE MODELS

Capture–recapture models play a vital role in population size estimation for many types of dynamic populations that are difficult to count; examples include fish, birds, and other types of wildlife [see, e.g., Pollock et al. (1990) and Seber (1982)], health applications (Laska 2002), estimating the number of Websites (Fienberg et al. 1999), and many other applications. In survey work, these models have found application in census counting error evaluations (Fienberg 1992; Bartolucci and Forcina 2001; Bartolucci and Pennoni 2007). In this section, we discuss the application of LCA in capture–recapture modeling for evaluating a population census. The use of LC models for capture–recapture applications was advocated by Pledger (2000) for wildlife population applications. Applying LC models to human population enumeration, Bartolucci and Forcina (2001) and Bartolucci and Pennoni (2007) showed how unobserved heterogeneity can be readily treated using an LC model framework. Biemer et al. (2002) also considered population heterogeneity using LC models and the conditions under which latent heterogeneity models are identifiable. Still, a number of issues of LC model identifiability that are unique to the capture–recapture framework have yet to be explored, and not much is known about the statistical properties of the LC model estimators in census coverage error evaluation applications.

The foundation of capture–recapture modeling is the creation of multiple lists (or *frames*) of the members of the target population (i.e., the population to be counted). A list may be created by an enumeration of the population or may be readily available from some source such as a population registry or administrative system. Each list may be imperfect in that it may miss some population units, may count other units more than once, or may contain non-population units that are erroneously counted. A determination is made as to whether a unit on some particular list is a member of the population (and should be counted) or is ineligible to be counted (sometimes referred to as an *erroneous enumeration*). In addition, the units must be matched across the lists so that it is possible to determine whether a particular unit on one list is included or excluded on the other lists.

In the LC model framework, the true status of a unit's membership in the target population is a latent variable. The process of listing population members is subject to classification error. Listing a non–population member is a false-positive error, and failing to list a population member is a false-negative error. Thus each listing process can be treated as an indicator of the latent variable. With a few modifications that will be discussed subsequently, the LC modeling framework can be applied to estimate N, the total number of persons in the target population. Although the methodology is theoretically sound, the modeling assumptions required for unbiased estimation of N are difficult to satisfy in practice unless at least three, and preferably four, lists are available. Some of these issues will be discussed in this section.

The basic methods for census count evaluation are well described in the classic text by Marks et al. (1974). The authors classify the methods into three types: national registration systems, administrative databases, and postenumeration surveys (PESs). Here we focus on the latter method, particularly in combination with the so-called *dual-system estimator* (Sekar and Deming 1949). In particular, we shall consider methods for three lists—the census, the PES, and an administrative records list (ARL)—in order to extend the dual-system methodology so that the assumptions made under than model can be evaluated to some extent.

The dual-system estimator (DSE) has been widely used throughout the world for evaluating the accuracy of the census counts. Countries that have used the method include the United States (Hogan and Wolter 1988), Canada (Statistics Canada 2001), England (UK) (Brown et al. 1999), Australia (Trewin 2006), India, Turkey, Pakistan, and a number of countries in Africa (Marks et al. 1974). With this approach, a PES is conducted and the persons in the PES are matched to persons in the census enumeration and vice versa to form a 2×2 table of counts cross-classifying the presence or absence of persons in the census enumeration with their presence or absence in the PES. The DSE approach provides an estimator of the number of persons in the fourth cell of this table, which corresponds to persons missed by both the census and the PES. The sum of the three observed and one estimated cells of the census–PES cross-classification table provides an estimate of the total population count.

Three key and rather strong assumptions are necessary in order for the DSE to produce consistent estimates of the population total N:

1. *Uncorrelated Error.* The probability of erroneously including/excluding an individual in the PES does not depend on whether the individual was erroneously included or excluded in the census. Failure of this assumption will induce correlations between the errors in the census and PES lists, sometimes referred to as *behavioral* (Wolter 1986) or *interlist correlation*. If a third list is available, the assumption can be tested [see, e.g., Bishop et al. (1975, Chapter 6)]. Zaslavsky and Wolfgang (1993) provide models for dealing with the interlist correlation in three systems.
2. *Homogeneity.* The probability of inclusion on a list does not vary from individual to individual. Although this assumption is known not to hold for the population as a whole, various strategies have been used to address the problem of heterogeneous enumeration probabilities, including poststratification (Sekar and Deming 1949) and logistic regression (Alho et al. 1993). Methods involving three systems have been explored by Darroch et al. (1993), Fienberg et al. (1999), and Chao and Tsay (1998). This is the correlation bias problem [see, e.g., Wolter (1986)]. Methods to adjust for this type of correlation involving the census and PES have been explored by Bell (1993) and Shores and Sands (2003).

3. *Perfect Enumeration and Matching.* Individuals on both lists are all population members that can be accurately matched between the two lists and any nonresidents who have been erroneously enumerated (EE) can be identified and eliminated. Matching errors can be fairly substantial. In 2000, the US Census Bureau Evaluation Followup (EFU) estimated that about 1.8 million enumerations in the PES were EEs (Fay 2002). On the basis of these results, the US Census Bureau concluded that the net undercount was overstated by three to four million persons and, thus, adjustments to the census count on the basis of the PES would substantially overcorrect the population counts in many areas. For the 1990 census, Biemer and Davis (1991) reported that the level of misclassified EEs in the 1990 PES exceeded 5% of the PES count for many areas of the country. In the worst cases, the Northeast urban and the Midwest noncentral city areas, the EE rate exceeded 20%.

The next section provides a general LC model framework for exploring models that relax these assumptions to some extent. For example, assumption 1 can be relaxed if a third list such as an ARL is available containing persons in the population. The additional degrees of freedom provided by the third system permits the estimation of correlations between the counting errors in the first two systems. Three systems are necessary and sufficient for relaxing assumption 3 by accounting for nonresidents in the census count.

6.3.1 Latent Class Capture–Recapture Models

For triple-system models, let U denote the union of persons included on at least one of the three lists plus all residents in the area who are not included on any of the lists. In addition, U contains nonresidents and fictitious persons who are erroneously included on the lists. Let P denote all persons who are true residents of the target population and should be counted, and let E denote the persons who are erroneous enumerations (nonresidents, fictitious persons, etc.) on at least one list and who should not be counted. By these definitions, U is the union of P and E. The number of persons in U will be denoted by M and the number of persons in P by N. Our goal is to obtain an accurate estimate of N on the basis of data from the census and two additional systems: the PES and the third list, which is assumed to be an ARL.

Let X_i denote a dichotomous latent variable defined for the ith person in U, where $X_i = 1$ if person i is in P (resident) and $X_i = 2$ if person i is in $E = U \sim P$ (i.e., nonresident or fictitious) where "$U \sim P$" denotes the set U after removing all units in the set P. In the triple-system framework, there are three manifest variables, one for each list, denoted A (census), B (PES), and C (ARL). For each variable, "1" denotes a person (or EE) in U who (or that) is on the list and "2" denotes a person in U who is not on the list. A correct enumeration for an indicator occurs when a manifest variable and X are both 1.

As an example, the probability of a correct enumeration in the census denoted by $\pi_{1|1}^{A|X}$, is referred to as the *correct enumeration probability*. Likewise, the probability a person is missed by the census is $\pi_{2|1}^{A|X}$. An EE (erroneous enumeration) occurs when a person in E is classified as in P. Thus, for example, the probability of an EE in the ARL is $\pi_{1|2}^{C|X}$. Note that, by these definitions, the 222 cell of the ABC table contains a structural zero since the number of persons in U missed by all three is unobserved. Indeed, the purpose of this analysis is to estimate $\pi_{222|1}^{ABC|X}$, that is, the proportion of true residents missed by all three systems, also known as the *undercount probability*.

Capture–recapture models that assume no EEs in any of the lists (i.e., $\pi_{1|2}^{A|X} = \pi_{1|2}^{B|X} = \pi_{1|2}^{C|X} = 0$) are equivalent to models specifying $\pi_1^X = 1$ (i.e., no nonresidents). In that case, no latent variable is required in the analysis, and traditional loglinear models can be used [see, e.g., Section 6.3 in Bishop et al. (1975)]. Our focus here is on models that allow for EEs on at least one list. The models to be considered invoke the homogeneity assumption (i.e., assumption 2 above). As described in Wolter (1986), this assumption is satisfied in practice to some extent by stratifying the sample according to geographic, demographic, and other characteristics related to enumeration probability. To simplify the exposition of the central ideas, the models considered below do not involve these homogenizing covariates, although including them in the models is quite straightforward. Analysis of triple-system data using meaningful grouping variables and other covariates is explored in Biemer et al. (2002).

6.3.2 Modeling Erroneous Enumerations

Brown and Biemer (2004) show that, with only three lists, the traditional LC model specifying local independence, namely, {XA XB XC} is nonidentifiable because of the structural zero in cell 222 of the ABC table. They further show that by adding a fourth list (say, D), a structural zero remains in the 2222 cell; still the model {XA XB XC XD} is identifiable. However, in the census evaluation context, obtaining four lists may be impractical. One approach for obtaining identifiability with three lists is to set any two of these three probabilities—$\pi_{1|2}^{A|X}$, $\pi_{1|2}^{B|X}$, and $\pi_{1|2}^{C|X}$—to 0. For illustration purposes, we shall assume

$$\pi_{2|1}^{A|X} = 0, \quad \pi_{2|1}^{B|X} = 0, \quad \pi_{2|1}^{C|X} > 0 \quad \text{and} \quad \pi_{222}^{ABC} = 0 \tag{6.15}$$

Biemer et al. (2002) refer to the model {XA XB XC} with these constraints as the L_t model.[1]

[1]The L indicates it is a LC model, and the subscript t indicates that capture–recapture probabilities vary by time or list. This is consistent with notational conventions in the literature on capture–recapture.

They note that local independence is often violated by capture–recapture models due to so-called *trap-happy* or *trap-shy* effects (Bishop et al. 1975, p. 230), which induces a type of behavior correlation between the manifest variables. For example, persons captured by the census are likely to have a higher probability of capture by the PES than are persons who are missed by the census (*trap-happy*). Similarly, persons who are missed by the census are more likely to be missed by the PES (*trap-shy*). One source of dependence between the census and the PES lists arises from the similarities in the list construction process. When both lists are generated by an enumeration process, errors due to enumerator judgments, enumerator–respondent interactions, common administration procedures and protocols, sample members' willingness to be located, and so on can cause the counting errors in the census and the PES to be correlated. Since the ARL is generated from noninterview sources (for, e.g., government program lists, employment roles, tax authority lists), it seems reasonable to expect errors in the C list to be less correlated with those in the census and the PES.

To model the potential interaction between A and B, the AB direct effect can be added to the model. Biemer et al. (2002) referred to this model as the L_{tAB} model, where the subscript AB indicates that the direct effect between A and B is added to the model (i.e., $\{XA\ XB\ XC\ AB\}$) with constraints in (6.15). The L_{tAB} model contains six parameters: π_1^X, $\pi_{2|1}^{A|X}$, $\pi_{2|1a}^{B|XA}$, $a = 1,2$, $\pi_{2|1}^{C|X}$, and $\pi_{1|2}^{C|X}$. They show that, for this model, the parameter $\pi_{1|2}^{C|X}$ is not identifiable and additional information must be provided in order to obtain meaningful inferences from the L_{tAB} model. Their solution is to fix one of the model parameters, which is sufficient to make the model identified. They do this by providing an estimate of $\gamma = \pi_{2|1}^{X|C}$ (i.e., the probability that a person on the C list is erroneous) obtained from an external source (e.g., a special study aimed at estimating EEs in the C list). With γ fixed to this estimate ($\hat{\gamma}$, say) in the model, the other parameters of L_{tAB} are identifiable and can be estimated conditionally on $\hat{\gamma}$.

6.3.3 Parameter Estimation

The method of moments and ML estimation methods have been used for parameter estimation in the literature for capture–recapture models. Method-of-moments estimates are typically easy to calculate but can have undesirable properties such as large variance or large bias. Consequently, MLEs are usually preferred. The standard ML estimation method consists of using the conditional likelihood (White and Burnham 1999). In this method, the model parameters are estimated by maximizing the conditional likelihood of the ABC table given $\pi_{222}^{ABC} = 0$. Then an estimate of the population size is derived via the Horvitz–Thompson estimator [see, e.g., Section 6.3 in Bishop et al. (1975)].

For the L_t or L_{tAB} models, the estimator of N has the general form

$$\hat{N} = \hat{\pi}_1^X \frac{n}{1-\hat{\pi}_{222|1}^{ABC|X}} \tag{6.16}$$

where n is the total number of persons on all three lists; $\hat{\pi}_1^X$ is an estimate of π_1^X, which, for the L_t model, is given by

$$\hat{\pi}_1^X = \frac{\sum_{abc} \exp(\hat{u}_{1a}^{XA} + \hat{u}_{1b}^{XB} + \hat{u}_{1c}^{XC})}{\sum_{xabc} \exp(\hat{u}_{xa}^{XA} + \hat{u}_{xb}^{XB} + \hat{u}_{xc}^{XC})} \tag{6.17}$$

and $\hat{\pi}_{222|1}^{ABC|X}$ is the estimate of the proportion of persons missed by all three systems given by

$$\hat{\pi}_{222|1}^{ABC|X} = \frac{\exp(\hat{u}_{12}^{XA} + \hat{u}_{12}^{XB} + \hat{u}_{12}^{XC})}{\sum_{abc} \exp(\hat{u}_{1a}^{XA} + \hat{u}_{1b}^{XB} + \hat{u}_{1c}^{XC})} \tag{6.18}$$

where $\hat{u}$ denotes the conditional likelihood estimate of the corresponding model parameter with $\pi_{222}^{ABC} = 0$.

In the next section, we compare the estimates of N from the L_{tAB} model and the so-called M_{tAB} model. The latter model is simply the nonlatent variable version of the L_{tAB} model. It assumes that no erroneous enumerations are possible in any lists (i.e., $\pi_2^X = 0$). Thus, M_{tAB} can be represented as $\{AB\ C\}$ (i.e., the model with effects A, B, C, and AB but without the X variable). Note its similarity to the L_{tAB} model given by $\{XA\ XB\ XC\ AB\}$. For the M_{tAB} model, an estimator of N is given by (6.16) with $\hat{\pi}_1^X = 1$ and replacing $\hat{\pi}_{222|1}^{ABC|X}$ by $\hat{\pi}_{222}^{ABC}$, the M_{tAB} model estimate of the proportion missed by all three systems given by

$$\hat{\pi}_{222}^{ABC} = \frac{\exp(\hat{u}_2^A + \hat{u}_2^B + \hat{u}_2^C + \hat{u}_{22}^{AB})}{\sum_{abc} \exp(\hat{u}_a^A + \hat{u}_b^B + \hat{u}_c^C + \hat{u}_{ab}^{AB})} \tag{6.19}$$

where, as before, $\hat{u}$ denotes the conditional likelihood estimate of the corresponding model parameter with $\pi_{222}^{ABC} = 0$.

6.3.4 Example: Evaluating the Census Undercount

To demonstrate the estimation process, L_{tAB} and M_{tAB} will be applied to data in Table 6.6, which are taken from a study conducted by Zaslavsky and Wolfgang (1993) for the 1988 US Census Dress Rehearsal in St. Louis, Missouri. The dataset for their analysis focused on 79 census blocks composed primarily of black renters. This group is among the hardest to enumerate, particularly black male renters in the 20–29 age group—historically one of the most seriously

Table 6.6 Triple System Data from the 1988 Dress Rehearsal Census in St. Louis, Missouri

A	*B*	*C*	Owners, 30–44 years	Renters, 20–29 years
1	1	1	91	58
1	1	2	262	144
1	2	1	13	12
1	2	2	77	73
2	1	1	10	11
2	1	2	69	70
2	2	1	35	43
2	2	2	0 (structural)	0 (structural)

undercounted subgroups in the United States. The estimates for this group will be compared with black owners in the 30–44 age group from the same areas.

Three sources of data in their study were collected—the census, the PES, and the ARL—to produce triple-system estimates of the test areas. The corresponding indicators will be denoted by A, B, and C, respectively. The ARL data were developed from precensus administrative records of state and federal government agencies, including state departments of employment security, state drivers' license bureaus, the US Internal Revenue Service, the US Selective Service System, and the US Veterans Administration.[2]

To use ℓEM to fit the L_{tAB} with $\gamma = \gamma_0$ for $0 < \gamma_0 < 1$, it was necessary to convert the constraint for γ to a constraint on π_2^X using the identity

$$\gamma = \pi_{2|1}^{X|C} = \frac{\pi_{12}^{CX}}{\pi_1^C} = \frac{\pi_{1|2}^{C|X}\pi_2^X}{\pi_1^C} \tag{6.20}$$

which follows from the assumptions of the L_{tAB} model. Since the model assumes that all EEs emanate from the C list, we have $\pi_{1|2}^{C|X} = 1$ and therefore

$$\pi_2^X = \gamma\pi_1^C \tag{6.21}$$

We can compute an estimate of π_1^C directly from the data in Table 6.6 for each stratum. For example, for owners, 30–44 years, $\pi_1^C = 0.268$. Therefore, fixing $\gamma = 0.1$ is equivalent to fixing $\hat{\pi}_2^X = 0.0268$. Therefore, L_{tAB} with $\gamma = 0.1$ can be fit by ℓEM with the constraint $\hat{\pi}_2^X = 0.0268$.

[2]The Selective Service System maintains information on those potentially subject to military conscription; the Internal Revenue Service maintains a database of all persons in the United States who are subject to income taxes; the Veterans Adminsitration maintains a database of persons serving in the US armed forces.

Figure 6.6 provides the ℓEM input statements to fit the M_{tAB} and the L_{tAB} models to the owners data. For the L_{tAB} model, γ is supplied as a constraint by using the γ-to-π_2^X conversion formula in (6.21). Note that $\gamma = 0.05$ is equivalent to $\pi_2^X = 0.013375$. Thus, the γ constraint is imposed by constraining π_2^X to this value using the `eq2` command (see Figure 6.6). Figure 6.7 presents the parameter estimates from the L_{tAB} model, where `beta` denotes the u parameters in our notation. The conditional probability estimates provided in ℓEM output should be ignored since they estimated with the restriction $\hat{\pi}_{222}^{ABC} = 0$ imposed. We require estimates after removing this constraint so that the entire population can be estimated, including persons in cell (2,2,2). To do this, we use the method shown in (6.17) and (6.18) and the estimates in Figure 6.7 to estimate the probability $\pi_{222|1}^{ABC|X}$ and subsequently $\hat{N}$.

To compute $\hat{\pi}_{222|1}^{ABC|X}$ we use the formula

$$\hat{\pi}_{222|1}^{ABC|X} = \frac{\exp(\hat{u}_2^A + \hat{u}_2^B + \hat{u}_{22}^{AB} + \hat{u}_{12}^{XC})}{\sum_{abc} \exp(\hat{u}_2^A + \hat{u}_2^B + \hat{u}_{22}^{AB} + \hat{u}_{12}^{XC})} \tag{6.22}$$

The numerator is given by

$$\exp(-0.2898 - 0.2247 + 0.4587 + 0) = 1.7891$$

For the denominator of (6.22), we repeat this calculation for the seven remaining ABC patterns: 111, 112, 121, 122, 211, 212, and 221 for $X = 1$ only. These calculations yield 1.3988, 5.0063, 0.3566, 1.2762, 0.3130, 1.1204, and 0.4999, respectively. The denominator is the sum of these seven numbers plus 1.7891 = 11.7603. This yields $\hat{\pi}_{222|1}^{ABC|X} = 1.7891/11.7603 = 0.1521$. Now using the constraint $\hat{\pi}_1^X = 0.9867$, we have

$$\hat{N} = \hat{\pi}_1^X \frac{n}{1 - \hat{\pi}_{222|1}^{ABC|X}} = 0.987 \frac{557}{1 - 0.1521} = 648.4$$

which rounded off is 648 as in Table 6.7.

Using this approach, Table 6.7 provides the estimates of $\hat{\pi}_1^X$, $\hat{\pi}_{222|1}^{ABC|X}$, and N for three estimators for M_{tAB}, and L_{tAB} with two values of γ_i: $\gamma = 0.05$ and $\gamma = 0.10$. From this table, we see that renters have a much higher undercount rate, $\hat{\pi}_{222|1}^{ABC|X}$, than do owners as hypothesized. In addition, we see that incorporating a nonzero level of EEs into the estimation process tends to reduce the estimates of N. This is also as expected since otherwise the EEs are counted as residents in the estimation of N rather than being deducted as false-positive counts.

This example illustrated the use of LC models estimating EEs in applications of capture–recapture models for census population evaluation. Agresti (2002, p. 544) uses an LC model to address the problem of unobserved

```
* MtAB Model with Owners Data
  man 3
  dim   2 2 2
  lab   A B C
  mod  ABC {AB C wei(ABC)}
  sta wei(ABC) [1 1 1 1 1 1 1 0] * STRUCTURAL 0 FOR CELL 222
  data [91 262 13 77 10 69 35 0] * OWNERS DATA
*******************************************************************************
*LtAB with Owners Data and Gamma = 0.05
  man 3
  lat 1
  dim  2 2 2 2
  lab  X A B C
  mod X      eq2
      ABC|X {AB XC wei(XABC)}
  des [0 -1]
  sta X [0.986625 0.013375]        * These constraints correspond to Gamma = 0.05
  sta wei(XABC) [1 1 1 1 1 1 1 0   * STRUCTURAL 0 AT 222
                 0 0 0 0 0 0 1 0] * SPECIFIES ALL EEs COME FROM THE C-LIST
  data [91 262 13 77 10 69 35 0]   * OWNERS DATA
```

Figure 6.6 ℓEM input statements for M_{tAB} and L_{tAB} ($\gamma = 0.05$).

```
*** LOG-LINEAR PARAMETERS ***
* TABLE XABC [or P(ABC|X)] * (WARNING: 1 fitted zero margins)
  effect           beta    exp(beta)
  A
   1             0.2898       1.3362
   2            -0.2898       0.7484
  B
   1             0.2247       1.2519
   2            -0.2247       0.7988
  C
   1            -0.3188       0.7270
   2             0.6375       1.8918
  AB
   1 1           0.4587       1.5820
   1 2          -0.4587       0.6321
   2 1          -0.4587       0.6321
   2 2           0.4587       1.5820
  XC
   1 1          -0.3188       0.7270
   1 2           0.0000       1.0000
   2 1           0.3188       1.3754
   2 2           ******   0.00E+0000
```

Figure 6.7 Model parameter estimates for the L_{tAB} model corresponding to the ℓEM input statements in Figure 6.6.

Table 6.7 Estimates of π_1^X, $\pi_{222|1}^{ABC|X}$, and $\hat{N}$ for L_{tAB} ($\gamma = 0.05$), L_{tAB} ($\gamma = 0.10$), and M_{tAB}, Respectively, for Owners, 30–44 years and Renters, 20–29 years

	$\hat{\pi}_1^X$	$\hat{\pi}_{222\|1}^{ABC\|X}$	$\hat{N}$
	Owners		
L_{tAB}, $\gamma = 0.05$	0.987	0.15	648
L_{tAB}, $\gamma = 0.10$	0.973	0.12	614
M_{tAB}	1.000	0.18	682
	Renters		
L_{tAB}, $\gamma = 0.05$	0.966	0.26	556
L_{tAB}, $\gamma = 0.10$	0.992	0.26	548
M_{tAB}	1.000	0.27	563

heterogeneity in a capture–recapture setting. The next example provides an illustration of the use of LCA for evaluating classification error in a census postenumeration survey (PES).

6.3.5 Example: Classification Error in a PES

Biemer et al. (2001) describe two studies conducted in 1995 and 1996 to evaluate the design of the 2000 Decennial Census PES [referred to as *integrated coverage measurement* (*ICM*) survey]. The PES was, of course, concerned with the coverage error associated with the census. The Biemer et al. studies were concerned with errors in the PES that may affect its ability to evaluate the census. To evaluate the PES, a second reinterview was conducted of a sample of PES households referred to as *evaluation interviews* (EIs). Thus, three indicators of census residency status were available for the LCA—the census pretest, the PES, and the EI.

As in the previous section, define the dichotomous latent variable X, which takes the value 1 for true residents and 2 for true nonresidents. Define A as the roster indicator for the test census; that is, $A = 1$ for an individual if the individual was "rostered" or listed by the test census and $A = 2$ otherwise. For the PES, Biemer et al. defined two indicators, P and B, where P is defined analogously to A for the PES; that is, $P = 1$ if the individual is listed by the PES and 2 otherwise. Initially, P is determined by the PES enumerator without regard to the census classification (i.e., A). However, following the determination of P, the PES employed a reconciliation methodology (see Section 2.5.2) that compared A and P to produce B, a postreconciliation roster. In other words, the PES enumerator compared A and P, and when they disagreed, asked a series of unstructured questions aimed at reconciling the discrepancy and determining the true residential status. This sometimes resulted in a revised (corrected) classification, which we denote by B.

For B, each person on the combined A–P list was classified into the following four categories

$$B = \begin{cases} 1 \text{ if resident and rostered by the ICM or census} \\ 2 \text{ if nonresident and rostered by the ICM or census} \\ 3 \text{ if undeterminable and rostered by the ICM or census} \\ 4 \text{ if not rostered by ICM or census} \end{cases} \tag{6.23}$$

where $B = 3$ is assigned when the PES enumerator cannot decide whether a person should be classified as a resident. In such cases, the individual is classified as "unresolved" or undetermined status. The fourth category applies to persons who are not rostered by the test census or the PES process but rather are later rostered in the EI reinterview. There were a substantial number of these persons in test censuses, and the US Census Bureau was quite interested in determining the true status of these persons. Ostensibly, it seems highly implausible that so many persons could be missed by both the census and the PES. The Census Bureau hypothesized that the persons classified as $B = 4$ where errorneous B-list enumerations.

The EI used procedures very similar to those of the PES interview. In the EI, a third, census-day, household roster was constructed independently of the previous two rosters. The EI roster was compared to the combined census–PES roster, and any differences were reconciled with the respondent. The reconciliation process used the following three classifications:

$$C = \begin{cases} 1 \text{ if resident} \\ 2 \text{ if nonresident} \\ 3 \text{ if undeterminable} \end{cases} \tag{6.24}$$

Note that, for the four-way $APBC$ table, there is no cell corresponding to persons missed by all three systems (or all four lists) since C is assumed to capture all residents. Therefore, there is no structural zero corresponding to this cell as there was in the capture–recapture setting. However, C is still considered a fallible indicator in the analysis since it is subject to misclassification of persons listed by at least one system. As Biemer et al. explain, the focus of the analysis is on persons missed and/or misclassified by the PES (rather than all systems combined) and therefore, persons missed by the EI can be ignored.

Nevertheless, there are three structural zero cells in the analysis. Note that, by design

$$\Pr(B = 4 \mid A = a, P = p) = 0 \quad \text{for} \quad (a,p) = (1,1),(1,2), \text{ or } (2,1) \tag{6.25}$$

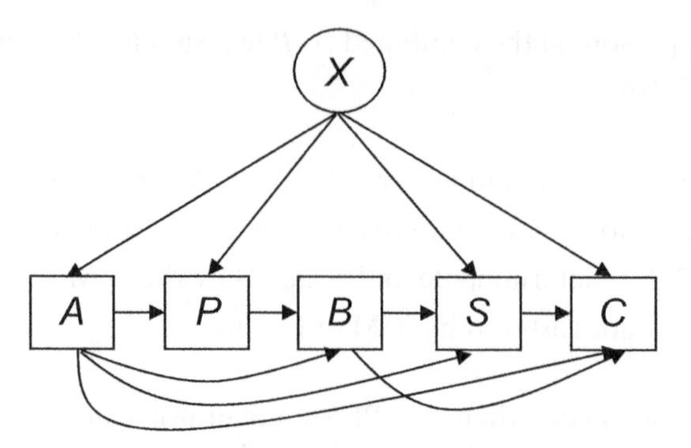

Figure 6.8 Path diagram for the PES evaluation model.

that is, persons who appear on the census roster or the PES independent roster (before reconciliation), cannot receive a "not rostered" classification in the reconciled PES (B). Thus, these constraints were added to the specifications of the models and 3 degrees of freedom were deducted from the model degrees of freedom in accordance with rules described in Section 4.4.5.

One aspect of this analysis that is quite unique for this book is that, for the first time, we consider indicators of X whose dimensionality differs from X, namely, B with four categories and C with three. Still, the dimensionality of X is well defined because eventually all persons in the census must be classified as either residents or nonresidents. In this formulation, the LC model will estimate the proportion of persons with response patterns that involve $B = 3$, $B = 4$, or $C = 3$ that are true residents or nonresidents.

To illustrate, we reanalyzed the 1996 data using a model slightly different from that of Biemer et al. (2001), whose model may have been overfitted. In this example, we used the procedures for modeling local dependence described in Section 5.2. This resulted in the model

$$\{XA\}\{XP\ AP\}\{AB\ \ PB\}\{XS\ SA\ SB\}\{SC\ XC\ AC\ BC\} \tag{6.26}$$

where S is a four-level stratification variable used in selecting the subsample for the EI. It is based on the degree of match between the census and the PES rosters. Since it depends on A and P, it is treated as an endogenous grouping variable in the analysis following A and P temporally in the modified path model. The corresponding path model is shown in Figure 6.8, and the ℓEM input statements for this model are shown in Figure 6.9.

This model, with 137 degrees of freedom, has a p value of 0.08 and a BIC of −1101, which is smaller than the best model used by Biemer et al. (2001). Note that the submodels for PES indicators (both B and P) include the interaction effects with the census. As the authors point out, the presence of this term in the model violates one of the key assumptions of the dual-system estimator (DSE) in use by the US Census Bureau—that the classifications error in the census and the PES are uncorrelated.

```
* 1996 Census Test Analysis using Data from Biemer et al (2001)
  man 5
  lat 1
  dim  2 4  2 4 2 3
  lab  X S  A B P C
  mod X
      A|X
      P|XA {XP PA}
      B|XPA {wei(PAB) PB AB}
      S|XAB {SX SA SB}
      C|XSPAB {SC XC AC BC}
  rec 384
  rco
  sta A|X [.75 .25  .3  .7]
  sta P|X [.9  .1 .5 .5]
  sta C|X [.95 .01 .04 .75 .10 .15]
  sta wei(PAB)  [1 1 1 1 1 1 0 0
                 1 1 1 1 1 1 0 1]
  dat census96.dat
```

Figure 6.9 ℓEM input statements for fitting the model in (6.26) and Figure 6.8.

Table 6.8 Comparison of PES Error Rate Estimates Using Three Methods

Parameter	$\hat{\pi}_{1\mid1}^{B\mid X}$	$\hat{\pi}_{2\mid1}^{B\mid X}$	$\hat{\pi}_{3\mid1}^{B\mid X}$	$\hat{\pi}_{4\mid1}^{B\mid X}$	$\hat{\pi}_{1\mid2}^{B\mid X}$	$\hat{\pi}_{2\mid2}^{B\mid X}$	$\hat{\pi}_{3\mid2}^{B\mid X}$	$\hat{\pi}_{4\mid2}^{B\mid X}$
New LC model	98.1	1.5	0.5	0.0	29.8	28.8	3.6	37.9
Biemer et al. (2001)	95.1	3.7	0.5	0.0	41.0	22.1	3.1	33.7
Traditional	79.1	4.0	1.3	15.1	58.5	34.8	1.5	5.2

Estimated error probabilities for the reconciled PES (indicator B) are shown in Table 6.8. The first row of the table shows the estimates under (6.26), the second from Biemer et al. (2001), and the third using the EI as a gold standard and the methods of Section 2.5 [see Raglin et al. (1998) for details]. The probability of correctly classifying a true resident is higher under both LC models than it is for the traditional analysis; it is in the range 95–98% for the LC models and less than 80% for traditional gold standard analysis. However, traditional analysis suggests the PES is better at correctly classifying nonresidents than the LC models suggest. The largest difference appears in the last column. This is the probability that nonresidents are persons discovered by the EI ($B = 4$). The LC model estimates a much higher probability for this than does traditional analysis.

It is interesting to estimate the probability that the new residents found in the EI that were somehow missed by the census and the PES are true residents. This probability is $\pi_{1\mid41}^{X\mid BC}$ and can be estimated from the XBC table available as optional output from ℓEM. According to the LC models, none of these newly found persons are true residents. This finding is consistent with the Census Bureau's initial hypothesis.

CHAPTER 7

Latent Class Models for Panel Data

Previous chapters have focused on the estimation of the LC model parameters for cross-sectional surveys. In that situation, the estimation of misclassification probabilities is not possible without replicate measurements of the same latent variable X. The replicate measurements could be either embedded within the same questionnaire or obtained from reinterviewing the same respondents at two or more timepoints. Still, the requirement of multiple indicators of X is rather restrictive because obtaining even two replicate measurements for each X of interest adds to the length of the interview and creates additional response burden. Reinterview surveys are costly as well. In addition, the assumption that each measurement is an indicator of the same latent variable can be problematic, particularly in reinterview surveys when characteristics may change since the original interview.

Panel surveys provide opportunities to model survey misclassification that are not available for cross-sectional surveys. In panel surveys, remeasurements are generated automatically across time in the normal wave-by-wave interviewing. However, the assumption that these remeasurements are indicators of the same latent variable must be relaxed for many characteristics of interest such as health, economic, and psychological variables that change with each panel wave. Since the primary purpose of longitudinal measurement is to study population dynamics and change, models describing interrelationships among the true characteristics and their indicators need to accommodate real changes in X at each panel wave.

This chapter considers a number of models for estimating classification errors in panel surveys. Unlike those previous chapters, the LC models considered in this chapter do not require multiple indicators referencing a single point in time. Typically in panel surveys, a single measurement of X is taken

Latent Class Analysis of Survey Error By Paul P. Biemer

at multiple timepoints called *panel waves*. Identifiable LC models can be fit to any panel survey with at least three panel waves. Panel surveys with only two waves may still require multiple (within-panel) indicators unless fairly strong parameter restrictions are imposed—restrictions that are often implausible for many practical applications. As the number of waves increases, more general and plausible models can be estimated. An important concern in LCA for panel surveys is data sparseness. For this reason, having more waves is not always better for modeling survey error.

Panel survey LC models attempt to account for the changes in X from wave-to-wave as well as the measurement error in the observations of X at each wave. The structural component of the LC model describes the way X changes over time, and the measurement component describes the error in the observations of X. As in previous chapters, both structural and measurement components are modeled simultaneously.

The next section considers a class of models that can be fit to dynamic measurements collected at three or more timepoints, referred to as *Markov latent class* (MLC) *models*. As the name suggests, these models assume the *Markov property* for changes in X described as follows. Let X_t denote the value of X at wave t, for $t = 1,\ldots,T$. The Markov property states that

$$\Pr(X_t = x_t \mid X_1 = x_1, \ldots, X_{t-1} = x_{t-1}) = \Pr(X_t = x_t \mid X_{t-1} = x_{t-1}) \qquad (7.1)$$

for $t \geq 2$. In words, it assumes the value of X at wave t only depends on the value of X at wave t-1 for $t = 2,\ldots,T$. Whether this is a plausible assumption depends upon the application. The assumption can be relaxed when four waves are available. Section 7.2 discusses some special cases and extensions of the MLC model. Section 7.3 considers some issues in the estimation MLC models including panel attrition and other types of nonresponse.

7.1 MARKOV LATENT CLASS MODELS

Markov latent class models were first proposed by Wiggins (1973) and refined by Poulsen (1982). Van de Pol and de Leeuw (1986) established conditions under which the standard MLC model is identifiable and gave other conditions of estimability of the model parameters. To begin this study of MLC models, we first consider Markov models without latent variables. The models, referred to as *manifest Markov* (MM) models, assume that the measurements of X are error-free (Goodman 1973a; Bartholomew 1982; Langeheine 1988). As we shall see, MM models represent the *structural component* of the MLC models. In fact, MLC models may be described as MM models that assume the X values are measured with error. The MM modeling framework is then extended to MLC models by adding a *measurement component* to the general MM model describing how the observations are related to the latent true values (i.e., the X values).

7.1.1 Manifest Markov Models

Suppose that some categorical variable of interest, say, X, is measured without error at each wave. For an arbitrary unit in the population, let X_t denote some characteristic of the unit at wave t and let x_t denote a specific value of X_t for $t = 1,\ldots,T$. In the literature of stochastic processes, the unit is said to *occupy state* x_t at wave t. Let K denote the number of *states* (or classes, in LCA parlance) that a unit can occupy at t that is the same for all t. Changes from state x_t to state x_{t+1} are called *transitions* and the probabilities $\Pr(X_{t+1} = x_{t+1}|X_t = x_t)$ are referred to as *transition probabilities* denoted by $\pi_{x_{t+1}|x_t}^{X_{t+1}|X_t}$. For example, $\pi_{1|1}^{X_{t+1}|X_t}$ is the probability that a unit in state 1 at time t remains in this state at $t + 1$; $\pi_{2|1}^{X_{t+1}|X_t}$ is the probability that a unit moves from state 1 at time t to state 2 at $t + 1$. If the transition probabilities satisfy the Markov property given in (7.1), then the sequence of random variables $X_1,\ldots,X_T$ is called a *Markov chain*, named after Andrey Markov, an early twentieth-century mathematician.

As an example, suppose X_t denotes an individual's employment status at wave t where $X_t = 1,2,3$ denotes employed (EMP), unemployed (UNE), and not in the labor force (NLF), respectively. Thus, an individual can occupy only one of $K = 3$ states at each wave. Suppose that three consecutive waves of the panel survey are analyzed (i.e., $T = 3$). Let $n_{x_1x_2x_3}$ denote the number of persons in the sample who occupied state x_1 at wave 1, x_2 at wave 2, and x_3 at wave 3. The data can be organized in a three-way table as in Table 7.1.

Let $\pi_{x_1x_2x_3}^{X_1X_2X_3}$ denote the probability that an observation falls in cell (x_1, x_2, x_3). The most general form of the model is

$$\pi_{x_1x_2x_3}^{X_1X_2X_3} = \pi_{x_1}^{X_1}\pi_{x_2|x_1}^{X_2|X_1}\pi_{x_3|x_1x_2}^{X_3|X_1X_2} \tag{7.2}$$

where (x_1, x_2, x_3) is an arbitrary cell in the $X_1X_2X_3$ table. Various conditional probabilities may be of interest. For example, it is often of interest to estimate the probability that an unemployed person at wave t finds employment at wave $t + 1$ ($\pi_{1|2}^{X_{t+1}|X_t}$) or conversely, that an employed person at wave t becomes unemployed at the wave $t + 1$ ($\pi_{2|1}^{X_{t+1}|X_t}$). The estimate of these conditional probabilities depends on what model is assumed for the relationships among the three waves.

Table 7.1 Cross-Classification for Three Waves

		Wave 3								
		EMP			UNE			NLF		
	Wave 2	EMP	UNE	NLF	EMP	UNE	NLF	EMP	UNE	NLF
Wave 1	EMP	n_{111}	n_{121}	n_{131}	n_{112}	n_{122}	n_{132}	n_{113}	n_{123}	n_{123}
	UNE	n_{211}	n_{221}	n_{231}	n_{212}	n_{222}	n_{232}	n_{213}	n_{223}	n_{233}
	NLF	n_{311}	n_{321}	n_{331}	n_{312}	n_{322}	n_{332}	n_{313}	n_{323}	n_{333}

The model in (7.2) is unrestricted and fully saturated having 26 (+1) parameters and 27 cells. Also, similar to multiple indicators in cross-sectional surveys, the temporal ordering of the measurements (wave 1, followed by wave 2 and followed by wave 3) provides a natural structure for specifying causal dependence among variables. For example, the form of (7.2) implies that wave 1 states do not depend upon the states occupied at waves 2 and 3. Further, the wave 2 state may depend on wave 1 but not wave 3, and wave 3 may depend on both prior waves' states. To obtain a more parsimonious model, various equality restrictions can be imposed. As an example, the last conditional probability in (7.2) represents 18 parameters. If the Markov[1] assumption is invoked, then $\pi_{x_3|x_1x_2}^{X_3|X_1X_2} = \pi_{x_3|x_2}^{X_3|X_2}$ for $x_1 = 1,2,3$, which eliminates 12 parameters, saving those degrees of freedom. Now, the model is

$$\pi_{x_1x_2x_3}^{X_1X_2X_3} = \pi_{x_1}^{X_1}\pi_{x_2|x_1}^{X_2|X_1}\pi_{x_3|x_2}^{X_3|X_2} \tag{7.3}$$

This model is referred to as the *standard MM model for three waves.*

In the present context, the Markov assumption states that the employment status of an individual at wave 3 depends only on their employment status at wave 2. Mathematically, this can be expressed as

$$\Pr(X_3 = x_3 \mid X_1 = x_1, X_2 = x_2) = \Pr(X_3 = x_3 \mid X_2 = x_2) \tag{7.4}$$

for any values of x_1, x_2, and x_3. For example, for a person who is unemployed at wave 2, the probability that this person is still unemployed at wave 3 is the same no matter what her or his labor force status was at wave 1. This assumption seems fairly restrictive and easily violated for panel surveys of frequent periodicity (such as a monthly survey). For example, a person who was unemployed in both months 1 and 2 would seem to have a different (perhaps higher) probability of being unemployed in month 3 than a person who was employed in month 1 and unemployed in month 2. Likewise, a person who is employed in both months 1 and 2 would seem to have a higher probability of being employed in month 3 than would a person who was unemployed in month 1 and employed in month 2. Despite its apparent implausibility, we show in Section 7.1.7 that the MLC model performs quite well for the US Current Population Survey (CPS), a monthly labor force panel survey.

Further reduction in the number of parameters can be achieved if the transition probabilities can be assumed to be *stationary* or *time-homogeneous*, that is, if the transition probabilities are the same for any two consecutive waves. This assumption would be tenable for stable processes. The Markov chain is stationary if

$$\pi_{x|x'}^{X_{t+1}|X_t} = \pi_{x|x'}^{X_{t+2}|X_{t+1}} \tag{7.5}$$

[1]Later, we shall define and discuss the second-order Markov assumption as well. When the order is not specified, the first-order Markov assumption is implied by default.

for all waves t and states x and x', which eliminates six parameters, that is, if the transition probabilities from wave 1 to wave 2 are equal to the corresponding transition probabilities from wave 2 to wave 3. Combined with the Markov assumption, the model can be written in terms of only 8 (+1) parameters, where the (+1) refers to the intercept (a constant): two parameters corresponding to $\pi_{x_1}^{X_1}$, $x_1 = 1,2$, three parameters corresponding to $\pi_{1|x_1}^{X_2|X_1}$ for $x_1 = 1,2,3$, and three parameters corresponding to $\pi_{2|x_1}^{X_2|X_1}$ for $x_1 = 1,2,3$.

For estimating the parameters of the MM model, the probabilistic parameterization in (7.2) can be written equivalently as a loglinear model for the expected frequencies, $m_{x_1x_2x_3}$ as follows

$$\log(m_{x_1x_2x_3}) = u + u_{x_1}^{X_1} + u_{x_2}^{X_2} + u_{x_3}^{X_3} + u_{x_1x_2}^{X_1X_2} + u_{x_2x_3}^{X_2X_3} + u_{x_1x_3}^{X_1X_3} + u_{x_1x_2x_3}^{X_1X_2X_3} \quad (7.6)$$

with the usual ANOVA-like restrictions on the u parameters:

$$\sum_{x_t} u_{x_t}^{X_t} = \sum_{x_t} u_{x_tx_{t'}}^{X_tX_{t'}} = \sum_{x_t} u_{x_1x_2x_3}^{X_1X_2X_3} = 0, \quad \text{for} \quad t = 1,2,3 \quad \text{and} \quad t \neq t' \quad (7.7)$$

This hierarchical model can be represented as $\{X_1X_2X_3\}$ in the standard shorthand notation and is a fully saturated model. Likewise, the unsaturated MM model in (7.3) can be represented simply by omitting the three-way interaction

$$\log(m_{x_1x_2x_3}) = u + u_{x_1}^{X_1} + u_{x_2}^{X_2} + u_{x_3}^{X_3} + u_{x_1x_2}^{X_1X_2} + u_{x_2x_3}^{X_2X_3} \quad (7.8)$$

or simply $\{X_1X_2 \quad X_2X_3 \quad X_1X_3\}$ with similar zero-sum restrictions. The stationarity assumption is imposed by setting

$$u_x^{X_2} = u_x^{X_3} \quad \text{and} \quad u_{x_1x_2}^{X_1X_2} = u_{x_2x_3}^{X_2X_3} \quad \text{for} \quad (x_1, x_2) = (x_2, x_3) \quad (7.9)$$

for all x, x_1, x_2, x_3. Note that simply equating the X_1X_2 and X_2X_3 interactions terms is not sufficient to impose the stationarity assumption. The restrictions in (7.9) saves 6 degrees of freedom. To estimate the relevant model probabilities, we can use these relationships

$$\begin{aligned}
\pi_{x_1}^{X_1} &= \frac{\exp(u_{x_1}^{X_1})}{\sum_{x_1} \exp(u_{x_1}^{X_1})} \\
\pi_{x_2|x_1}^{X_2|X_1} &= \frac{\exp(u_{x_2}^{X_2} + u_{x_1x_2}^{X_1X_2})}{\sum_{x_2} \exp(u_{x_2}^{X_2} + u_{x_1x_2}^{X_1X_2})} \\
\pi_{x_3|x_2}^{X_3|X_2} &= \frac{\exp(u_{x_3}^{X_3} + u_{x_2x_3}^{X_2X_3})}{\sum_{x_3} \exp(u_{x_3}^{X_3} + u_{x_2x_3}^{X_2X_3})}
\end{aligned} \quad (7.10)$$

replacing the u parameters with their corresponding MLEs.

7.1.2 Example: Application of the MM Model to Labor Force Data

To illustrate the model, consider the data in Table 7.2, which are unweighted frequencies from three consecutive months of the CPS. A more appropriate analysis would acknowledge the complex sample design of the CPS and incorporate survey weights. However, unweighted data will be used for the purposes of this illustration. Only persons who responded in all three waves (months) of the CPS are included in the table—a total of 41,751 respondents. Alternative methods for dealing with wave attrition will be considered later in the chapter.

We use ℓEM to estimate the initial probability vector $\boldsymbol{\pi}^{X_1} = [\pi_1^{X_1}, \pi_2^{X_1}, \pi_3^{X_1}]'$, that is, the prevalence of the three states in month 1, and the transition matrices $\boldsymbol{\pi}^{X_2|X_1} = [\pi_{j|i}^{X_2|X_1}]$ with rows representing month 1 states ($i = 1,2,3$) and the columns the month 2 states ($j = 1,2,3$). The transition matrix for month 2–month 3, denoted $\boldsymbol{\pi}^{X_3|X_2}$ is defined analogously. Using ℓEM (see the input statements in Figure 7.1), we fit the standard MM model to these data with 12 degrees of freedom. The chi-square likelihood ratio is $L^2 = 2178.4$, which is highly significant, indicating model lack of fit. But given the extremely large sample size, the fit of the model may still be adequate since the power of this test is nearly 1. In fact, the dissimilarity index is small (0.03), suggesting that the model fits well. The parameter estimates are

$$\hat{\boldsymbol{\pi}}^{X_1} = [0.604 \quad 0.038 \quad 0.359]'$$
$$\hat{\boldsymbol{\pi}}^{X_2|X_1} = \begin{bmatrix} 0.961 & 0.013 & 0.026 \\ 0.292 & 0.473 & 0.235 \\ 0.046 & 0.025 & 0.929 \end{bmatrix} \tag{7.11}$$
$$\hat{\boldsymbol{\pi}}^{X_3|X_2} = \begin{bmatrix} 0.967 & 0.012 & 0.021 \\ 0.254 & 0.514 & 0.232 \\ 0.039 & 0.026 & 0.935 \end{bmatrix}$$

Thus, in month 1, 60.4% of the sample is EMP, 3.8% is UNE, and 35.9% is NLF. The transition probability matrices for months 1–2 and 2–3 are quite

Table 7.2 Cross-Classification of 3 Months of 1996 CPS Data

		Month 3								
		EMP			UNE			NLF		
	Month 2	EMP	UNE	NLF	EMP	UNE	NLF	EMP	UNE	NLF
Month 1	EMP	23,661	140	211	238	136	26	326	49	413
	UNE	379	150	55	45	472	107	38	125	209
	NLF	505	78	316	20	135	252	164	162	13,339

```
* X1 is the labor force status for Month 1 with
* X1 = 1 for Employed
* X1 = 2 for Unemployed
* X1 = 3 for NLF
* X2 and X3 are defined similarly for Months 2 and 3
man 3
dim  3  3  3
lab  X1 X2 X3
mod X1
    X2|X1
    X3|X2
rec 27
rco
data table7-2.dat
cri 0.00000001
ite 10000
```

Figure 7.1 ℓEM input statements for the Markov manifest model.

similar. Considering the month 1–2 transitions ($\hat{\pi}^{X_2|X_1}$), the probability that an EMP person stays EMP is 0.961, moves to UNE is 0.013, and moves to NLF is 0.026. The probability of staying UNE from month to month is about 50% and the probability of staying NLF is about 93%.

Since the two transition matrices are very similar, we can explore the fit of the stationary MM model. This model was fit using the code in Figure 7.1 by substituting the instruction `X3|X2 eql X2|X1` for the `X3|X2` submodel in Figure 7.3. In doing so, the estimate of π^{X_1} is unchanged while the estimate of the lone transition matrix is

$$\hat{\pi}^{X_2|X_1} = \hat{\pi}^{X_3|X_2} = \begin{bmatrix} 0.964 & 0.012 & 0.023 \\ 0.274 & 0.492 & 0.233 \\ 0.043 & 0.025 & 0.932 \end{bmatrix} \tag{7.12}$$

Although the differences in the estimates from the standard MM model are small, the stationary MM model is rejected by the nested chi-square test. Recall that the nonstationary MM model L^2 was 2178 while the MM model with stationary transitions imposed has an L^2 of 2209. The difference in L^2 values of 31 with 6 degrees of freedom is highly statistically significant, rejecting the restricted model. The BIC for the stationary model is smaller, however: 108,350 compared with 108,383 for the nonstationary model. Therefore, under the BIC criterion, the stationary MM model is preferred.

The analysis of transitions from one labor force state to another across time is an important component of econometric analysis referred to as *gross-flow analysis* [see, e.g., Barkume and Horvath (1995)]. Labor analysts are concerned with changes and shifts in the population from employment to unemployment and vice versa. Growth in unemployment can signal a recession, while growth in employment can signal the end of a recessionary period. It is well known that gross-flow estimates are biased, sometimes severely so, when a large number of respondents are misclassified as to their labor force status by the survey [see, e.g., Abowd and Zellner (1985) and Biemer and Bushery (2001)].

Suppose that instead of observing X_t, the true labor force category, we observe A_t, the survey classification subject to measurement error. The MLC model can be viewed as an MM model where X_t is treated as a latent variable representing the true "state" (now referred to as the *latent class*) that is measured with error by the observed state, A_t. The next section considers this situation in some detail.

7.1.3 Markov Latent Class Models

The MLC model was originally proposed by Wiggins (1973), although Poulsen (1982), Van de Pol and de Leeuw (1986), and Van de Pol and Langeheine (1990) made a number of important contributions that facilitated practical application of the methodology. Markov latent class analysis (MLCA) is a set of procedures and methods based on the MLC model that can be used for separating real changes from spurious changes in gross-flow analysis and other longitudinal survey applications. More recently [see, e.g., Biemer and Bushery (2001)], it has been exploited as a method for analyzing the measurement error in categorical panel survey measurements.

Markov LCA assumes that observations on X, the categorical variable of interest, are subject to classification errors. Like LCA, MLCA treats X as a latent variable in the modeling. The classification error probabilities affecting the measurements of X at each wave can be estimated if two or more replicate measurements (denoted by A, B, C, etc.) are available at one or more waves. Like the LC model, the MLC model contains two components: (1) the structural component that describes the interdependencies between the X_t, $t = 1,\ldots,T$ and the model covariates (grouping variables) and (2) the error component describing the interdependences among the observations $A_t, B_t, C_t, \ldots$, at each wave $t = 1,\ldots,T$ as well as their interactions with X_t and other model covariates. However, as we shall see, MLC models are much more flexible than LC models because estimation of error probabilities does not require (within-wave) replicate measurements for identifiability. The basic assumptions of the MLC model include Markov transitions, independent between-wave classification errors, and time-homogeneous error probabilities. With a minimum of three panel waves, MLCA can provide estimates of the wave 1 prevalence probabilities, waves t to $t + 1$ transition probabilities, and the classification error probabilities associated with A_t, the single measurement of X_t.

An important related analysis technique in the fields of public health, psychology, and sociology is *latent transition analysis* (LTA) [see, e.g., Collins and Lanza (2010)]. Like MLCA, LTA is also a longitudinal extension of LCA; however, the focus of LTA is somewhat different from that of MLCA. In LTA, the number of states underlying a set of panel measurements is unknown and must be determined as part of the analysis. For example, states may be stages in the growth or development of an individual that are indicated by responses to questions designed to uncover the dynamic structure of some underlying, immeasurable processes. In MLCA, the number of states is known and, except in rare cases, correspond to the number of categories in the indicator variables. MLCA latent states are measurable quantities rather than theoretical constructs. Thus, LTA may be considered "exploratory" analysis while MLCA is "confirmatory." In addition, LTA focuses on the estimation of the transitions between the hypothesized states and the internal and external factors affecting these transitions. This can also be a focus of MLCA. However, most of the applications of MLCA in this chapter focus on the error probabilities associated with the indictors of the latent variable. This focus is motivated by the desire of the survey methodologist to study the sources and root causes of survey error. These analyses can lead to discoveries about the measurement process useful for error prevention, reduction and control.

A typical LTA application might consider the various stages of drug use and addiction such as: no use of drugs or alcohol, alcohol use, use of marijuana and "softer" drugs, experimentation with harder drugs, and finally, drug addiction. The exact number of stages and their precise definitions are unknown. Rather, determining the number of stages, how each stage should be interpreted, and the correlates of an individual's transition from stage to stage are key objectives of LTA. Models postulating different numbers of stages (or latent classes) are specified and compared as a means of developing and testing theories about this *dynamic stage-sequential process*. The latent stages are purely theoretical and are usually immeasurable by direct observations. Instead, a series of questions are designed to estimate the number of stages as well as the probability an individual resides in any particular stage and at various points in time.

By contrast, in our applications, the latent variable is measurable albeit with classification error. In our applications, as in previous chapters, the number of latent classes (K) will be known or prespecified by an accepted theory of the measurement process. The researcher designs one or more survey questions to measure X at each wave, t. These measurements are called *indicators* of X_t and will be denoted by A_t, B_t, C_t, and so on. In most cases, the indicators will also have K categories, although this is not required for MLCA. In our initial treatment of MLCA, only one indicator of X_t, denoted by A_t, will be available at each wave. Section 7.2.7 considers the case where multiple indicators of X_t are available at each wave.

The discussion of MLCA in this chapter is somewhat specialized, and the models we consider are special cases of the general Markov latent class model. For example, MLC models can be applied in situations where K is unknown as in LTA and when the number of categories of the manifest variables A_t, B_t, C_t, ... is different for each variable. For our purposes, this more general model is not needed and will not be considered further in this book.

The standard MLC model assumes that some characteristic of interest, say, X, is measured for $T \geq 3$ panel waves. Initially, only the case of three waves will be considered in detail. Later, straightforward extensions to four or more waves will also be considered. The assumptions for the standard MLC model are as follows:

1. *Markov Property.* Assume that

$$\pi_{x_3|x_1x_2}^{X_3|X_1X_2} = \pi_{x_3|x_2}^{X_3|X_2}$$

that is, that a unit's latent state at wave 3 (X_3), given its state at wave 2 (X_2), is independent of its state at wave 1 (X_1)

2. *Independent Classification Error (ICE).* This assumption may be regarded as an extension of the assumption of local independence to longitudinal data. ICE assumes that

$$\pi_{a_1a_2a_3|x_1x_2x_3}^{A_1A_2A_3|X_1X_2X_3} = \pi_{a_1|x_1}^{A_1|X_1}\pi_{a_2|x_2}^{A_2|X_2}\pi_{a_3|x_3}^{A_3|X_3}$$

This essentially means that classification errors for the three indicators are mutually independent across waves (Singh and Rao 1995).

3. *Time-Homogeneous Classification Error.* Classification errors for the indicator A_t are assumed to be the same for all waves $t = 1,2,3$:

$$\pi_{a_t|x_t}^{A_t|X_t} = \pi_{a|x}^{A|X} \quad \text{for} \quad a = a_t, x = x_t, t = 1,2,3$$

4. *Homogeneous Error Probabilities.* This is equivalent to the homogeneity assumption made for LC models. Assume that all individuals within the same latent class have the same probability of being misclassified; that is, $\pi_{a_t|x_t}^{A_t|X_t}$ for $a_t = 1,\ldots,T$ is the same for all units in class $X_t = x_t$. As for LC models, this assumption is unlikely to be satisfied for most MLC models, and the addition of grouping variables may be required to model the heterogeneity.

Let $A_1A_2A_3$ denote the $K \times K \times K$ cross-classification table for the observations for n persons interviewed at waves 1, 2, and 3. Wave nonresponse is ignored at this point but will be considered in Section 7.3.2. Thus, the joint probability that an individual is classified in cell (a_1,a_2,a_3) in the table is

$$\pi_{a_1a_2a_3}^{A_1A_2A_3} = \sum_{x_1}\sum_{x_2}\sum_{x_3}(\pi_{x_1x_2x_3}^{X_1X_2X_3})(\pi_{a_1a_2a_3|x_1x_2x_3}^{A_1A_2A_3|X_1X_2X_3}) \tag{7.13}$$

Applying the Markov property, the structural component of the model may be rewritten as

$$\pi_{x_1x_2x_3}^{X_1X_2X_3} = \pi_{x_1}^{X_1}\pi_{x_2|x_1}^{X_2|X_1}\pi_{x_3|x_2}^{X_3|X_2} \tag{7.14}$$

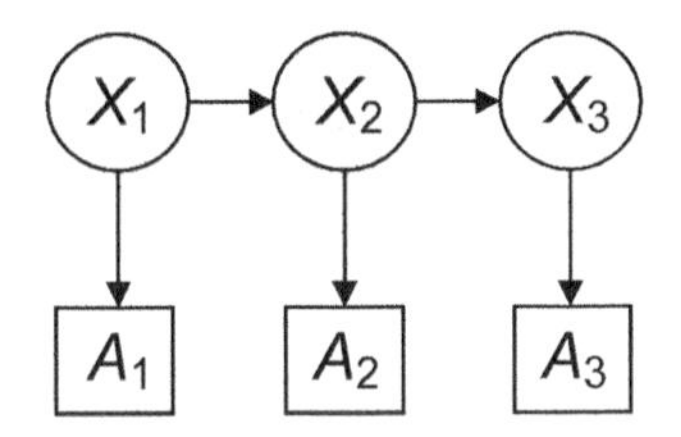

Figure 7.2 Path diagram for the markov latent class model.

Applying the ICE and homogeneous error probability assumptions, the measurement component can be rewritten as

$$\pi_{a_1a_2a_3|x_1x_2x_3}^{A_1A_2A_3|X_1X_2X_3} = \pi_{a_1|x_1}^{A_1|X_1}\pi_{a_2|x_2}^{A_2|X_2}\pi_{a_3|x_3}^{A_3|X_3} \tag{7.15}$$

which can be further simplified under assumption 3 to

$$\pi_{a_1a_2a_3|x_1x_2x_3}^{A_1A_2A_3|X_1X_2X_3} = \pi_{a_1|x_1}^{A|X}\pi_{a_2|x_2}^{A|X}\pi_{a_3|x_3}^{A|X} \tag{7.16}$$

Note that the same superscript ($A|X$) is used for each wave's classification probabilities to represent time homogeneity. This leads to the following expression for the likelihood kernel of the standard MLC model

$$\pi_{a_1a_2a_3}^{A_1A_2A_3} = \sum_{x_1}\sum_{x_2}\sum_{x_3}(\pi_{x_1}^{X_1}\pi_{x_2|x_1}^{X_2|X_1}\pi_{x_3|x_2}^{X_3|X_2})(\pi_{a_1|x_1}^{A|X}\pi_{a_2|x_2}^{A|X}\pi_{a_3|x_3}^{A|X}) \tag{7.17}$$

where x_1, x_2, x_3, a_1, a_2 and a_3 assume the values 1, 2,..., K. The path diagram for this model is shown in Figure 7.2. Time homogeneity is not represented in the diagram.

For T waves, (7.17) is easily extended by simply adding multiplicative terms and writing

$$\pi_{a_1a_2\cdots a_T}^{A_1A_2\cdots A_T} = \sum_{x_1}\sum_{x_2}\cdots\sum_{x_T}\pi_{x_1x_2\cdots x_T}^{X_1X_2\cdots X_T}\pi_{a_1a_2\cdots a_T|x_1x_2\cdots x_T}^{A_1A_2\cdots A_T|X_1X_2\cdots X_T} \tag{7.18}$$

Extending the assumptions of the standard MLC model, the structural component is

$$\pi_{x_1x_2\cdots x_T}^{X_1X_2\cdots X_T} = \pi_{x_1}^{X_1}\pi_{x_2|x_1}^{X_2|X_1}\pi_{x_3|x_2}^{X_3|X_2}\cdots\pi_{x_t|x_{t-1}}^{X_T|X_{T-1}} \tag{7.19}$$

and the measurement component is

$$\pi_{a_1a_2\cdots a_T|x_1x_2\cdots x_T}^{A_1A_2\cdots A_T|X_1X_2\cdots X_T} = \pi_{a_1|x_1}^{A|X}\pi_{a_2|x_2}^{A|X}\cdots\pi_{a_T|x_T}^{A|X} \tag{7.20}$$

The combination of the structural and measurement model components is referred to as the *first-order MLC model.* When $T > 3$, assumptions 1 and 3 above can be relaxed in a number of ways. For example, with four waves, the first-order Markov property can be replaced by the second-order Markov property, which assumes

$$\pi_{x_1x_2x_3x_4}^{X_1X_2X_3X_4} = \pi_{x_1}^{X_1}\pi_{x_2|x_1}^{X_2|X_1}\pi_{x_3|x_1x_2}^{X_3|X_1X_2}\pi_{x_t|x_2x_3}^{X_4|X_2X_3} \tag{7.21}$$

Now, the current wave's latent state is allowed to depend on latent states of the previous two waves. In general, for any $T > 2$, the Markov property may be relaxed to order $T - 2$ if desired and the model will still be identifiable.

In addition, with $T > 3$, one can relax assumption 3 as an alternative to relaxing the first order Markov assumption. Van de Pol and Langeheine (1990) show that the error probabilities for the endpoints of the time series (i.e., A_1 and A_T) are not separately estimable in the standard MLC model. The error probabilities become identifiable only if they are equated to the corresponding error probabilities for the adjacent occasions. However, for the intermediate timepoints, the error probabilities can be separately estimated. For example, with four waves, two sets of error probabilities can be estimated after setting $\pi_{a_1|x_1}^{A_1|X_1} = \pi_{a_2|x_2}^{A_2|X_2}$ and $\pi_{a_3|x_3}^{A_3|X_3} = \pi_{a_4|x_4}^{A_4|X_4}$, where $\pi_{a_1|x_1}^{A_1|X_1} \neq \pi_{a_3|x_3}^{A_3|X_3}$. With five waves, three sets of probabilities can be estimated: one set for waves 1 and 2, another for wave 3, and a third set for waves 4 and 5. With six waves, four sets of probabilities can be estimated and so on. In addition, with four or more waves, no restrictions on the error probabilities are required if wave-to-wave transitions are restricted to be stationary. Additional options are avaialable when there is more than one indicator at each timepoint (see Section 7.2.7) or when grouping variables are added to the models as discussed in Section 7.1.6.

7.1.4 Example: Application of the MLC Model to Labor Force Data

For this illustration, the data in Table 7.2 was reanalyzed under the assumption that X_1, X_2, and X_3 are subject to misclassification. The classification error probabilities for the observations A_1, A_2, and A_3 were estimated under the stationary MLC model using the ℓEM (input statements in Figure 7.3). The error probability estimates are shown in Table 7.3. The first column of this table provides the true classification according to the model (i.e., X_1). Reading across the columns, we find the estimates of error probabilities, $\pi_{a_1|x_1}^{A_1|X_1}$. The smallest *accuracy rate* (defined as the estimate of the probability of a correct classification or $\hat{\pi}_{a_1|x_1}^{A_1|X_1}, a_1 = x_1$) is obtained for unemployed persons where only 74.6% are classified correctly. Of those misclassified, about two-thirds are classified as NLF and about one third as EMP. By contrast, true EMP and NLF persons are seldom misclassified. These findings are similar to those in Section 3.4.2 for the Hui–Walter analysis of the CPS data.

Finally, the transition probability estimates, corrected for classification error, are

```
* Definition of latent variables is the same as Figure 7.1
* A1, A2 and A3 are indicators of X1, X2 and X3, respectively.
* Categories of indicators corresponds to their respective latent variables
lat 3
man 3
dim  3  3  3  3  3  3
lab  X1 X2 X3 A1 A2 A3
mod X1
    X2|X1
    X3|X2 eq1 X2|X1     * Stationary transition probabilities
    A1|X1
    A2|X2 eq1 A1|X1     * Time homogeneous error probabilities
    A3|X3 eq1 A1|X1     * Time homogeneous error probabilities
rec 27
rco
sta A1|X1 [.9 .05 .05 .1 .8 .1 .05 .05 .9] *Added for flippage control
data table7-2.dat
cri 0.00000001
ite 10000
```

Figure 7.3 ℓEM input statements for the standard MLC with stationary transition probabilities.

Table 7.3 Estimated Labor Force Classification Probabilities for the Stationary MLC Model

	Observed Classification		
True Classification	EMP	UNE	NLF
EMP	0.988	0.004	0.008
UNE	0.084	0.752	0.168
NLF	0.012	0.009	0.980

$$\hat{\pi}^{X_2|X_1} = \hat{\pi}^{X_3|X_2} = \begin{bmatrix} 0.986 & 0.007 & 0.008 \\ 0.188 & 0.762 & 0.050 \\ 0.014 & 0.008 & 0.978 \end{bmatrix} \tag{7.22}$$

Comparison with the corresponding estimates from the stationary MM model (7.12) reveals that a much larger proportion of the unemployed remain unemployed from month to month: about 76% for the MLC model compared with 49.2% for the corresponding MM model. All transition probabilities are also somewhat greater for the MM model. These results illustrate the tendency for misclassification to exaggerate wave-to-wave changes.

7.1.5 The EM Algorithm for MLC Models

The EM algorithm described in Sections 4.2.3 and 4.4.1 for estimating LC model parameters can also be employed for estimating MLC model parameters. As originally shown by Poulsen (1982), the same basic principles apply. A number of software packages are available for estimating MLC models, including ℓEM, PANMARK, Latent GOLD, and MPlus. (See Section 4.1.3 for a general description of these packages.) For three waves and K latent classes, the number of parameters of the MLC model is $(K-1)(3K+1)$. Thus, for binary data, the model contains exactly seven parameters (plus 1 for the overall mean) and is saturated. Otherwise the model degrees of freedom are positive and model fit can be tested.

Consider the application of the EM algorithm for a simple MLC model for three waves under multinomial sampling. The observed data are the cross-classification of the variable A at three waves, denoted by $A_1A_2A_3$. This table may be referred to as the *incomplete* data table. Now let X_t denote the observable true value of A_t at wave t. If X_t were known for each individual at each wave, we could form the *complete* data table $X_1X_2X_3A_1A_2A_3$ which contains K^6 cells, where K is the number of latent classes. Let $(x_1, x_2, x_3, a_1, a_2, a_3)$ denote an arbitrary cell in the complete data table, and let $\pi_{x_1x_2x_3a_1a_2a_3}^{X_1X_2X_3A_1A_2A_3}$ denote the probability of a sample unit being classified into this cell. The loglikelihood kernel for the incomplete data table is therefore

$$\ell = \sum_{a_1,a_2,a_3} n_{x_1x_2x_3a_1a_2a_3} \log \sum_{x_1,x_2,x_3} \pi_{x_1x_2x_3a_1a_2a_3}^{X_1X_2X_3A_1A_2A_3} \tag{7.23}$$

where $n_{x_1x_2x_3a_1a_2a_3}$ is the observed count in cell $(x_1,x_2,x_3,a_1,a_2,a_3)$. Under the standard MLC model assumptions, we can write

$$\pi_{x_1x_2x_3a_1a_2a_3}^{X_1X_2X_3A_1A_2A_3} = (\pi_{x_1}^{X_1}\pi_{x_2|x_1}^{X_2|X_1}\pi_{x_3|x_2}^{X_3|X_2})(\pi_{a_1|x_1}^{A_1|X_1}\pi_{a_2|x_2}^{A_2|X_2}\pi_{a_3|x_3}^{A_3|X_3}) \tag{7.24}$$

and $\pi_{a_1|x_1}^{A_1|X_1} = \pi_{a_2|x_2}^{A_2|X_2} = \pi_{a_3|x_3}^{A_3|X_3}$ for all $a_1 = a_2 = a_3$ and $x_1 = x_2 = x_3$, due to the restriction of time homogeneous error probabilities.

As noted in Section 4.2.3, the EM algorithm (Dempster et al. 1977) consists of two separate steps at each iteration cycle: an E(xpectation)-step and an M(aximization)-step. At the E-step of a cycle, the algorithm computes new entries for the $X_1X_2X_3A_1A_2A_3$ table, based on estimates of $\pi_{x_1x_2x_3|a_1a_2a_3}^{X_1X_2X_3|A_1A_2A_3}$ computed at the previous cycle. Recall that at the first cycle, the values of $\pi_{x_1x_2x_3|a_1a_2a_3}^{X_1X_2X_3|A_1A_2A_3}$ are based on user-provided and/or randomly generated starting values. Then

$$\hat{n}_{x_1x_2x_3a_1a_2a_3} = n_{a_1a_2a_3}\hat{\pi}_{x_1x_2x_3|a_1a_2a_3}^{X_1X_2X_3|A_1A_2A_3} \tag{7.25}$$

can be computed, where $\hat{n}_{x_1x_2x_3a_1a_2a_3}$ denotes the estimated count for cell $(x_1, x_2, x_3, a_1, a_2, a_3)$. These counts are then used to obtain new estimates of $\pi_{x_1x_2x_3|a_1a_2a_3}^{X_1X_2X_3|A_1A_2A_3}$ in the M-step as follows:

$$
\begin{aligned}
\hat{\pi}_{x_1}^{X_1} &= \frac{n_{x_1+++++}}{n_{++++++}} \\
\hat{\pi}_{x_2|x_1}^{X_2|X_1} &= \frac{n_{x_1x_2++++}}{n_{x_1+++++}} \\
\hat{\pi}_{x_3|x_2}^{X_3|X_2} &= \frac{n_{+x_2x_3+++}}{n_{+x_2++++}} \\
\hat{\pi}_{a_1|x_1}^{A_1|X_1} &= \frac{n_{x_1++a_1++} + n_{+x_2++a_2+} + n_{++x_3++a_3}}{n_{x_1+++++} + n_{+x_2++++} + n_{++x_3+++}}
\end{aligned}
\tag{7.26}
$$

where the "+" subscript indicates summation over the subscript it replaces. These new estimates are used via (7.25) to produce new estimates of the cells in the complete data table (E-step), which, in turn, produces another cycle of parameter estimates via (7.26) (M-step). The E–M cycle is repeated until the largest difference between parameter estimates at two consecutive cycles is less than the convergence criterion (usually, 10^{-6} or a smaller number).

Instead of the probabilistic model, a loglinear model can be specified for $m_{x_1x_2x_3a_1a_2a_3}$, the expected frequency in cell $(x_1,x_2,x_3,a_1,a_2,a_3)$. The loglinear model producing estimates equivalent to those from (7.24) is

$$
\begin{aligned}
\log m_{x_1x_2x_3a_1a_2a_3} = {} & u + u_{x_1}^{X_1} + u_{x_2}^{X_2} + u_{x_3}^{X_3} + u_{a_1}^{A_1} + u_{a_2}^{A_2} + u_{a_3}^{A_3} \\
& + u_{x_1x_2}^{X_1X_2} + u_{x_2x_3}^{X_2X_3} + u_{x_1a_1}^{X_1A_1} + u_{x_2a_2}^{X_2A_2} + u_{x_3a_3}^{X_3A_3}
\end{aligned}
\tag{7.27}
$$

or, in shorthand notation, $\{X_1X_2\ X_2X_3\ X_1A_1\ X_2A_2\ X_3A_3\}$. However, it is often more intuitively appealing to specify an MLC model as a modified path model. For example, (7.27) may be specified as: $\{X_1\}\{X_1X_2\}\{X_2X_3\}\{X_1A_1\}\ \{X_2A_2\}\{X_3A_3\}$. This model essentially specifies a logit model for each probability in (7.24). For example, $\pi_{x_2|x_1}^{X_2|X_1}$ is specified as

$$
\pi_{x_2|x_1}^{X_2|X_1} = \frac{\exp(u_{x_2}^{X_2} + u_{x_1x_2}^{X_1X_2})}{\sum_{x_2} \exp(u_{x_2}^{X_2} + u_{x_1x_2}^{X_1X_2})}
\tag{7.28}
$$

Likewise, $\pi_{a_1|x_1}^{A_1|X_1}$ is specified as

$$
\pi_{a_1|x_1}^{A_1|X_1} = \frac{\exp(u_{a_1}^{A_1} + u_{x_1a_1}^{X_1A_1})}{\sum_{a_1} \exp(u_{a_1}^{A_1} + u_{x_1a_1}^{X_1A_1})}
\tag{7.29}
$$

and so on. The assumption of time-homogeneous error probabilities translates as

$$\frac{\exp(u_{a_1}^{A_1} + u_{x_1 a_1}^{X_1 A_1})}{\Sigma_{a_1} \exp(u_{a_1}^{A_1} + u_{x_1 a_1}^{X_1 A_1})} = \frac{\exp(u_{a_2}^{A_2} + u_{x_2 a_2}^{X_2 A_2})}{\Sigma_{a_2} \exp(u_{a_2}^{A_2} + u_{x_2 a_2}^{X_2 A_2})} = \frac{\exp(u_{a_3}^{A_3} + u_{x_3 a_3}^{X_3 A_3})}{\Sigma_{a_3} \exp(u_{a_3}^{A_3} + u_{x_3 a_3}^{X_3 A_3})} \tag{7.30}$$

or, equivalently,

$$u_{a_1}^{A_1} = u_{a_2}^{A_2} = u_{a_3}^{A_3} \quad \text{and} \quad u_{x_1 a_1}^{X_1 A_1} = u_{x_2 a_2}^{X_2 A_2} = u_{x_3 a_3}^{X_3 A_3} \tag{7.31}$$

for $a_1 = a_2 = a_3$ and $x_1 = x_2 = x_3$.

Estimation via the EM algorithm can still use the same E-step as in (7.25); however, the M-step differs somewhat. Estimating the structural components proceeds by applying the M-step separately to tables X_1X_2 and X_2X_3 to estimate the parameters of $\{X_1X_2\}$ and $\{X_2X_3\}$, respectively. Estimation for the measurement components is somewhat more complicated due to the equality restrictions across submodels imposed by the time-homogeneous error probability assumption. In that case, the likelihood functions for tables X_1A_1, X_2A_2, and X_3A_3 must be maximized simultaneously using the procedure described in Vermunt (1997a, pp. 309–311).

7.1.6 MLC Model with Grouping Variables

In this section we consider MLC models that include one or more grouping variables. As in the case of LC models, grouping variables are important in Markov latent class modeling for several reasons:

1. For the measurement component, they allow us to better satisfy the assumption of homogeneous error probabilities with latent classes (assumption 4 above). By incorporating grouping variables, this assumption can be relaxed to one of *conditional* homogeneity—error probabilities that are homogeneous within each *grouping variable by latent class cell.*
2. For the structural component, grouping variables are often needed to the explain variation in latent classes prevalence probabilities as well as variations in the interactions of the latent variables over time.
3. The researcher may be interested in comparing transition or error probabilities for various population subgroups or domains. Indeed, examining the error structure as a function of various demographic and economic variables such as age, race, gender, education, and income level can provide important clues as to the causes of classification error in panel surveys.

To simplify the discussion, consider a single grouping variable G with L levels or categories. Assume that an individual's classification by G does not change over the waves considered in the analysis. For example, G may be gender or age at wave 1. *Time-varying* grouping variables, such age at the time of interview, interview mode, and proxy/self-response are considered in Section 7.2.5. Let π_g^G denote the proportion of the population belonging to group g, for $g = 1,\ldots,L$. Let $GA_1A_2A_3$ denote the LK^3 cross-classification table for the n sample members classified by G as well as their observed states for three panel waves. Then, in its most general form, the first-order MLC model specification of joint probability for cell (g,a_1,a_2,a_3) is

$$\pi_{ga_1a_2a_3}^{GA_1A_2A_3} = \pi_g^G \sum_{x_1}\sum_{x_2}\sum_{x_3}(\pi_{x_1x_2x_3g}^{X_1X_2X_3|G})(\pi_{a_1a_2a_3|gx_1x_2x_3}^{A_1A_2A_3|GX_1X_2X_3}) \tag{7.32}$$

The structural component can be rewritten as

$$\pi_{x_1x_2x_3|g}^{X_1X_2X_3|G} = \pi_{x_1|g}^{X_1|G}\pi_{x_2|gx_1}^{X_2|GX_1}\pi_{x_3|gx_2}^{X_3|GX_2} \tag{7.33}$$

and the measurement component as

$$\pi_{a_1a_2a_3|gx_1x_2x_3}^{A_1A_2A_3|GX_1X_2X_3} = \pi_{a_1|gx_1}^{A|GX_1}\pi_{a_2|gx_2}^{A|GX_2}\pi_{a_3|gx_3}^{A|GX_3} \tag{7.34}$$

where now the time homogeneous error probability assumption takes the form

$$\pi_{a_1|gx_1}^{A|GX_1} = \pi_{a_2|gx_2}^{A|GX_2} = \pi_{a_3|gx_3}^{A|GX_3} \tag{7.35}$$

Note that this model, represented schematically in Figure 7.4, is equivalent to fitting the standard MLC model in Figure 7.2 (i.e., the model $\{X_1A_1\ X_2A_2\ X_3A_3\}$) separately in each group. As before, the arrows emanating from G that point to other arrows represent the three-way interactions.

The model may be written in the log-linear formulation as

$$\{X_1G\ \ X_1X_2G\ \ X_2X_3G\ \ X_1GA_1\ \ X_2GA_2\ \ X_3GA_3\}$$

For the modified path model specification, the structural component is $\{X_1G\}$ $\{X_1X_2G\}$ $\{X_2X_3G\}$, and the measurement component is

$$\{X_1GA_1\}\ \{X_2GA_2\}\ \{X_3GA_3\}$$

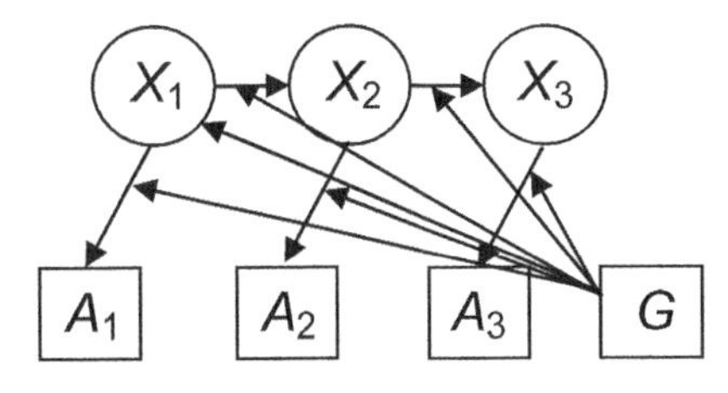

Figure 7.4 Path model diagram for an MLC model with one grouping variable.

Analysis usually proceeds by testing the significance of higher-order interactions involving G using the Wald statistic or the conditional likelihood ratio test for nested models. For an MLCA involving multiple waves and grouping variables, data sparseness can be a problem for estimation. As an example, a trichotomous characteristic measured at four waves produces a table containing $3^4 = 81$ cells. Adding three trichotomous grouping variables to the model, the number of cells grows to 2187. Thus a sample size of more than 2000 is needed merely to achieve an average cell count of unity. The good news is that the EM algorithm is quite robust even in situations where the expected cell size is very small [see, e.g., Collins et al. (1996) and Collins and Tracy (1997)].

A more serious problem with sparse data is model selection and testing since, as noted in Sections 4.4.2 and 5.1.4, the usual chi-square tests are not be valid in these situations. The methods suggested there for addressing this issue for LCA can also be applied in MLCA. Perhaps the best solution is to simulate the actual distribution of L^2 using pseudoreplication methods such as the bootstrap (see Section 5.1.4) This approach has been adopted for PANMARK (Langeheine et al. 1996), MPlus, and Latent GOLD, but is not available in the ℓEM package. In addition, using the traditional tests of model fit, determining the appropriate degrees of freedom for the reference chi-square distribution can be complicated as a result of zero cells and expected frequencies. This problem is circumvented when using bootstrap simulation.

Finally, the problems of model nonidentifiability and local maxima are even greater in MLCA than in LCA because of the larger number of parameters that are typical in MLC models. As noted in Section 5.1.2, rerunning the models with multiple starting values and comparing the results offer some protection against these problems. However, larger models often require a much larger number of runs using differing starting values to find the global maximum.

7.1.7 Example: CPS Labor Force Status Classification Error

A common application of the MLC model is to model the classification error in labor force survey panel data. For example, Van de Pol and Langeheine (1997) applied these models to The Netherlands Labor Market Survey; Vermunt (1996), to the US Survey of Income Program Participation (SIPP) labor force series, and Biemer and Bushery (2001) and Biemer (2004a), to the CPS. The latter two articles also evaluated a number of the MLC model assumptions specifically for the CPS, including the Markov assumption, and provided evidence of the empirical, theoretical, and external validity of the MLC model estimates of CPS classification error. The authors used MLCA to evaluate the accuracy of labor force status classification process for the revised CPS questionnaire that was introduced in 1994 and compared it with the corresponding accuracy of the classification process used for the questionnaire that had been in use before 1994.

To illustrate the general approach, we use the three labor force categories defined in Section 7.1.2 (i.e., EMP, UNE, and NLF) and consider any three consecutive months of the CPS, say January, February, and March. Let X_1 denote an individual's true labor force status in January: $X_1 = 1$ for EMP, $X_1 = 2$ for UNE, and $X_1 = 3$ for NLF.[2] We shall define X_2 and X_3 analogously for February and March. Similarly, A_1, A_2 , and A_3 will denote the observed labor force statuses for January, February, and March, respectively, with the same categories as their corresponding latent counterparts. Similar models can be defined and fit for February, March, and April; March, April, and May; April, May, and June; and so on. The resultant estimates can be compared and/or averaged for increased estimator stability.

For grouping variables, Biemer and Bushery (2001) and Biemer (2004a) considered age, race, gender, education, income, and self- or proxy response. The self-proxy variable is an indicator of whether a subject's labor force status was obtained from the subject him herself or from another person in the household (i.e., a proxy). In the CPS, the interview informant can change from month to month and, thus, self-proxy can be regarded as a time-varying covariate (to be discussed in Section 7.2.5). However, in doing so, Biemer and Bushery found the models to be quite unstable because of the data sparseness issues and the large number of parameters that were needed. They instead opted for a time-invariant self-proxy variable, denoted by G, where $G = 1$ if the interview was by taken by self-response in all 3 months, $G = 2$ if taken by self-response in exactly 2 months, $G = 3$ taken by proxy response in exactly 2 months, and $G = 4$ if taken by proxy response in all 3 months. This variable was highly explanatory of the variation in the labor force classification probabilities.

Biemer and Bushery analyzed data from 3 years—1993, 1994, and 1996—of the CPS, fitting models to each year separately. The data were weighted and rescaled as described in Section 5.3.4, and the ℓEM software was used to fit the models. A wide range of models were fit to the data including the following five essential models:

Model 0 — Homogeneous and stationary transition probabilities and homogeneous error probabilities: $\{X_1\}\{X_1X_2\}\{X_2X_3\}\{X_1A_1\}\{X_2A_2\}\{X_3A_3\}$, under constraints given by $X_1X_2 = X_2X_3$ and $X_1A_1 = X_2A_2 = X_3A_3$

Model 1 — Nonhomogeneous but stationary transition probabilities and homogeneous error probabilities: $\{X_1G\}\{X_1X_2G\}\{X_2X_3G\}\{X_1A_1\}\{X_2A_2\}\{X_3A_3\}$, with constraints $X_1X_2G = X_2X_3G$ and $X_1A_1 = X_2A_2 = X_3A_3$

[2] Biemer (2004a) used four categories corresponding to employed, unemployed (laid off), unemployed (looking for work), and not in the labor force.

Model 2 Homogeneous and nonstationary transition probabilities and homogeneous error probabilities: same as model 0 without constraint $X_1X_2 = X_1X_2$

Model 3 Nonhomogeneous and nonstationary transition probabilities and homogeneous error probabilities: same as model 1 with constraint $X_1X_2G = X_2X_3G$

Model 4 Nonhomogeneous and nonstationary transition probabilities and nonhomogeneous response probabilities: the model $\{X_1G\}\{X_1X_2G\}\{X_2X_3G\}\{X_1GA_1\}\{X_2GA_2\}\{X_3GA_3\}$, with constraints $X_1GA_1 = X_2GA_2 = X_3GA_3$

Table 7.4 shows the fit diagnostics for these five models by year. As shown, only model 4 provided an acceptable fit when the p-value criterion is used. Model 4 is the most general model in the table and allows the January–February and February–March transition probabilities to vary independently across the four self-proxy groups. The model further specifies that the error probabilities are the same for January, February, and March, but may vary by self-proxy group. This latter specification is consistent with the survey methods literature [see, e.g., O'Muircheartaigh (1991) and Moore (1988)]. In addition, the dissimilarity index D for model 4 is 0.3%, which indicates a very good model fit. Thus, the authors used model 4 to generate the estimates of

Table 7.4 Model Diagnostics for Alterative MLCA Models by Year

Model	Year	df[a]	npar[b]	L^2	p Value	BIC	D
Model 0	1993	90	17	645	0.000	–320	0.048
	1995	—	—	697	0.000	–275	0.044
	1996	—	—	632	0.000	–325	0.045
Model 1	1993	84	23	632	0.000	–269	0.047
	1995	—	—	668	0.000	–240	0.043
	1996	—	—	585	0.000	–308	0.044
Model 2	1993	66	41	99	0.006	–609	0.007
	1995	—	—	146	0.000	–567	0.008
	1996	—	—	159	0.000	–543	0.010
Model 3	1993	42	65	64	0.016	–386	0.005
	1995	—	—	82	0.000	–372	0.005
	1996	—	—	83	0.000	–364	0.010
Model 4	1993	24	83	23	0.501	–234	0.002
	1995	—	—	25	0.410	–234	0.002
	1996	—	—	39	0.026	–216	0.003

[a]Degrees of freedom.
[b]Number of free (unconstrained) parameters π or u.

Table 7.5 Estimated Labor Force Classification Probabilities by Group and Year[a]

True Classification	Observed Classification								
	EMP			UNE			NLF		
	1993	1995	1996	1993	1995	1996	1993	1995	1996
EMP	98.8 (0.1)	98.7 (0.1)	98.8 (0.1)	0.3 (0.11)	0.5 (0.1)	0.4 (0.1)	0.9 (0.1)	0.8 (0.1)	0.8 (0.1)
UNE	7.1 (0.7)	7.9 (0.9)	8.6 (1.0)	81.8 (0.9)	76.1 (1.2)	74.4 (1.2)	11.1 (0.9)	16.0 (1.2)	17.0 (1.2)
NLF	1.4 (0.1)	1.1 (0.1)	1.1 (0.1)	0.8 (0.1)	0.7 (0.1)	0.9 (0.1)	97.8 (0.1)	98.2 (0.1)	98.0 (0.1)

[a]Standard error values are shown in parentheses.

labor force classification error. These estimates appear in Table 7.5 along with their standard errors (in parentheses). As noted in Section 5.3.2, the standard errors are likely to be understated since the effects of unequal weighting and clustering were not taken into account in the analysis.

For the true EMP and true NLF, the probability of a correct response is quite high: 98% and 97%, respectively. This result is similar to what we found in Section 3.4.2 in the Hui–Walter LCA of the CPS reinterview. Also, consistent with prior results, the probability of a correct response for UNE varies across years from 72% to 84%. However, a surprising result from Table 7.5 is the magnitude of reporting accuracy for 1994 and 1995 compared to 1993. As Biemer and Bushery note, the CPS questionnaire was substantially redesigned in 1994 to increase the accuracy of the labor force status classifications as well other survey improvements. The results in Table 7.4 suggest that reporting accuracy is higher for the year prior to the major redesign (i.e., in 1993) than for the years following the redesign.

Subsequently, Biemer (2004a) applied the MLC model to further explore this anomaly. His analysis considered the error associated with the two primary subclassifications of the unemployed: (1) persons who are unemployed and *on layoff* and (2) persons who are unemployed and *looking for work*. His analysis suggests that the primary cause of the anomaly in Table 7.4 was a reduction in classification accuracy of persons who are on layoff. Using MLCA, the analysis revealed that two questions on the revised questionnaire appear to be the cause. Considerable error was associated with the revised question: "LAST WEEK, did you do ANY work (either) for pay (or profit)?" More that half of the error in the revised layoff classification may be attributed to the error in this question. In addition, there is considerable classification error in determining whether individuals reporting some type of layoff have a date or indication of a date to return to work. This question alone appeared to explain between 30% and 40% of the layoff classification error. Biemer concluded that the combination of these two questions explains the reduction in accuracy in UNE in the revised questionnaire.

Biemer and Bushery also compared the MLCA estimates of the CPS classification probabilities with similar estimates from the literature, including Fuller and Chua (1985), Chua and Fuller (1987); Poterba and Summers (1995), and the CPS reconciled reinterview program (Biemer and Bushery 2001). Unlike the MLCA estimates, these other estimates relied on reconciled reinterview data, which was considered a gold standard. Biemer and Bushery found that the relative magnitude of the MLCA estimates across the labor force categories agree fairly well with the previous estimates. The largest differences occurred for the true unemployed population where the estimates of response accuracy from the literature were 3–7 percentage points higher than the corresponding MLCA estimates. One explanation for this difference is that the comparison estimates are biased upward as a result of correlations between the errors in interview and reinterview. Another explanation is that the MLCA estimates are biased downward as a result of the failure of the Markov assumption to hold. Both explanations may be true to some extent. However, later in this section we provide evidence that failure of the Markov assumption is likely to have a small effect on estimates of classification error.

The next example shows how MLCA can be applied to a quarterly panel survey of consumer expenditures in order to evaluate the error in questions about the types of expenditures that household members may have had. Of particular interest in this analysis were underreporting error (i.e., not reporting an actual expenditure) and its causes. In this analysis, underreporting is modeled as a false-negative error and overreporting, as a false-positive error. The analysis examines whether certain types of households have higher levels of error than others, whether the use of household receipts and other records reduces the reporting errors, and whether items that are asked later in the hour-long interview are more prone to error than are items that are asked earlier. The analysis also shows that while some items seem to satisfy the Markov assumption, other items do not. Therefore, the first-order MLC model must be used cautiously and interpreted under the threat of model misspecification.

7.1.8 Example: Underreporting in U.S. Consumer Expenditure Survey

The US Consumer Expenditure Interview Survey (CEIS), sponsored by the US Bureau of Labor Statistics (BLS), is a quarterly panel survey that collects data on the buying habits of American consumers, including data on their expenditures, income, and consumer unit[3] (CU) characteristics. The survey consists of five interview waves spaced approximately 3 months apart. The initial interview is a bounding interview and is not reported. A bounding interview is the CU's first interview that is used primarily to establish the beginning of the recall period for the second interview. For example, although respondents are only supposed to report expenditures occurring within the

[3]A *consumer unit* pertains to household members who may be part of the same family as well as other persons who share the household and who are financially dependent for housing, food, or other living expenses.

past 3 months, many may erroneously report expenditures that occurred during the past 4 or more months (referred to as *external telescoping*). The bounding interviews are used to identify telescoped reports in the second interview, that is, reports in the second interview that were already reported in the bounding interview. Then the second interview is used in the same way for the third interview, and so forth. This is intended to rid the 2 through 5 interviews of external telescoping errors. Thus, only waves 2, 3, 4, and 5 are actually used in the CU estimates. For the present analysis, only these waves were analyzed.

At each wave, respondents are asked a number of so-called screening questions to determine what consumer items within a comprehensive list of items they purchased in the previous quarter. The question stated somewhat like the following: "Since the first of [month], have you or has any member of your CU purchased any clothing for persons age 2 and over either for members of your CU or for someone outside your CU?" If the respondent responds positively to a screening question (indicating at least one purchase) for an item, detailed questions concerning when the purchases were made, description of each item, and the amounts paid are then asked. If the response is negative, the detailed questions are skipped. Since the list of consumer items included in the survey is quite extensive, interviews can last 2 hours or longer. Thus, there is a high risk of false-negative responses to the screening questions once respondents learn that denying their purchases will shorten the interview. This phenomenon is a form of *satisficing* (Krosnick 1991), which is a type of respondent behavior during the interview that is intended to reduce the respondent's cognitive burden, but usually with adverse effects for response accuracy.

In a series of articles (Biemer 2000; Biemer and Tucker 2001; Tucker et al. 2005), BLS has sought to estimate the magnitude of the error in the screening questions, particularly the underreporting of purchases. The goals of the research are threefold:

- To estimate the level of reporting error (particularly, underreporting) in the CEIS
- To identify factors that influence the errors; for example, household characteristics such as income, family size, and other demographic characteristics; item characteristics such as the frequency with which an item is purchased, whether a major or minor expenditure and whether the item was asked earlier or later in the interview; and interview characteristics such as length of the interview, use of household records, and the amount of item nonresponse in the questionnaire
- To generate hypotheses regarding the causes of the errors so that future research could explore these potential causes and develop improved methods to reduce the errors

Traditional methods for evaluating survey error are quite difficult to apply in this setting. For example, methods that rely on gold standard measurements

Table 7.6 A Selection of Consumer Items Analyzed

Variable	Description
CABLE	Cable TV, satellite services, or community antenna
GAS	Bottled or tank gas
SPORTS	Combined sports, recreation, and exercise equipment
FURN	Combined furniture
CLOTH	Combined clothing
SHOES	Footwear (including athletic shoes not specifically purchased for sports)
DENTAL	Dental care
DRUGS	Combined medicine and medical supplies
EYES	Eye exams, eyeglasses, other eye services
PETS	Pet supplies services, and including veterinary expenses

such as sales receipts and household purchase records are very difficult to obtain and are often unavailable. For this reason, BLS has used MLCA for evaluating the magnitude of the reporting error in the CEIS. In this example, some of the various MLC models that have been applied to the CEIS will be illustrated. The focus of the analyses is on the false-positive and false-negative probabilities for the consumer item screening questions.

For illustration purposes, only one year of the CEIS (1997) will be analyzed here—a total of 2189 cases. However, in the referenced studies, Biemer and his colleagues analyzed multiple years of CEIS data. Further, the referenced studies examined 19 items or more, while the present analysis will be confined to just 10 items selected to be exemplary of the key findings. These items are shown in Table 7.6. Only MLC models for three consecutive quarters will be considered, although some of the parent studies considered all four quarters of data. The ideas contained in this illustration will be generalized for multiple years and all four quarters in Section 7.2.4.

Biemer (2000) considered a number of possible definitions of the dependent variable in the analysis. Since the CEIS screening questions are asked each month of a quarter, three quarterly interviews would produce nine dichotomous responses—one for each month. Thus, for each consumer item, one could define a dichotomous latent variable X_t for each month denoting a true purchase ($X_t = 1$) or true nonpurchase ($X_t = 2$) at month t for $t = 1,\ldots,9$. The corresponding dichotomous response variable, A_t, could be defined analogously. However, as reported in Biemer (2000), MLC models fit to these nine time points did not perform well because of their complexity; in particular, too many parameters, data sparseness issues, and size constraints imposed by the available software for the analysis. In addition, the ICE and Markov assumptions did not hold for the 3 months within a quarter. Adding model terms to address these model failures added even greater complexity to the models.

Instead, the author defined a dependent variable for each CU that summarized the purchases over all three months of a quarter as follows:

$$X_t = \begin{cases} 1 & \text{if CU purchased the item each month of quarter } t \text{ (\textit{purchaser})} \\ 2 & \text{if CU did not purchase the item each month of quarter } t \text{ (\textit{nonpurchaser})} \\ 3 & \text{if CU purchased 1 or 2 months during quarter } t \text{ (\textit{mixed purchaser})} \end{cases} \tag{7.36}$$

Here X_t is subject to misclassification and, therefore, is treated as a latent variable in the analysis. The indicator of X_t is A_t which is defined analogously:

$$A_t = \begin{cases} 1 & \text{if CU reported purchasing the item each month of quarter } t \text{ (\textit{purchaser})} \\ 2 & \text{if CU reported not purchasing the item each month of quarter } t \text{ (\textit{nonpurchaser})} \\ 3 & \text{if CU reported purchasing the item 1 or 2 months during quarter } t \text{ (\textit{mixed purchaser})} \end{cases} \tag{7.37}$$

The author believes that the ICE and Markov assumptions are more likely to hold with these definitions. With regard to the ICE assumption, it seems plausible that errors are less likely to be correlated across the 3-month reference periods (referred to as *quarters*) than across the 3 months within a quarter. Within a quarter, expenditure reports are all collected in the same interview, and it is easy for respondents to provide erroneous albeit consistent information as a result of memory effects, acquiescent behavior, and other cognitive phenomena [see, e.g., Tourangeau et al. (2000, p. 125)]. However, for the definition in (7.37), A_1, A_2, and A_3 correspond to three separate interviews spaced 3 months apart, and thus there is less opportunity for these types of effects to contaminate the reports.

Likewise, regarding the Markov assumption, an individual's purchasing behavior during a 3-month period is likely to be more intercorrelated than his or her purchases over three quarters. Therefore, defining X as a quarterly, rather than a monthly, indicator affords greater opportunity for the Markov assumption to hold as well.

To address the heterogeneity issues as well as to test for group differences in underreporting rates, the authors defined a large number of grouping variables for the analysis that were potentially related to the structural and/or measurement components of the model. An extensive examination of these grouping variables was conducted, and three variables emerged as particularly explanatory: CU size (F), average interview duration (L), and use of household expenditure records during the interview (R). These time-constant grouping variables are defined as follows: F with three levels—single-person CU, two or three persons, four or more persons; L with four levels—very short, short, medium, or long average interview duration; and R with three levels—always, sometimes, and seldom or never used records.

In the initial stages of the analysis, the author explored a number of MLC models from the simplest model with no grouping variables to models that incorporated up to three grouping variables and their interactions. Including more than three grouping variables in the models was somewhat problematic because of increased data sparseness and the number of zero cells. Because of the number of runs that were needed for fitting a model to all 19 consumer items, a single-model structure was identified and applied to all the items. The risk of this "one model fits all" approach is over fitting for some items and misspecifying for others. Nevertheless, for the purposes of estimating misclassification, this approach seems to work fairly well, based on the author's subsequent efforts to check the robustness of the estimates to model misspecification.

The model that emerged as the overall best across the 19 consumer items contained 98 parameters and contained the following terms:

Grouping variable structure	$\{FLR\}$
Latent variable structure	$\{X_1F\ X_1L\ X_1R\}\ \{X_1X_2\ X_2F\ X_2L\ X_2R\}$ $\{X_2X_3\ X_3F\ X_3L\ X_3R\}$
Measurement component	$\{X_1A_1\ FA_1\ LA_1\ RA_1\}\ \{X_2A_2\ FA_2\ LA_2\ RA_2\}$ $\{X_1A_3\ FA_3\ LA_3\ RA_3\}$ including time-homogeneous error probability constraints: $\pi_{a_1\mid x_1 flr}^{A_1\mid X_1FLR}=\pi_{a_2\mid x_2 flr}^{A_2\mid X_2FLR}=\pi_{a_3\mid x_3 flr}^{A_3\mid X_3FLR}$ or, equivalently, $A_1\mid X_1FLR=A_2\mid X_2FLR=A_3\mid X_3FLR$

The grouping variable structure and the latent variable structure constitute the structural component. They were separated in the specification shown above and in the ℓEM input statements for convenience and clarity. The model fit the data for all 19 consumer items quite well with p values very close to 1. Estimates based on both unweighted and weighted data were compared. Since few differences were found, the unweighted data were used in the analyses. The ℓEM input code for this model is provided in Figure 7.5, and the corresponding path model is shown in Figure 7.6. For the sake of clarity, the path diagram is decomposed into structural and measurement components.

The estimates of reporting error from model 1 ($\pi_{a_1\mid x_1}^{A_1\mid X_1}$, $a_1 \neq x_1$) as well as the true prevalence of each purchaser status ($\pi_{x_1}^{X_1}$) are provided in Table 7.7 for all 19 consumer items considered. A number of points can be made from this table as well as other results reported in Biemer (2000) as follows:

```
lat 3
man 6
dim  3  3  3 3 4 3  3  3  3
lab X1 X2 X3 F L R A1 A2 A3
mod F.L.R       * Dot separators required when multi-character variable
                  names are used
    X1|F.L.R     {X1.F X1.L X1.R}
    X2|X1.F.L.R {X1.X2 X2.F X2.L X2.R}
    X3|X2.F.L.R {X2.X3 X3.F X3.L X3.R}
    A1|X1.F.L.R {A1.X1 A1.F A1.L A1.R}
    A2|X2.F.L.R eq1 A1|X1.F.L.R
    A3|X3.F.L.R eq1 A1|X1.F.L.R
dat ceis.dat
cri 0.00000001
ite 10000
```

Figure 7.5 ℓEM input statements for the CEIS MLC model.

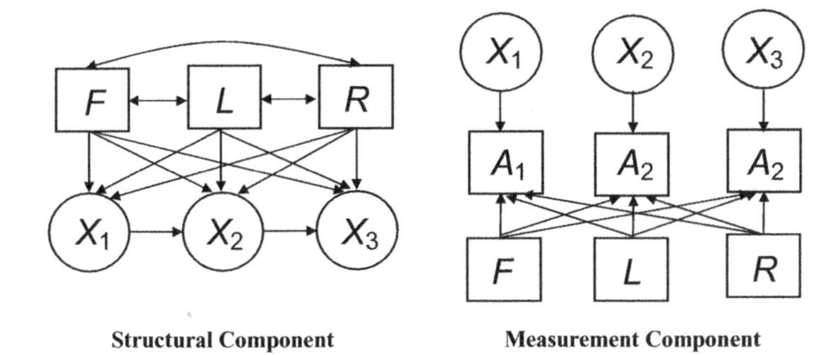

Figure 7.6 Path diagrams for the fitted CEIS MLC model: structural and measurement components.

Table 7.7 True Prevalence of Consumer Class and the Probability of Correct Classification by Consumer Item[a]

Variable	Purchaser ($X_1 = 1$) $\pi_1^{X_1}$	Purchaser ($X_1 = 1$) $\pi_{1\|1}^{A_1\|X_1}$	Nonpurchaser ($X_1 = 2$) $\pi_2^{X_1}$	Nonpurchaser ($X_1 = 2$) $\pi_{2\|2}^{A_1\|X_1}$	Mixed Purchaser ($X_1 = 3$) $\pi_3^{X_1}$	Mixed Purchaser ($X_1 = 3$) $\pi_{3\|3}^{A_1\|X_1}$
CABLE	22.50	98.42	47.62	97.19	29.88	78.63
GAS	5.57	93.84	86.12	99.86	8.31	64.65
SPORTS	5.04	5.25	86.33	100.00	8.63	97.85
FURN	4.52	5.73	90.52	92.77	4.96	58.94
CLOTH	11.29	55.74	56.10	85.29	32.61	82.31
SHOES	5.34	10.02	85.80	100.00	8.86	70.37
DENTAL	2.89	53.40	74.41	97.41	22.70	59.66
DRUGS	8.56	86.41	67.24	87.19	24.20	77.43
EYES	12.20	2.02	76.66	98.69	11.14	73.33
PETS	5.34	10.02	85.80	100.00	8.86	70.37
Average	*8.33*	*42.09*	*75.67*	*95.84*	*16.02*	*73.35*

[a]Cell entries are percentages.

- Across all items, the probability of correctly classifying a true nonpurchaser is high, averaging about 96%. *Nonpurchaser* is also the most prevalent consumer status among the three status groups.
- The accuracy with which true purchasers are classified varies considerably across items and is quite low, about 42% on average. The true prevalence rates for this type of purchaser are also quite low for the items considered.
- The accuracy with which mixed consumers are classified is about 73% on average with considerable variability across items. With few exceptions, the prevalence of mixed consumers in the sample is higher, often considerably so, than the prevalence of purchasers.
- Regular purchases such as CABLE, GAS, and DRUGS tend to have the smallest levels of underreporting error.

The study also found that underreporting tended to increase as the interview progressed. For example, true purchasers were misclassified as mixed purchasers or nonpurchasers 50% of the time for items appearing in the first half of the questionnaire. For items appearing in the second half of the questionnaire, this rate rises to 95%.

Finally, the error probabilities for true purchasers and true mixed consumers by item are shown in Table 7.8. The probability that a true purchaser is misclassified as a nonpurchaser is shown in column 2 and the probability that a true purchaser is misclassified as a mixed purchaser is shown in column 3. Averaged across the items, both types of misclassification appear equally

Table 7.8 Probability of an Error[a]

	Misclassification of True Purchasers ($X_1 = 1$)		Misclassification of True Mixed Purchaser ($X_1 = 3$)	
Variable	$\pi_{2\mid 1}^{A_1\mid X_1}$	$\pi_{3\mid 1}^{A_1\mid X_1}$	$\pi_{1\mid 3}^{A_1\mid X_1}$	$\pi_{2\mid 3}^{A_1\mid X_1}$
CABLE	0.25	1.33	16.36	5.01
GAS	0.95	5.22	12.03	5.39
SPORTS	46.52	48.23	2.15	0.00
FURN	60.73	50.16	0.00	10.92
CLOTH	1.93	42.32	5.49	12.19
SHOES	32.29	57.69	0.39	29.24
DENTAL	17.75	28.86	0.00	40.34
DRUGS	2.91	10.68	16.87	5.70
EYES	82.76	15.22	0.00	26.67
PETS	32.29	57.69	0.39	29.24
Average	*27.84*	*31.74*	*5.37*	*16.47*

[a]Cell entries are percentages.

likely; however, as before, the story varies considerably by item. For example, for EYES, true purchasers are more likely to be misclassified as nonpurchasers than as mixed purchasers. However, for CLOTH, the opposite appears to be true.

Column 4 shows the probability that a true mixed purchaser is classified as a purchaser, and the last column shows the probability that a true mixed purchaser is classified as a nonpurchaser. Here, there appears to be a stronger tendency to classify mixed purchasers as nonpurchasers rather than purchasers—leading to greater underreporting of their expenditures. However, for the items CABLE, GAS, and DRUGS, there is a higher risk that mixed purchasers are classified as purchasers—possibly leading overreporting for these items. Thus, although underreporting appears to be the dominate type of screener error, the MLCA indicates that overreporting can also be a problem for some items.

The study also found that, in general, infrequent purchases tend to be underreported more often than frequent purchases; length of interview implies higher accuracy, and CU size is not related to reporting accuracy.

7.2 SOME NONSTANDARD MARKOV MODELS

In this section, we discuss some models that are quite useful in the evaluation of panel survey error but whose forms and assumptions deviate from the standard MLC models considered thus far. The first of these is the so-called mover–stayer model, whose manifest and latent class forms will be considered. This model is very useful in situations where some latent group in the population has essentially no chance of moving from its initial state. Then the second-order Markov model will be considered, which may provide an improved fit for some dynamic phenomena. Next we consider models that incorporate grouping variables (such as income, family size, and mode of interview) that may vary across the survey waves. Finally, we consider models appropriate for panel survey designs employing multiple indicators of the latent variable at one or more waves.

7.2.1 Manifest Mover–Stayer Model

The manifest mover–stayer (MMS) model [see, e.g., Goodman (1961) and Blumen et al. (1966)] is an extension of the MM model described in Section 7.1.1. For some applications, it may be plausible to assume the population consists of two types of individuals or units: those having a positive probability of changing from wave to wave and those having zero probability of changing. As an example, in a survey of smokers, the population may consist of persons who never have smoked and have zero probability of ever starting to smoke (at least for the T panel waves in the analysis). Likewise, there may be persons who have always smoked and have essentially a zero probability of stopping this habit during the T waves. There may also be persons whose smoking status may vary—smoking at some waves, not smoking at others. The first two groups are referred to as *stayers*, so called because with probability 1, they always stay in the state occupied at the initial wave. The latter group are the *movers* because they can move in and out (transition) between states across the T waves. For dichotomous (1–2) variables, there may be *negative stayers* and *positive stayers*; for example, nonsmoking stayers and smoking stayers, respectively. The former occupies state 2 at wave 1 and does not change for T waves, while the later occupies state 1 at wave 1 and does not change for T waves.

Movers and stayers create heterogeneity in the transition probabilities that may be difficult to model by simply adding more grouping variables to the structural component of the model. Unless the grouping variables do a good job of isolating the stayers in the population into homogeneous cross-classification cells or the stayer subpopulation is quite small, the model will not fit the data adequately. An innovative approach for dealing with this type of model misspecification is to define a latent grouping variable, denoted by H, which captures this cause of unobserved heterogeneity in transition probabilities.

To illustrate, let $H = 1$ denote the stayers and $H = 2$ denote the movers. For the binary latent variable X, stayers may be positive or negative depending on their state at time $t = 1$. Thus, for positive stayers, $\pi_{1|1}^{X_t|X_1} = 1$ and for negative stayers, $\pi_{2|2}^{X_t|X_1} = 1$ for $t > 1$. In other words, stayers occupy the same state over time with probability 1. On the other hand, movers can occupy either positive or negative states at each wave. The Markov property is assumed for their wave-to-wave transitions.

For example, for $T = 3$ waves, only two wave-to-wave transition patterns are possible for stayers: either 111 or 222. Thus, stayers can occupy only two cells of the $X_1X_2X_3$ table—cell 111 for positive stayers and cell 222 for negative stayers. Movers can occupy any cell of the table; however, in the observed data, they are indistinguishable from stayers in the two so-called *nontransitory* cells. Fortunately, the proportion of stayers of each type in the nontransitory cells can be estimated using ML estimation.

To understand how, we first consider the purely manifest form of the mover–stayer model (i.e., the MMS model) shown in Figure 7.7 and then generalize this model to the latent class form, which assumes misclassification in the observed states. We Write the MMS model as

$$\pi_{x_1x_2x_3}^{X_1X_2X_3} = \pi_1^H \pi_{x_1x_2x_3|1}^{X_1X_2X_3|H} + \pi_2^H \pi_{x_1|2}^{X_1|H} \pi_{x_2|x_1 2}^{X_2|X_1H} \pi_{x_3|x_2 2}^{X_3|X_2H} \tag{7.38}$$

where $\pi_{x_1x_2x_3|1}^{X_1X_2X_3|H} = 1$ if $x_1 = x_2 = x_3$ and 0 otherwise. The last term in (7.38), apart for the conditioning on H, is essentially a manifest Markov (MM) model. Generalizing (7.38) to $K > 2$ classes is straightforward. Note that the mover–stayer indicator variable H is still dichotomous so that (7.38) is essentially

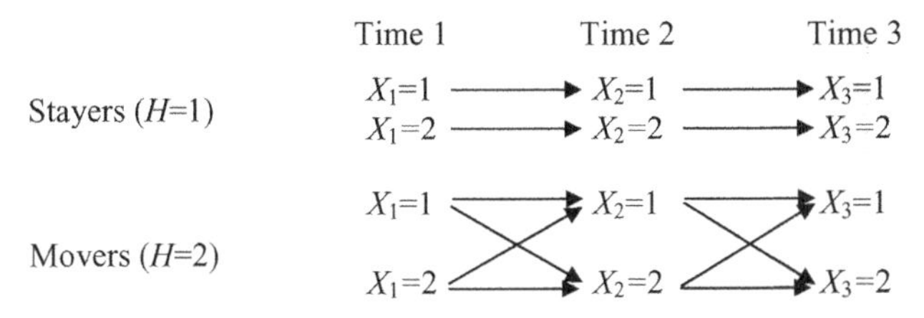

Figure 7.7 Graphical illustration of an MMS model for two states and three timepoints.

unchanged. However, now there are K groups of stayers corresponding to the K initial states, that is, $\pi_{x_1x_2x_3|1}^{X_1X_2X_3|H}=1$ for patterns 111, 222,..., KKK.

A large number of observations in the nontransitory cells of the $X_1X_2X_3$ table does not necessarily suggest a large stayer subpopulation since, as mentioned earlier, movers can also occupy these cells. Rather, the question is whether the number of observations in these cells is much larger than what would be predicted from an MM model. For example, the probability the pattern 111 for an MM model is $\pi_{111}^{X_1X_2X_3}=\pi_1^{X_1}\pi_{1|1}^{X_2|X_1}\pi_{1|1}^{X_3|X_2}$. An excess of observations in cell 111 results when the model predicted value of the expected number of observations (viz., $\hat{m}_{111}=n\pi_{111}^{X_1X_2X_3}$) is significantly less than n_{111}, the observed number. If this is also true for the other nontransitory cells, then the MM model will not provide an adequate fit to the data. In that case, the MMS model may provide a better fit since, as seen from (7.38), it supplements the MM model estimate for these cells by a factor representing the stayer group in those cells. If the size of the stayer group is small, then the fit of the MMS model will show little improvement over the MM model. However, if the stayer group is large enough, the MMS model will fit the data better.

For dichotomous characteristics, the model contains seven (+1) parameters, two more than the MM model. With only eight cells, the model is fully saturated. The seven parameters are π_1^H, $\pi_{111|1}^{X_1X_2X_3|H}$, $\pi_{1|2}^{X_1|H}$, $\pi_{1|x_1 2}^{X_2|X_1H}$ for $x_1=1{,}2$; and $\pi_{1|x_2 2}^{X_3|X_2H}$ for $x_2=1{,}2$. For trichotomous variables, the model contains 17 (+1) parameters (three more than the MM model) leaving 9 degrees of freedom for model testing. These parameters are π_1^H, $\pi_{111|1}^{X_1X_2X_3|H}$, $\pi_{222|1}^{X_1X_2X_3|H}$, $\pi_{x_1|2}^{X_1|H}$ for $x_1=1{,}2$; $\pi_{x_2|x_1 2}^{X_2|X_1H}$ for $x_2=1{,}2$; and $x_1=1{,}2{,}3$ (representing six parameters) and $\pi_{x_3|x_2 2}^{X_3|X_2H}$ for $x_3=1{,}2$ and $x_2=1{,}2{,}3$ (representing six parameters).

A number of variations of the mover–stayer formulation can also be considered. One such variation is the *partial* mover–stayer model. For some applications, it might be preferable to assume that movers and stayers coexist within some but not all latent classes. As an example, consider the smoker–nonsmoker variable described above, where $X=1$ if an individual never smoked and $X=2$ if otherwise. Suppose for some population, that the nonsmokers ($X=1$) consists of persons who have zero probability of ever starting to smoke (nonsmoking stayers) and others who may decide to start and stop smoking (nonsmoking movers). Within the smoker group ($X=2$), there are no stayers; that is, all smokers have some positive probability of stopping. Thus, the manifest

```
* H is the latent indicator variable for stayers (H=1) and movers (H=2)
* X1 is the labor force status for Month 1 with
* X1 = 1 for Employed
* X1 = 2 for Unemployed
* X1 = 3 for NLF
* X2 and X3 are defined similarly for Month 2 and 3
lat 1
man 3
dim 2  3  3  3
lab H  X1 X2 X3
mod H
   X1|H
   X2.X3|X1.H {H.X1.X2 H.X2.X3 wei(H.X1.X2.X3)}
rec 27
rco
sta wei(H.X1.X2.X3) [1 0 0 0 0 0 0 0 0   * stayer constraint for pattern 111
                     0 0 0 0 1 0 0 0 0   * stayer constraint for pattern 222
                     0 0 0 0 0 0 0 0 1   * stayer constraint for pattern 333
                     1 1 1 1 1 1 1 1 1   * movers are unconstrained
                     1 1 1 1 1 1 1 1 1
                     1 1 1 1 1 1 1 1 1]

data table7-2.dat
cri 0.00000001
ite 10000
```

Figure 7.8 ℓEM input statements for MMS model.

cell 111 contains both movers and stayers, while the cell 222 contains only movers. This can be reflected in the model by lifting the constraint $\pi_{222|1}^{X_1X_2X_3|H} = 1$ representing the smoker stayers and leaving the constraint $\pi_{111|1}^{X_1X_2X_3|H} = 1$ representing the nonsmoker stayers.

Another variation of the mover–stayer model is the *black–white model* (Converse 1964; Taylor 1983; Langeheine and Van de Pol 1990). This model essentially assumes that transitions are not dependent on prior states. Rather, at each wave, units are assumed to be assigned to states at random. This model can be combined with a mover–stayer model, where a typical assumption is that transitions between states are equiprobable for movers regardless of their prior state and the states a stayer may occupy at time 1 are decided completely at random. The black–white model has limited applicability for our applications other than as a starting point for fitting more elaborate models. As such, it will not be considered further in the sequel.

Illustration of MMS Model. We fit an MMS model to the labor force data in Table 7.2 using the ℓEM software and the input statements in Figure 7.8. Other packages (e.g., PANMARK, Latent GOLD, and MPlus) could also be used. One caution in using ℓEM is that the degrees of freedom, reported for the model are not calculated correctly in some cases where model constraints are imposed. For these data, the MMS model has 9 degrees of freedom, and the model L^2 is 325.96, which is highly significant, indicating the MMS model does not fit the data. However, compared to the MM model fit in Section 7.1.1, the fit has been greatly improved. To see this, note that the nested chi-square test statistic is computed as 2178.4 – 325.6 = 1852.8 which, for a chi-square with 3 degrees of freedom is highly significant. According to the model, the stayer domain represents 87.2% of the population. The vast majority of the stayers are either EMP (63%) or NLF (36%). The UNE state accounts for only ~1% probably because, unlike the other two labor force states, unemployment is a state that most individuals want to leave quickly. Perhaps a better way to view this effect is to compute $\pi_{1|2}^{H|X_1}$—that is, the proportion of the population that is unemployed at wave 1 and who are stayers. The model estimate of this parameter is about 19%. Among movers, 47% are employed, 18% are unemployed, and 35% are not in the labor force.

7.2.2 Latent Class Mover–Stayer Model

When the observations are subject to misclassification, the MMS model can be extended in a straightforward manner to obtain the latent class mover–stayer (LCMS) model. Again we consider the case of three waves, which is easily generalized to $T > 3$ waves. Suppose that instead of observing X_t, we observe A_t, which is assumed to be an indicator of X_t. The structural component of the LCMS model follows an MMS model while the measurement component is identical to the corresponding component for the standard MLC model. Thus, following (7.38), we can write the following model for an arbitrary cell of the observed $A_1A_2A_3$ table

$$\begin{aligned}\pi_{a_1a_2a_3}^{A_1A_2A_3} &= \sum_{x_1}\sum_{x_2}\sum_{x_3}(\pi_{x_1x_2x_3}^{X_1X_2X_3})(\pi_{a_1a_2a_3|x_1x_2x_3}^{A_1A_2A_3|X_1X_2X_3}) \\ &= \sum_{x_1}\sum_{x_2}\sum_{x_3}(\pi_1^H\pi_{x_1x_2x_3|1}^{X_1X_2X_3|H} + \pi_2^H\pi_{x_1|2}^{X_1|H}\pi_{x_2|x_1 2}^{X_2|X_1H}\pi_{x_3|x_2 2}^{X_3|X_2H})\pi_{a_1|x_1}^{A_1|X_1}\pi_{a_2|x_2}^{A_2|X_2}\pi_{a_3|x_3}^{A_3|X_3}\end{aligned} \tag{7.39}$$

with MMS model constraints for the structural component and the time-homogeneity error probability constraints for the measurement component. With only three waves, the model is not identifiable for dichotomous measures unless additional constraints are imposed. For example, assuming stationary transition probabilities for movers saves 2 degrees of freedom, and the model is just identified. For trichotomous data, the measurement component adds six new parameters to the MMS model, leaving 3 degrees of freedom to test model fit. No additional constraints are necessary to achieve an identifiable model. An important extension of this model is to allow the error probabilities for stayers to differ from those of movers. This is accomplished by replacing the probabilities $\pi_{a_t|x_t}^{A_t|X_t}$ by $\pi_{a_t|hx_t}^{A_t|HX_t}$ for $t = 1,2,3$, or equivalently, by adding the three 3-way interaction terms, HX_tA_t, $t = 1,2,3$, to the model. However, this addition may encounter additional identifiability and/or convergence problems even when $T > 3$ panel waves are available. A common solution is to

```
* H is the latent indicator variable for stayers (H=1) and movers (H=2)
* X1 is the true labor force status for Month 1 with
* X1 = 1 for Employed
* X1 = 2 for Unemployed
* X1 = 3 for NLF
* X2 and X3 are defined similarly for Months 2 and 3
* A1, A2 and A3 are the observed classifications corresponding to X1, X2 and X3
* A1 is the observed labor force status for Month 1 with A2 and A3 defined
* analogously for Months 2 and 3, respectively.
lat 1
man 3
dim 2  3  3  3
lab H  X1 X2 X3
mod H
    X1|H
    X2.X3|X1.H {H.X1.X2 H.X2.X3 wei(H.X1.X2.X3)}
    A1|X1
    A2|X2 eq1 A1|X1
    A3|X3 eq1 A2|X2
rec 27
rco
sta wei(H.X1.X2.X3) [1 0 0 0 0 0 0 0 0  * stayer constraint for pattern 111
                     0 0 0 0 1 0 0 0 0  * stayer constraint for pattern 222
                     0 0 0 0 0 0 0 0 1  * stayer constraint for pattern 333
                     1 1 1 1 1 1 1 1 1  * movers are unconstrained
                     1 1 1 1 1 1 1 1 1
                     1 1 1 1 1 1 1 1 1]
sta A1|X1 [.8 .1 .1 .1 .8 .1 .1 .1 .8] * added for flippage control
data table6-2.dat
cri 0.00000001
ite 10000
```

Figure 7.9 ℓEM input statements for LCMS model.

Table 7.9 Estimated Labor Force Classification Probabilities for the LCMS Model

	Observed Classification		
True Classification	EMP	UNE	NLF
EMP	0.992	0.001	0.007
UNE	0.042	0.786	0.172
NLF	0.010	0.009	0.981

constrain the error probabilities for stayers to zero, which may be plausible in some applications. Since stayers have no chance of transitioning to other states, it seems reasonable that their chances of being misclassified into other states would also be zero. As an example, it may be plausible to assume that nonsmokers who are stayers have zero probability of being misclassified as smokers.

Illustration of the LCMS Model. To illustrate the LCMS model, we apply (7.39) to the data in Table 7.2 using ℓEM (input statements in Figure 7.9). Of particular interest in this illustration is the comparison of the error probability estimates for the MLC and the LCMS models. The MLC model estimates were reported in Table 7.3, while the estimates from the LCMS model are reported in Table 7.9. In this comparison, the maximum difference occurs for the classification accuracy rate for the UNE. Under the MLC model, the accuracy rate is about 0.75, while for the LCMS model it is about 0.79. Adding the mover–stayer constraints also improved the fit of the model. One could be tempted to compute the conditional L^2 test by subtracting $L^2 = 17.3$ for the LCMS model from $L^2 = 29.4$ for the MLC model, resulting in 12.1 with 3 degrees of freedom. Unfortunately, this test is not recommended for comparing these models because of the zero-probability constraints for stayers in the LCMS model. As noted in Section 5.1.5, zero-probability constraints are boundary values that invalidate all chi-square testing. Instead, we use the BIC criterion. For the MLC model, BIC = 106,298, while it is 106,445 for the LCMS model. Thus, the MLC model is preferred.

7.2.3 Second-Order MLC Model

Recall that for $T > 2$ waves, the maximum order of the Markov property is $T - 1$ for an identifiable model assuming time-homogeneous classification error probabilities. It may also be possible to fit even higher-order Markov models when grouping variables are added to the model. Here we consider the case of $T = 4$ waves where the first-order Markov assumption is replaced by the second-order Markov assumption. The path diagram for the second-order MLC model is shown in Figure 7.10. The probability for an arbitrary cell in the $A_1A_2A_3A_4$ cross-classification table is

$$\pi_{a_1a_2a_3a_4}^{A_1A_2A_3A_4} = \sum_{x_1}\sum_{x_2}\sum_{x_3}\sum_{x_4}(\pi_{x_1}^{X_1}\pi_{x_2|x_1}^{X_2|X_1}\pi_{x_3|x_1x_2}^{X_3|X_1X_2}\pi_{x_4|x_2x_3}^{X_4|X_2X_3})(\pi_{a_1|x_1}^{A_1|X_1}\pi_{a_2|x_2}^{A_2|X_2}\pi_{a_3|x_3}^{A_3|X_3}\pi_{a_4|x_4}^{A_4|X_4}) \tag{7.40}$$

This model is obtained by simply extending the four-wave, first-order MLC model to include two second-order interaction terms—namely, $X_1X_2X_3$ and

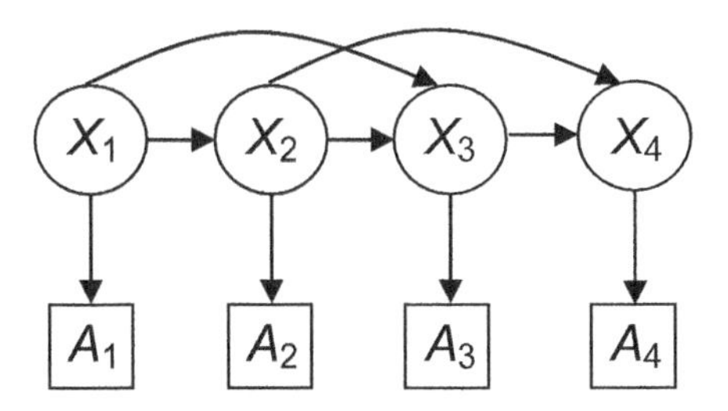

Figure 7.10 Path diagram for the second-order MLC model.

$X_2X_3X_4$. For binary latent variables, the first-order Markov model contains 9 (+1) parameters with 6 degrees of freedom for testing fit. Extending this model to the second-order formulation increases the number of parameters by 4 because each three-way interaction adds two parameters. This leaves 2 degrees of freedom for testing model fit.

It could be important to test the second-order MLC model against its first-order counterpart in situations where this test is possible, even when the first-order MLC model fits the data well, particularly if the focus of the analysis is on transition probabilities. Very different transition probabilities can be obtained by the addition of the second-order interaction terms. However, the effect of these terms on the error probability estimates may be less important. For example, Biemer and Bushery (2001) found estimates of the error probabilities for the CPS application to be quite robust to violations of the first-order Markov assumption. However, for the CEIS, this was not the case, as the following example shows.

7.2.4 Example: CEIS Analysis with Four Timepoints

In Section 7.1.8, data from the 1997 CEIS were analyzed to illustrate the MLC model for three panel waves. In this example, we revisit the CEIS to illustrate the application of the second-order MLC and the LCMS models, this time using all four waves of the CEIS (i.e., all four quarters of data for a given CU). In addition, data from 1997, 1998, and 1999 are combined for a total of 8817 CUs. Only CUs responding to all four waves are retained in the analysis.

For household expenditures, a *stayer* can be defined as a CU whose purchase status has essentially no chance (zero probability) of changing over the four quarters in the survey. Note that persons who did not purchase a particular commodity for any quarter (i.e., a 2222 purchase pattern) or who purchased the commodity in all four quarters (i.e., a 1111 purchase response pattern) may be still be movers because movers have a nonzero or nonunity probability of purchase. For example, a CU that owns no pet would have essentially no chance of purchasing petcare supplies. Likewise, a CU that has no automobile has essentially zero probability of purchasing automobile services. These are true nonpurchasing stayers. On the other hand, persons who own pets and/or automobiles may still not have purchased either item in any quarter just by chance. Of course, these are not stayers since there is a nonzero probability (albeit small) that they would purchase the consumer item.

As another example, CUs that are located in areas where cable TV is not offered have zero probability of purchasing cable services and are true nonpurchasing stayers. CUs that have cable TV, like it, and can afford it will have essentially zero probability of canceling their service during the four-quarter period and are true purchasing stayers. One objective of mover–stayer modeling is to divide these CUs who ostensibly look like stayers (so-called *nontransitional* CUs) into latent stayer ($H = 1$) and mover ($H = 2$) domains.

For some consumer items, a full or partial mover–stayer model could provide a better fit than could a mover-only model (i.e., the standard MLC model). For items having a small stayer subpopulation, the mover–stayer formulation may not offer any improvement. We shall apply the LCMS model to all 10 items considered in Section 7.1.8 and test whether the model fit is improved and the effect that this model has on the key parameter estimates.

Alternatively, the second-order Markov property might also be preferred for some consumer items, particularly items whose purchase in one quarter likely precludes their purchase in subsequent quarters. As an example, a CU that purchased a new car in the first quarter could have a much different probability of purchasing a car in subsequent quarters. The same might be said of other major purchases such as furniture and large kitchen appliances. A model postulating that the current quarter's purchasing status depends on the two previous quarters' purchases (nor nonpurchases) may provide a better fit to the data than might a model that considers only the previous quarter's purchases. The second-order MLC model has the advantage of a longer "memory" (one wave longer) than either the standard MLC model or the LCMS model for capturing influential, prior transitions. This longer memory could be advantageous for modeling the purchasing and reporting behaviors for some consumer items.

For this illustration, all three models were fit to the CEIS data. In the other studies [see, e.g., Tucker et al. (2003)], extensive model fitting was conducted to customize a model for each consumer item in the analysis. Their analysis investigated a number of grouping variables related to consumption and reporting accuracy, including CU size, income, age of CU informant, use of records, and interview length. For the present analysis, only two grouping variables are considered: the frequency with which the CU informant used records (R with three levels) and the length of the interview (L with two levels). Further, only three simple forms of each model will be fitted and compared as follows:

- MLC model (52 degrees of freedom)
  - Structural component: $\{X_1RL\}\{X_1X_2R\ X_1X_2L\}\{X_2X_3R\ X_2X_3L\}$ $\{X_3X_4R\ X_3X_4L\}$
  - Measurement component: $\{X_1RA_1\ X_1LA_1\}\{X_2RA_2\ X_2LA_2\}$ $\{X_3RA_3\ X_3LA_3\}$
- Second-order MLC model (44 degrees of freedom):
  - Structural component: MLC model + $\{X_1X_3R\ X_1X_3L\}$ $\{X_2X_4R\ X_2X_4L\}$
  - Measurement component: (same as MLC model)
- LCMS model (33 degrees of freedom):
  - Mover–stayer component: $\{HRL\}$
  - Structural component (stayers, $H = 1$) $\{X_1X_2X_3X_4R\ X_1X_2X_3X_4L\}$
  - Structural component (movers, $H = 2$) (same as MLC model}
  - Measurement component (stayers and movers)

Table 7.10 Fit Statistics for the Three Alternative Longitudinal Models

	MLC (52 df)				Second-order MLC (44 df)				LCMS (33 df)			
Variable	L^2	$D \times 100$	$p(L^2)$	BIC	L^2	$D \times 100$	$p(L^2)$	BIC	L^2	$D \times 100$	$p(L^2)$	BIC
CABLE	104.29	1.00	0.00	−368.1	73.9	0.81	0.003	−325.8	32.7	0.47	0.48	−267.1
GAS	123.61	0.84	0.00	−348.8	104.5	0.74	0.000	−295.2	29.4	0.35	0.65	−270.4
SPORTS	83.22	1.77	0.00	−389.2	67.1	1.44	0.014	−332.6	38.3	0.99	0.24	−261.5
FURN	70.52	1.46	0.04	−401.9	55.7	1.21	0.111	−344.0	51.1	1.04	0.02	−248.7
CLOTH	113.65	3.53	0.00	−358.7	101.2	3.24	0.000	−298.5	27.9	1.30	0.72	−271.8
SHOES	80.36	2.77	0.01	−392.0	63.5	2.26	0.029	−336.2	27.8	1.27	0.73	−272.0
DENTAL	93.60	2.37	0.00	−378.8	81.7	2.07	0.000	−318.0	30.4	1.04	0.60	−269.4
DRUGS	193.01	4.74	0.00	−279.4	167.3	3.97	0.000	−232.4	33.6	1.17	0.44	−266.2
EYES	66.29	1.55	0.09	−406.1	50.6	1.24	0.229	−349.1	35.9	0.91	0.33	−263.8
PETS	117.56	2.18	0.00	−354.8	108.2	1.93	0.000	−291.5	31.9	0.77	0.52	−267.9

$$\{X_1HRA_1 \;\; X_1HLA_1\}\{X_2HRA_2 \;\; X_2HLA_2\}\{X_3HRA_3 \;\; X_3HLA_3\}$$

All three models assumed time-homogeneous error probabilities. Obviously, the MLC model is nested within the second-order MLC model. Less obvious is that it is also nested within the LCMS model since setting $\pi_1^H = 0$ (i.e., setting the probability of a stayer to zero) will produce the first-order MLC model.

Table 7.10 displays some of the fit statistics produced by ℓEM for the three models and ten consumer items. For the MLC model, the p values are quite small which, as noted in previous analyses, does not necessarily indicate inadequate fit since the sample size (8817 observations) is quite large. On the other hand, the dissimilarity index D indicates the model fits the data adequately for all 10 items. It is particularly interesting to compare the fit for alternative models. The second-order MLC model shows some improvement in fit for two criteria (the dissimilarity index D and the model p value) but not for the BIC. For the LCMS model, all except one consumer item (FURN) show an acceptable fit by the p value and dissimilarity index criteria. However, the BIC is never smaller than BIC for the MLC model, which suggests that the standard MLC model may be the preferred model when fit and parsimony are balanced. Under the p-value criterion, the LCMS model is preferred for all consumer items except FURN, for which the second-order MLC model is preferred.

The choice of model can make an important difference in the estimates of indicator accuracy, as Table 7.11 shows. In this table the accuracy rates for true purchasers and nonpurchasers are shown for all three models. The differences between the first- and second-order MLC models are fairly minor, although there is a slight tendency for the second-order model to provide higher accuracy rates. The LCMS model, on the other hand, shows somewhat larger differences, particularly for SPORTS, EYES, and PETS. There is no consistent pattern regarding the direction of the differences. About half of the items have larger and half have smaller accuracy rates compared to the MLC model estimates. The last column in this table shows the estimated proportion of stayers in the population. The analysis indicates the existence of a substantial

Table 7.11 Selected Parameter Estimates for the Three Alternate Models

	MLC		Second-Order MLC		LCMS		
Variable	$\pi_{1\|1}^{A_1\|X_1}$	$\pi_{2\|2}^{A_1\|X_1}$	$\pi_{1\|1}^{A_1\|X_1}$	$\pi_{2\|2}^{A_1\|X_1}$	$\pi_{1\|1}^{A_1\|X_1}$	$\pi_{2\|2}^{A_1\|X_1}$	π_1^H
CABLE	0.989	0.990	0.992	0.996	0.991	0.990	0.556
GAS	0.986	0.995	0.989	0.996	0.990	0.997	0.885
SPORTS	0.557	0.944	0.595	0.955	0.739	0.967	0.639
FURN	0.403	0.931	0.434	0.936	0.347	0.932	0.694
CLOTH	0.861	0.791	0.890	0.835	0.852	0.819	0.545
SHOES	0.632	0.821	0.651	0.835	0.578	0.654	0.478
DENTAL	0.600	0.946	0.663	0.970	0.551	0.915	0.337
DRUGS	0.909	0.919	0.945	0.919	0.928	0.891	0.458
EYES	0.440	0.940	0.440	0.940	0.353	0.963	0.634
PETS	0.625	0.974	0.642	0.980	0.753	0.980	0.670

stayer group for all items. GAS has the highest proportion (almost 90%) of stayers and DENTAL has the lowest (about 34%).

All three models suggest that the most underreported items are SPORTS, SHOES, DENTAL, EYES, and PETS. Items with the best reporting accuracy include CABLE, GAS, CLOTH, and DRUGS. Surprisingly, false positives (overreporting) may be a concern for several items, especially SHOES. It is not well understood why the estimated reporting accuracies vary so much across consumer items and whether this variation indicates real error in the CEIS or is an artifact of the modeling process. For further details and discussion of the use of these models for the CEIS, see Tucker et al. (2003, 2004, 2005) and Biemer and Tucker (2001).

7.2.5 MLC Model with Time-Varying Grouping Variables

The previous sections of this book incorporated grouping variables in an LCA for three reasons: (1) to address the issues of heterogeneity; (2) to add degrees of freedom to the model, thus allowing more complex models to be fit; and (3) for their own intrinsic interest. The use of grouping variables was extended to MLC models in the same way that they were added to LC models in Section 7.1.6. However, special issues arise when using grouping variables in MLCA since, like the other variables in the model, grouping variables can also change values over time. Such grouping variables are referred to as *time-varying* as opposed to *time-constant* or *time-invariant* grouping variables. As an example, mode of interview, proxy/self response, and use of household records are just a few of the variables that have been used in prior analyses that could be redefined as time-varying.

There are several ways to handle time-varying covariates in a panel analysis. One way is to create a time-invariant variable that is equal to the value of the time-varying variable at one particular wave. As an example, an individual's age may be defined for all waves as "age at wave 1." This is adequate for most analyses, especially if waves are equally spaced in time. Another approach is to define a time-invariant grouping variable that summarizes the value of the time-varying variable over all waves. As an example, for the CPS labor force analysis of Section 7.1.7, a time-invariant grouping variable was introduced for proxy/self interviewing. This variable assumed four values according to whether proxy interviewing was used for (1) all of the waves, (2) most of the waves, (3) only one of the waves, or (4) none of the waves. Likewise, for the CEIS analysis, we defined a time-invariant variable summarizing the use of records at each interview.

A third way of dealing with time-varying characteristics that will be introduced in this section is to model changes in the value of the grouping variable at each wave. For T waves, this would involve introducing in the model T separate grouping variables, one for each wave. To illustrate, consider the proxy/self variable for a three-wave panel survey. Define a dichotomous grouping variable F_t, where $F_t = 1$ for self response and $F_t = 2$ for proxy response at wave t, for $t = 1,2,3$. Using these three variables instead of a time-constant proxy/self summary variable could produce a better fitting model but at the cost of using more degrees of freedom.

Another drawback of using time-varying covariates is that it increases the problems of data sparseness, which is already a problem for MLC models, especially for more than four waves. As noted in Section 5.1.4, data sparseness can render the usual chi-square fit statistics meaningless for assessing model fit. Convergence issues can also arise for some software packages. In addition, the problems of estimated zero frequencies and zero margins are also increased, making it difficult to compute the appropriate degrees of freedom for a model. Nevertheless, for some applications, the use of time-varying covariates can be important.

The path diagram in Figure 7.11 shows an MLC model for three waves with time-varying grouping variables, F_1, F_2, F_3, and a single time-constant grouping variable, G. For clarity, the path diagram is shown separately for the structural component and the measurement component. The structural component is further divided into a component that just involves the grouping variables and another that shows the interactions of the grouping variables with the latent variables. These can be written in shorthand notation as follows:

Grouping variable structure	$\{GF_1F_2F_3\}$
Latent variable structure	$\{X_1G\ X_1F_1\}\{X_1X_2G\ X_1X_2F_2\}\{X_2X_3G\ X_2X_3F_3\}$
Measurement component	$\{X_1GA_1\}\{X_2GA_2\}\{X_3GA_3\}$ where $X_1GA_1 = X_2GA_2 = X_3GA_3$

Mathematically the model can be written as

$$\pi_{gf_1f_2f_3a_1a_2a_3}^{GF_1F_2F_3A_1A_2A_3} = (\pi_{gf_1f_2f_3}^{GF_1F_2F_3})(\pi_{x_1|gf_1}^{X_1|GF_1}\pi_{x_2|gf_2x_1}^{X_2|GF_2X_1}\pi_{x_3|gf_3x_2}^{X_3|GF_3X_2})(\pi_{a_1|gx_1}^{A_1|GX_1}\pi_{a_2|gx_2}^{A_2|GX_2}\pi_{a_3|gx_3}^{A_3|GX_3}) \quad (7.41)$$

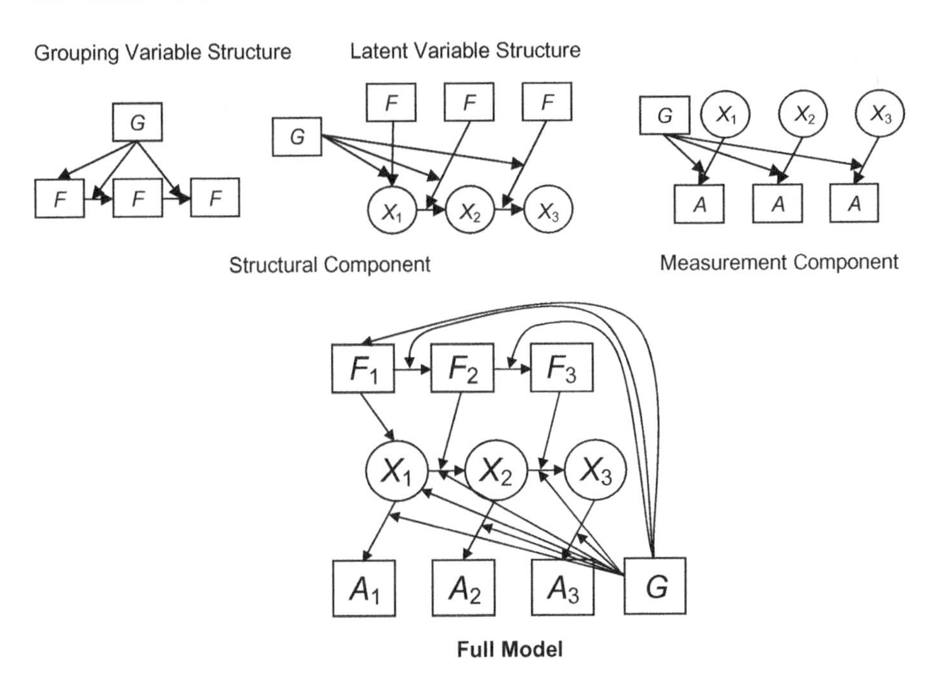

Figure 7.11 Path diagram for the MLC model with Markov time-varying covariates model specification: $\{GF_1F_2F_3\}\{X_1G\ X_1F_1\}\{X_1X_2G\ X_1X_2F_2\}\{X_2X_3G\ X_2X_3F_3\}\{X_1GA_1\}\{X_2GA_2\}\{X_3GA_3\}$ with constraint $X_1GA_1 = X_2GA_2 = X_3GA_3$.

with constraints on the latent parameters given by

$$
\begin{aligned}
\pi_{x_1|gf_1}^{X_1|GF_1} &= \frac{\exp(u + u_{x_1}^{X_1} + u_{x_1g}^{X_1G} + u_{x_1f_1}^{X_1F_1})}{\sum_{x_1}\exp(u + u_{x_1}^{X_1} + u_{x_1g}^{X_1G} + u_{x_1f_1}^{X_1F_1})} \\
\pi_{x_2|gf_1x_1}^{X_2|GF_2X_1} &= \frac{\exp(u + u_{x_2}^{X_2} + u_{x_1x_2}^{X_1X_2} + u_{x_2g}^{X_2G} + u_{x_2f_2}^{X_2F_2} + u_{x_1x_2g}^{X_1X_2G} + u_{x_1x_2f_2}^{X_1X_2F_2})}{\sum_{x_2}\exp(u + u_{x_2}^{X_2} + u_{x_1x_2}^{X_1X_2} + u_{x_2g}^{X_2G} + u_{x_2f_2}^{X_2F_2} + u_{x_1x_2g}^{X_1X_2G} + u_{x_1x_2f_2}^{X_1X_2F_2})} \\
\pi_{x_3|gf_3x_3}^{X_3|GF_3X_2} &= \frac{\exp(u + u_{x_3}^{X_3} + u_{x_2x_3}^{X_2X_3} + u_{x_3g}^{X_3G} + u_{x_3f_3}^{X_3F_3} + u_{x_2x_3g}^{X_2X_3G} + u_{x_2x_3f_3}^{X_2X_3F_3})}{\sum_{x_3}\exp(u + u_{x_3}^{X_3} + u_{x_2x_3}^{X_2X_3} + u_{x_3g}^{X_3G} + u_{x_3f_3}^{X_3F_3} + u_{x_2x_3g}^{X_2X_3G} + u_{x_2x_3f_3}^{X_2X_3F_3})}
\end{aligned}
\tag{7.42}
$$

and where it is assumed that $\pi_{a_1|gx_1}^{A_1|GX_1} = \pi_{a_2|gx_2}^{A_2|GX_2} = \pi_{a_3|gx_3}^{A_3|GX_3}$ (i.e., time-homogeneous error parameters).

The model specifies a saturated model for the component of the model that involves only grouping variables: $GF_1F_2F_3$. The number of parameters in this component can be reduced by considering only three-way or two-way interactions for the manifest component. As an example, one could replace the manifest component in (7.41) by $\{GF_1F_2\ GF_2F_3\}$. However, this reduction in complexity is unlikely to affect the structural or measurement components, which are usually the focus of the MLCA.

The three submodels representing the latent variable part of the structural component specifies that the initial state and the wave-to-wave transitions both depend on the time-constant variable G as well as the time-varying variable F. These components can also be further reduced if desired to achieve greater parsimony. However, when the focus of the analysis is on the measurement component, it is seldom necessary to optimize the structural component.

For the measurement component, the error parameters are allowed to depend on G but not F at each wave. A more elaborate structure would allow the error parameters to depend also on the current value of F by substituting $\{X_1GF_1A_1\}$ or, perhaps $\{X_1GA_1\ X_1F_1A_1\}$, for $\{X_1GA_1\}$. For example, if G is gender and F_t is mode of interview at wave t, then either of the former models would allow the error probabilities to differ by mode of interview. Further, the effects of the interview mode on misclassification may be different for males and females. Many other model specifications are possible and can be explored. However, the risk of model nonidentifiability increases as the models become more elaborate, so evaluating model identifiability is particularly important for these analyses.

7.2.6 Example: Assessment of Subject Interests

For this example, data from Vermunt et al. (1999) will be used to illustrate the use of time-varying covariates. These data are from a study of secondary school students in Germany who were interviewed annually for 3 years regarding their interests in various subjects in school and their grades in these subjects. Vermunt et al. conducted an MLCA for the variable X defined as "interest in physics" in order to test the hypothesis that interest in physics as a course of study declines at a faster rate for females than for males. Since X is likely to be measured with error, an MLCA was conducted for this variable.

Unlike the examples considered thus far, this example focuses primarily on the structural component of the model rather than the measurement component. The major hypotheses to be tested concerns whether males and females differ in their interests in physics as they progress through the grades, rather than whether their interest in physics is being measured accurately by the survey questions. For that reason, the authors gave much attention to correctly specifying the structural component of the model.

Using our notation, let X_t denote a dichotomous latent variable "at least some interest" ($X_t = 1$) and "little or no interest" ($X_t = 2$) for $t = 1,2,3$ representing the interviews at grades 7, 8, and 9, respectively. Correspondingly, A_t denotes the survey observation of X_t. Define two grouping variables: a time-constant variable, G for gender where $G = 1$ for males and $G = 2$ for females and a time-varying covariate representing a student's "grade in physics" in

Table 7.12 Test Results for the Estimated Models

Model	X^2	L^2	df	$p(X^2)$	$p(L^2)$	BIC
1. Basic	139.45	142.94	99	0.005	0.003	4466
2. Basic + $\{X_1X_3\}$	118.35	127.27	98	0.079	0.025	4556
3. Basic + $\{X_1X_2F_2\}$	141.35	140.69	97	0.002	0.003	4476
4. Basic + $\{X_1X_2G\ X_2X_3G\}$	137.99	142.69	97	0.004	0.002	4478
5. Basic + $\{X_2F_1\ X_3F_2\}$	140.08	142.52	97	0.003	0.002	4478
6. Basic + $\{X_1F_2\ X_2F_3\}$	119.28	123.08	97	0.062	0.038	4459
7. Basic + $\{X_1X_3\ X_1F_2\ X_2F_3\}$	95.23	107.88	96	0.503	0.192	4450

Source: Vermunt et al. (1999), Table 1.

year t denoted by F_t, where $F_t = 1$ for "low" grade and 2 for "high" grade. Of particular interest in the study was the change in student interest in physics (X) over the 3 years as a function of their gender and their grades in physics. Only students who completed all three interviews were included in this analysis—a total of 541 students split approximately evenly between boys and girls. Vermunt et al. (1999) also considered the effect of panel attrition on the results. Panel attrition effects for this example are discussed in Section 7.3.3.

The authors fit a number of models ranging from the standard MLC model to various second-order Markov models using the ℓEM software. Table 7.12 summarizes their results. Their model-building strategy was to begin with a plausible yet highly restricted model and then to consider progressively more complex models, improving model fit with each enhancement. The "basic" or starting model was a first-order MLC model that assumed that students' interests in physics at time t depend only on their interests at time $t-1$, their gender, and their grades at time t. Measurement error was assumed to be independent of gender and grade, which seems untenable but still a reasonable place to start the model-building process. This is precisely the model in (7.41) after setting $GA_1X_1 = 0$; that is, the model with $\{A_1X_1\}$ as the measurement component. This model specifies that the classification error is unrelated to the grouping variables.

Vermunt et al. make several points about these results, which can be summarized as follows:

- According to the values of the p values [i.e., $p(X^2)$ for the X^2 criterion and $p(L^2)$ for the L^2 criterion], models 1, 3, 4, and 5 do not fit the data.
- Models 2 and 6 are accepted at the 5% level by the X^2 criterion but not by the L^2 criterion. The reason for the conflicting test results is due to data sparseness. Conservatively, they decide to reject both models.
- Model 7 fits the data under both the X^2 and L^2 criteria and is therefore accepted. This model suggests that the first-order Markov assumption

does not hold for these data. Further, students' interests at time t seem to affect their grades at time $t + 1$.

The authors further explored the seemingly oversimplified structure of the measurement component. They imposed the restriction that $\pi_{2|1}^{A_1|X_1} = \pi_{1|2}^{A_1|X_1} = 0$ (i.e., the hypothesis of no classification error) and accepted this restriction, concluding that there was no evidence of measurement error in these data. Doubting this result, they explained that the time-varying covariates may have caused the estimates of measurement error to be attenuated. The reason, according to Vermunt et al., is that F_t probably absorbed some of the variation in the measurement of the latent states as a result of classification error. (See Section 8.3 for further discussion of this phenomenon.) Likewise, they concluded that when F_t was not in the model, the classification error terms may have absorbed some of the unobserved heterogeneity in the latent states that could be explained by F_t. This suggests that in Markov models with a single indicator per wave, it can be difficult to distinguish measurement error from unobserved homogeneity. The next section offers one remedy for this problem: the multiple indicators per wave.

7.2.7 Multiple Indicators at One or More Waves

In this section, we consider the situation where there are multiple indicators of the latent variable at one or more waves. The path diagrams in Figures 7.12 and 7.13 depict two possible scenarios that will be considered. Many additional designs are feasible. Figure 7.12 depicts a two-wave panel survey with two indicators per wave. A good example of this scenario is the CPS, where, at each wave, a subsample of the original survey respondents is reinterviewed. Each

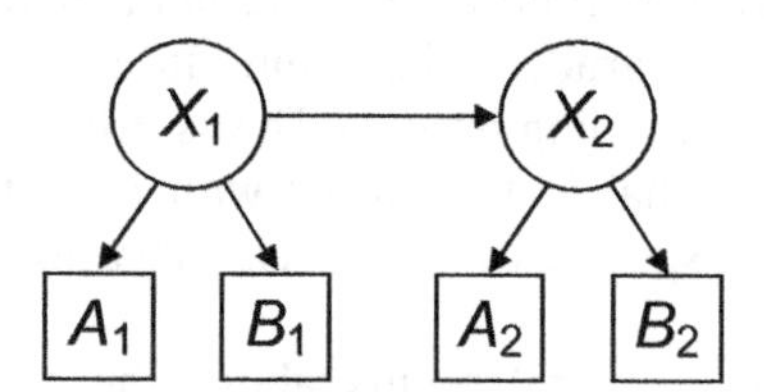

Figure 7.12 Path diagram for an LC model for two waves and two indicators per wave.

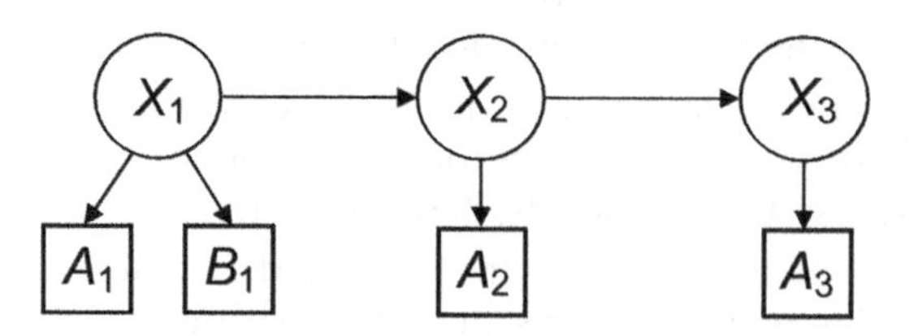

Figure 7.13 MLC model with two indicators at wave 1 only.

reinterview person is asked about his or her labor force status at the time of the original interview. Thus, A_t denotes the original interview response and B_t denotes the reinterview response at wave t. Both A_t and B_t are indicators of X_t. Now the data to be analyzed are organized in the K^4 cross-classification table $A_1B_1A_2B_2$ with the likelihood kernel given by

$$\pi_{a_1b_1a_2b_2}^{A_1B_1A_2B_2} = \sum_{x_1}\sum_{x_2}(\pi_{x_1x_2}^{X_1X_2})(\pi_{a_1b_1a_2b_2|x_1x_2}^{A_1B_1A_2B_2|X_1X_2}) \tag{7.43}$$

The structural component for this model can be written as

$$\pi_{x_1x_2}^{X_1X_2} = \pi_{x_1}^{X_1}\pi_{x_2|x_1}^{X_2|X_1} \tag{7.44}$$

while the measurement component takes the form

$$\pi_{a_1b_1a_2b_2|x_1x_2}^{A_1B_1A_2B_2|X_1X_2} = \pi_{a_1|x_1}^{A_1|X_1}\pi_{b_1|x_1}^{B_1|X_1}\pi_{a_2|x_2}^{A_2|X_2}\pi_{b_2|x_2}^{B_2|X_2} \tag{7.45}$$

which assumes local independence within waves and independent errors between waves (i.e., ICE). Bassi (1997) showed that, for $K = 2$ classes, the model is identifiable without further restrictions. Note that for $K = 2$ there are 16 cells and 11 (+1) parameters so the model has 4 degrees of freedom left for testing. For $K = 3$ there are 81 cells and 32 (+1) parameters for a net of 48 model degrees of freedom.

Bassi (1997) also showed that it is possible to relax the ICE assumption for the model in Figure 7.12. For $K = 2$ classes, Bassi showed that the model including a direct effect (i.e., A_1A_2) in the submodel for $A_2|A_1X_2$ is globally identifiable without further constraints on the model specified in (7.43)–(7.45). Testing the effects of correlated errors between waves may be critical for panel surveys where the time between panel interviews is short (say, one month or less) and the risk of memory effects is high.

By extension, MLC models for three or more waves and having two indicators per wave are also identifiable without imposing the time-homogeneous error probability assumption. This could be important for panel surveys were the error probabilities at the first wave are quite different from those for subsequent waves. As an example, in surveys like the CEIS, respondents who respond positively to an expenditure screening question are asked a series of additional questions about that expenditure. These additional questions can be burdensome to some respondents. Therefore, after the initial interview, some respondents may be reluctant to report their expenditures in order to shorten the interview [i.e, they satisfice; see Krosnick (1991)]. Consequently, error probabilities for the first and second interviews could be quite different.[4]

[4]This may not be an important problem for the CEIS analysis since the first interview is a bounding interview and consequently was omitted from the MLCA as well as from the CEIS estimates.

Bassi (1997) showed that the unconstrained, three wave first-order MLC model given by

$$\{X_1\}\{X_2|X_1\}\{X_3|X_2\}\{A_1|X_1\}\{B_1|X_1\}\{A_2|X_2\}\{B_2|X_2\}\{A_3|X_3\}\{B_3|X_3\}$$

is identifiable for $K = 2$. The identifiability of more complex models involving, for instance, time-constant and time-varying grouping variables, second-order Markov properties, and correlated errors both between and within waves have not been explored in the literature. Therefore, carefully and thoroughly checking for identifiability of such models is especially important to ensure model validity.

In some situations, multiple indicators are not available at all waves. For example, a special study to evaluate measurement error—say, a test–retest reinterview survey—may have been conducted only at wave 1 of a panel survey. Thus, at wave 1, two replicate measurements of the latent variable are available whereas subsequent waves have only one (see Figure 7.13). In this scenario, it is possible to estimate the classification probabilities separately for wave 1. However, time homogeneity must be assumed for waves 2 and 3. The identifiable model takes the form

$$\{X_1\}\{X_2|X_1\}\{X_3|X_2\}\{A_1|X_1\}\{B_1|X_1\}\{A_2|X_2\}\{A_3|X_3\}$$

with the constraint $A_2|X_2 = A_3|X_3$ (i.e., $\pi_{a_2|x_2}^{A_2|X_2} = \pi_{a_3|x_3}^{A_3|X_3}$) imposed.

7.3 FURTHER ASPECTS OF MARKOV LATENT CLASS ANALYSIS

7.3.1 Estimation Issues with MLCA

In Chapter 5, a number of issues in the estimation of LC models were discussed, including nonidentifiability, data sparseness, boundary estimates, local maxima, and latent class flippage. These problems are also encountered in MLCA since the standard LC model is just a special case of the standard MLC model (Langeheine and van de Pol 1993). To understand why, consider the standard ML model in Figure 7.2. If we constrain the transition probabilities so that

$$\pi_{x_2|x_1}^{X_2|X_1} = \pi_{x_3|x_2}^{X_3|X_2} = 1 \quad \text{for all} \quad x_1 = x_2 = x_3 \tag{7.46}$$

or, equivalently, if we constrain the transition matrix between any two successive timepoints to be the identity matrix, the model is reduced to an LC model for three indicators. Thus, any LC model can be expressed as an MLC model with certainty constraints and, therefore, the estimation issues affecting LC models also affect MLC models. Moreover, the approaches for addressing these issues described in Chapter 5 can also be applied to MLC models with similar results.

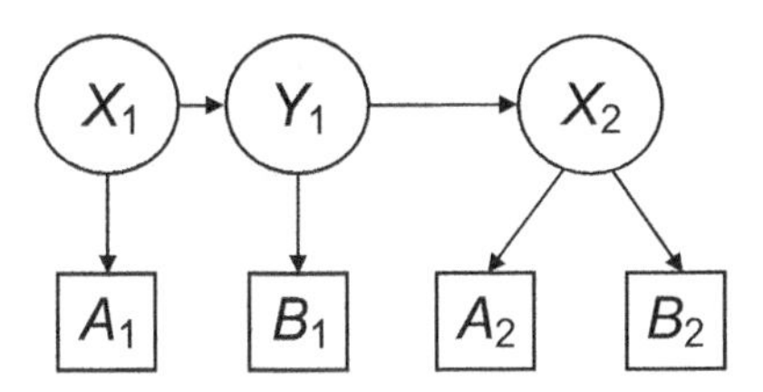

Figure 7.14 MLC model with two indicators at time 1 expressed as a single indicator MLC model with constrained transition probabilities between X_1 and Y_1.

This is not only a clever way of viewing LC models as MLC models with constrained transitions; it is also very useful technique that allows us to deal with many of the issues common to both LCA and MLCA simultaneously. The approach can also be applied to any subset of successive panel waves to create an MLC model with multiple indicators with waves. As an example, note that the model in Figure 7.14 is equivalent to the model in Figure 7.13 when $\pi_{y_1|x_1}^{Y_1|X_1} = 1$ for $x_1 = y_1$. In that situation, $\Pr(X_1 = Y_1) = 1$; that is, X_1 and Y_1 are the same construct.

The estimation issues for LC models are often compounded in MLC models because of their increased complexity. Like LC models, MLC models with multiple indicators per wave are subject to local dependence issues caused by heterogeneity, within-wave error correlations, and bivocality. Moreover, the additional assumptions of Markov transitions, time-homogeneous error, and ICE can also be violated in MLCA. Addressing these issues usually requires adding more parameters to the model, thus increasing model complexity.

As an example, to address heterogeneity, both time-constant and time-varying grouping variables should be added to the MLC model. However, this solution will exacerbate the problem of data sparseness and over-parameterization. Fortunately, as Collins and Tracy (1997) and Collins et al. (1996) show, data sparseness is not an important concern for obtaining model-unbiased estimates of the MLC model parameters. However, as noted in Chapter 5, model testing and selection can be challenging because of distributional problems with the chi-square statistics. Software packages such as PANMARK, Latent GOLD, and MPlus aid the analyst in these situations through the use of empirical bootstrapped distributions for the chi-square statistics. However, ℓEM has no provision for this. Moreover, ℓEM can handle a maximum of only eight panel waves for dichotomous indicators and four waves for indicators having five or more categories (Vermunt et al. 1999).

Panel survey samples are typically selected with unequal probability, multistage cluster sampling designs. Also, as noted in Section 5.3, it is important to take into account the complex sample design in the analysis using the methods described in that section. The bias due to survey nonresponse is also an important issue for both LCA and MLCA. However, the problems of missing data can be exacerbated for panel surveys since the response burden on respondents can be much greater. The increased respondent burden affects both unit nonresponse and item nonresponse. Especially for panel surveys with many waves, the combined effects of nonresponse at the baseline interview, at each subsequent interview wave, and at each questionnaire item can be considerable, requiring different modeling strategies. This topic is addressed in the next section.

7.3.2 Methods for Panel Nonresponse

For almost all sample surveys, nonresponse is inevitable and unavoidable. Sampling units may be missing for a variety of reasons [see Groves and Couper (1998) for a comprehensive study of household survey nonresponse]. One type of nonresponse is at the unit (e.g., individual, household, establishment) level (referred to as *unit nonresponse*). This arises as a result of some sample members not being interviewed and, thus, the entire interview is missing. However, nonresponse also occurs at the question or item level (referred to as *item nonresponse*) when an interviewed sample member fails to respond to an item. Unit nonresponse may be the result of refusals, noncontacted or unlocatable units, language barriers, illnesses, temporary absences, or even lost or destroyed questionnaires. Item nonresponse can take the form of a refusal to answer a question, a "don't know" response, or an inadvertently skipped question.

The problems of missing data are further exacerbated in panel surveys, particularly for longitudinal analysis. In addition to unit and item nonresponse, a panel member may drop out of the panel, never to return; this is referred to as *panel attrition*. In other cases, panel members may be interviewed in some waves and not in others in more or less random patterns. Of course, this type of attrition is possible only if the survey specifies that previously noninterviewed panel members are followed up in subsequent panel waves. Not all panel surveys use this followup approach. Panel nonresponse shares some characteristics of both unit and item nonresponse since there may be extensive data on panel nonrespondents that were acquired in prior panel waves. These data may be used to impute the missing panel interviews instead of using survey weighting adjustments for unit nonresponse. In what follows, panel nonresponse will be treated as a type of item nonresponse for the purposes of data analysis.

Regardless of the type, level, or reason, the consequences of nonresponse are the same—missing data limits data analysis. In this section, we will briefly discuss some techniques for compensating for missing data in loglinear modeling with latent variables with particular emphasis on MLCA. However, as discussed previously, LCA can be viewed as a special case of MLCA, so the techniques discussed can be applied to LCA as well. Our discussion is not intended to be a comprehensive review of the methods; rather, it is intended to cover a few commonly used methods. For example, the problems of unit nonresponse are not discussed. For a general treatment of the methodology for dealing with missing data in survey data analysis, including latent variable analysis, see Little and Rubin (1987), Allison (2001), and Schafer (1997).

One common method for dealing with missing values in data analysis due to item nonresponse is *case deletion*. With this approach, cases that have a missing value on one or more of the items in analysis are simply removed from the analysis. What are left are cases that have complete data on all the analysis variables. Many statistical software packages automatically perform case deletion for any case having a missing value for one or more variables. In fact, case deletion was the default method used in the previous examples throughout this book. When only a small fraction (say, 5% or less) of the cases are incomplete, case deletion can be reasonable a way to deal with nonresponse because the bias in the estimates caused by nonresponse is probably not important. However, when item missingness or panel nonresponse is extensive, the issue of analysis bias due to nonresponse should be considered.

As an example, for an MLCA of a three-wave survey, only persons who responded at all three waves would be retained using the case deletion approach. Moreover, persons having a missing value on either A_1, A_2, A_3, or any grouping variable in the analysis would also be removed from the analysis dataset even if they otherwise responded at all three waves. When the data analysis involves many variables, each having an appreciable level of item nonresponse, case deletion will result in discarding a substantial portion of the entire dataset. This not only increases the standard errors of the estimates due to the reduction of n but can also impart biases in the resulting parameter estimates to the extent that the deleted cases differ systematically from the retained ones.

An alternative to case deletion is *imputation*. This method fills in the missing data with plausible values. Imputation is usually done for a limited number of variables deemed most important for general data analysis; however, in panel surveys, it is also possible to impute an entire missing panel interview. In using imputed variables, analysis proceeds as though the filled-in data were actually observed. Imputation methods can also create problems because the joint distributions of the variables may be distorted. As Schafer (1997) notes, estimates of variances and covariances will tend to be attenuated and standard errors, and p values and other measures of uncertainty may also be quite biased. Methods for multiply imputing missing values [see, e.g., Rubin (1987)] are designed to compensate for imputation uncertainty, but these methods can be quite complex to use without the proper software. In addition, methods for multiple imputations for loglinear models with latent variables have yet to be developed. Furthermore, retaining the imputed data in MLCA would seem to bias estimates of classification error in ways that are not well understood. It may be preferable to remove the imputed data records from the analysis, especially if the primary goal of the MLCA is the estimation of misclassification probabilities. Clearly, the use of imputed data in latent variable analysis raises many issues that require further research.

Fortunately, for MLCA (as well as other types of categorical data analysis), there is an alternative methodology for handling missing data that does not involve case deletion or imputation. This is referred to as the *maximum-likelihood approach.* With this approach, a model is specified for the *response* (or *missing data*) *mechanism*—namely, the underlying process that generates or *causes* the missing data. To specify an appropriate model for the response mechanism, a new type of grouping variable is created called a *response indicator*, which is 1 if the variable, say, A, is observed and 2 if it is missing. The response indicator behaves somewhat like a grouping variable in that it identifies a group of sample members who have a value for A and another group that does not. It is different from a grouping variable because it is defined at the sample level, not the population level. A submodel is specified for the response indicator to describe its variation in the sample in terms of the other variables in the model. In this way, the response indicator is treated like another dependent variable in the MLC model. Using familiar methods from modified path modeling, the functional form of the *incomplete data likelihood*, that is, the likelihood associated with the respondent only data table, can be specified. Maximizing this likelihood using the approaches discussed previously in this book produces the estimates of all model parameters, including those for the response mechanism. The EM algorithm is ideally suited for this estimation process. Before examining the maximum-likelihood approach in greater detail, some additional terminology must be introduced.

Little and Rubin (1987) developed a widely used terminology for describing the statistical assumptions underlying the response mechanism. They define three types of missing data: data that are *missing completely at random* (MCAR), *missing at random* (MAR), or neither MCAR nor MAR referred to as *not missing at random* (NMAR). MCAR means that the response mechanism does not depend on any of the variables in the model. MCAR means that the probability of a missing data value is the same for every cell of the unobserved, complete manifest data table. This assumption is equivalent to selecting the missing observations by SRS.

Missing at random, which can be a much weaker assumption than MCAR, implies that the probability of a particular pattern of nonresponse depends only on variables that are fully observed for every sample unit. Obviously, the plausibility of this assumption depends on which fully observed variables are available for modeling the nonresponse mechanism. If the response mechanism is neither MCAR nor MAR, then it is NMAR by definition. The response mechanism is said to be *ignorable* either if it is MAR or if is not MAR, but the joint likelihood for the incomplete data table is still factorable into two components: one that does not depend on the response indicators and another that depends *solely* on the response indicators. In can be shown [see, e.g., Vermunt (1997a)] that when the response mechanism is ignorable, the model parameters estimates, apart from those that describe the response mechanism, will not depend on the specification of the response model. If the response mechanism does not satisfy these two conditions, it is said to be *nonignorable* (Rubin 1987, p. 51). It is important that the model take into account nonresponse when it is nonignorable since, otherwise, the model parameter estimates could be biased, sometimes substantially so.

For a three-way panel survey with indicators A_1, A_2, and A_3, define the response indicators R_1, R_2, and R_3, respectively, as indicator variables for "not missing" (denoted by $R_t = 1$ whenever A_t is observed) and "missing" (denoted by $R_t = 2$ whenever A_t is not observed). Note the dual use of the word "indicator" here. To avoid confusion, R_t will always be referred to as a "response indicator," while A_t will be referred to simply as an "indicator" (as in "an indicator of the latent variable X_t"). Suppose that the model contains a single grouping variable G. If R_t is independent of G and A_t, $t = 1,2,3$, then nonresponse to A_t is said to be MCAR. If R_t and G are correlated, then nonresponse to A_t is said to be MAR. In other words, the missing data mechanism is explained by G. If R_t depends on A_t, nonresponse is nonignorable. Vermunt (1996) discusses other possibilities with the MLC model that will result in either MAR or nonignorable nonresponse. For example, nonresponse at one wave may influence nonresponse at a later wave. Either ignorable or nonignorable nonresponse may result in MLC models depending on the circumstances and whether the response indicator component of the incomplete data likelihood can be factored as discussed above. For our purposes, the distinction is not particularly cogent because both ignorable and nonignorable response mechanisms can be modeled in ℓEM and other software packages using Fay's technique, which will be discussed next.

Fay (1986) showed how the parameters of the response model can be estimated for manifest path models and loglinear models. Under Fay's approach, various types of ignorable and nonignorable response models may be estimated and the assumptions of nonignorability, MCAR and MAR can all be tested. Vermunt (1996) extended Fay's approach to loglinear models with latent variables and implemented his methodology in the ℓEM software. As a requirement of both Fay's approach and Vermunt's extension, the structural and measurement components of the model must not contain any response indicators; that is, the response indicators must have their own submodel, and they cannot appear as independent variables in a submodel for any variable other than another response indicator.

In addition to occurring for the dependent variables, nonresponse can also occur in the covariates. Covariate nonresponse can be handled in the same manner as nonresponse in the independent variables, specifically, using response indicator variables. For example, suppose that a time-varying covariate F_t is added to the three-wave model, and let S_t, defined in analogy to R_t, denote the response indicator for F_t. Now the incomplete data likelihood will be quite complex, involving 13 variables: G, F_1, F_2, F_3, A_1, A_2, A_3, R_1, R_2, R_3, S_1, S_2, S_3. Obviously, data sparseness issues are at the forefront of such an analysis since, even for dichotomous variables, the number of cells in the complete data tables will be $2^{13} = 8192$ cells! With each new wave, the number of cells increases exponentially. For this reason, a more prudent approach might involve case deletion for variables having a small or moderate amount of missing data (e.g., less than 10% of the cases are missing the variable) and the maximum-likelihood approach for variables subject to more substantial amounts of missing data.

A somewhat simpler method for handling nonresponse in the indicator variables proposed by Langeheine and van de Pol (1994) deserves mention. For this method, a "missing" category is added to each indicator variable to represent their missing values while their corresponding latent variables are unchanged. For example, if A_t is a dichotomous indicator of X_t, a third category would be added to A_t, and all cases missing an A_t classification would be assigned to this third category. However, X_t is still dichotomous. This is essentially equivalent to assuming the MAR mechanism for A_t. An MCAR mechanism can be specified by setting the $\Pr(A_t = 3)$ equal across all variables in the model.

7.3.3 Example: Assessment of Subject Interests with Nonresponse

To illustrate the maximum-likelihood approach, the data used in Section 7.2.6 will be reanalyzed. Recall that these data are from a panel study of secondary school students in Germany analyzed in Vermunt et al. (1999). As before, let A_t denote interest in science for three grades, $t = 1,2,3$, where G denotes gender and F_t is a time-varying covariate denoting "grade in physics" for year t. All variables are dichotomous.

For this analysis, gender (G) is a fully observed variable and there is very little missing data for the A variables. Therefore, case records that were missing either A_1, A_2, or A_3 were deleted in the analysis. Thus, only the response mechanisms associated with the time-varying covariates need be modeled in this analysis. Of the 637 retained cases, 49 cases were missing F_1, 19 were missing F_2, and 38 were missing F_3. Thus, let R_1, R_2, and R_3 denote the response indicators for F_1, F_2, and F_3, respectively. There are eight groups of respondents defined by the eight possible response patterns $R_1R_2R_3$ as follows: 111, 112, 121, 122, 211, 212, 221, and 222. For example, the pattern 111 denotes a group of sample members for whom F_1, F_2, and F_3 were all observed; the pattern 122 denotes a group for whom only F_1 was observed; the pattern 222 denotes a group for whom F_1, F_2, and F_3 are all missing; and so on.

The best and final model considered in Section 7.2.6 for these data (viz., model 7 in Table 7.12) was respecified with the addition of three submodels corresponding to the three response mechanisms for F_1, F_2, and F_3. Figure 7.15 provides the ℓEM input statements for fitting this model under the MCAR assumption for the response mechanism. The MAR assumption could be imposed for R_1 by replacing $\{R_1\}$ in the MCAR model by $\{GR_1\}$ (written as

```
man 7
res 3
lat 3
dim  2  2  2  2  2  2 2  2  2  2  2  2  2
lab R1 R2 R3 X1 X2 X3 G F1 A1 F2 A2 F3 A3
sub G.F1.A1.F2.A2.F3.A3  *R1.R2.R3 = 111
    G.F1.A1.F2.A2.A3     *R1.R2.R3 = 112
    G.F1.A1.A2.F3.A3     *R1.R2.R3 = 121
    G.F1.A1.A2.A3        *R1.R2.R3 = 122
    G.A1.F2.A2.F3.A3     *R1.R2.R3 = 211
    G.A1.F2.A2.A3        *R1.R2.R3 = 212
    G.A1.A2.F3.A3        *R1.R2.R3 = 221
    G.A1.A2.A3           *R1.R2.R3 = 222
mod G.F1.F2.F3    {G.F1.F2.F3}
    X1|G.F1.F2    {G.X1 F1.X1 F2.X1}
    X2|G.F2.F3.X1 {G.X2 F2.X2 X1.X2 F3.X2}
    X3|G.F3.X1.X2 {G.X3 F3.X3 X1.X3 X2.X3}
    A1|X1
    A2|X2 eq1 A1|X1
    A3|X3 eq1 A1|X1
    R1|X1.F1  {R1}  * MCAR
    R2|X2.F2  {R2}  * MCAR
    R3|X3.F3  {R3}  * MCAR
rec 432
rco
dat vlb99.dat
cri 0.00000001
ite 10000
```

Figure 7.15 ℓEM input statements for model 7 in Table 7.9 with response indicators under MCAR response mechanism.

`{G.R1}` in the ℓEM syntax). Likewise, a nonignorable assumption could be imposed substituting `{F1.R1}` for `{R1}` in the ℓEM syntax. The other two response indicators could be handled similarly. Many other possibilities could be explored and tested.

Unfortunately, Vermunt et al. note that the small dataset did not permit the formal testing of alternative assumptions for the response mechanism. Problems with data sparseness forced the authors to resort to more ad hoc methods for exploring the effects of nonresponse on the estimates. After exploring both ignorable and nonignorable response models and noting that none of them improved the model fit appreciably, the authors concluded that the data are very nearly MCAR and, thus, that the effects of nonresponse on the estimates are negligible.

CHAPTER 8

Survey Error Evaluation: Past, Present, and Future

The previous chapters of the book traced the remarkable evolution of survey error evaluation since the 1950s. The current methodology for evaluating survey error has evolved from simple concepts established in the midtwentieth century to the complex, loglinear, latent variable models that have been implemented in modern software packages. This progress has been aided to a large extent by similar sweeping advances in categorical data analysis more generally. Even after this incredible evolution, the conceptual foundations of the early models are still ever-present in contemporary model parameterizations. As an example, the two-stage response process described in Chapter 2 still forms the basic foundation of all survey error evaluation. Chapters 3, 4, and 5 explicated and exploited these conceptual and technical linkages from early to current models. The current chapter highlights some of the major innovations encountered in this evolution and reflects on the relationships between early models and modern models. It also briefly examines the current state of the art and considers what the future holds for this important area of survey methodology and data analysis. The chapter concludes with some ideas for future research for continuing the evolution.

8.1 HISTORY OF SURVEY ERROR EVALUATION METHODOLOGY

8.1.1 The US Census Bureau Model for Survey Error

In the early 1930s and 1940s, Jerzy Neyman's groundbreaking work on randomization (Neyman 1934) provided the theoretical foundations for survey sampling. Survey error was essentially sampling error, reliability referred to

---

sampling variance. Little was known about nonsampling errors. But in the early 1940s, statisticians at the US Census Bureau (referred to as the US Bureau of the Census at that time) began to realize the importance of nonsampling error in surveys (Deming 1944). Many of the essential concepts for survey error modeling that we use today were developed at the Census Bureau in the 1950s and 1960s by Morris Hansen and his colleagues (Hansen et al. 1951, 1961, 1964). Hansen and his colleagues at the Census Bureau considered the following simple model for an observation y_i on the ith unit

$$y_i = Y_i + e_i \tag{8.1}$$

where $Y_i = E(y_i|i)$ is the mean of the ith individual's response distribution and e_i represents the departure of a single response from Y_i. From a sampling statistician's perspective, the response process can be viewed as a two-stage sample where the primary-stage sampling unit (PSU) is the individual or respondent and the secondary-stage unit (SSU) is the response from the individual. In that sense, e_i is a type of sampling error associated with the secondary-stage selection that is implemented by the act of responding to a question. Expressions for the variance of common survey statistics follow immediately by applying the usual two-stage (or multistage) sampling formulas.

In two-stage sampling, the variance of the usual estimator of the population mean consists of two variance components: between-PSU variance and the within-PSU sampling variance. For the Census Bureau model, these variance components become the sampling variance (SV) and the simple response variance (SRV), respectively (Section 2.1.1). These components can also be derived by the usual decomposition of variance formula, which equates the variance of the observation to the expected value of the conditional variance plus the variance of the conditional expected value

$$Var(y_i) = E\, Var(y_i \mid i) + Var\, E(y_i \mid i) \tag{8.2}$$

where the conditional variance and expectation fix the sampling unit and the unconditional variance and expectation vary the sampling unit across all possible samples from the population. It is shown in Section 2.1.1 that SV $= E\,Var(y_i \mid i)$ and SRV $= Var\,E(y_i \mid i)$. For dichotomous variables, Y_i can be rewritten as $P_i = \Pr(y_i = 1 \mid i)$; that is, the mean of the response distribution for unit i is the probability of a positive response for unit i. It is then shown that the SRV component has the simple form

$$\text{SRV} = E[P_i(1 - P_i)] \tag{8.3}$$

Working in parallel and seemingly independently of the Census Bureau statisticians, psychometricians were developing a theory of measurement known as *classical test theory*. Classical test theory shares many concepts with the Census Bureau model but with its own distinct terminology. For example, the test theory concept of a true score for unit i, denoted by τ_i, is equivalent to Y_i in the Census Bureau model. True score variance is equivalent to sampling variance and error variance is equivalent to simple response variance. The main difference between the two theories is that, while the Census Bureau model is concerned with the effects of measurement errors on the survey estimators and summary statistics, classical test theory is more focused on their effects on individual measurements. Interestingly, both models lead to the development of the concept of reliability (in psychometrics) or response inconsistency (in the Census Bureau paradigm). The latter approach defines the inconsistency ratio as

$$I = \frac{\text{SRV}}{\text{SV} + \text{SRV}} \tag{8.4}$$

This ratio is equivalent to one minus the reliability ratio R, defined in classical test theory as

$$R = \frac{Var(\tau_i)}{Var(y_i)} \tag{8.5}$$

In a typical survey, each survey item is measured only once, which is analogous to the case of a two-stage sample where only one secondary unit is obtained within each PSU. It is well-known from sampling theory that a single observation within each PSU is not sufficient for estimating the within PSU variance. Likewise, for the typical survey situation of one realization of each characteristic per respondent, separate estimation of SRV and SV is not possible, and, therefore, R and I also cannot be estimated. Only when at least one replicate measurement is obtained for two or more sample units is SRV estimable. In the minimally sufficient case, the estimability criteria require the assumptions of parallel and locally independent measurements for unbiased estimation.

Both classical test theory and the Census Bureau's theory are concerned with the estimation of reliability; however, the Census Bureau theory goes further to evaluate the effects of inconsistency on statistical inference. Hansen, Hurwitz, and their colleagues derived expressions for the bias and variance of common descriptive statistics such as means, totals, and proportions. On the other hand, classical test theorists invented the concept of validity, including theoretical validity, construct validity, convergent validity, and predictive validity. As shown in Chapter 2, the concept of theoretical validity and bias are similar in that both require the assumption of a true value. But the former is a correlation between unit-level measurements, while the latter is a difference between aggregate statistics. In Chapter 2, the Census Bureau model was applied in order to derive expressions for the bias and variance estimators of the sample mean under SRS. These expressions suggest that measurement errors tend to bias survey estimators and increase their standard errors. These concepts and results can be easily extended to more complex sample designs [see, e.g., Appendix D in Wolter (2007)].

8.1.2 From Bross' Model to the Standard LC and MLC Models

Chapter 3 revealed some important limitations of both classical test theory and the Census Bureau model for evaluating misclassification. These limitations become quite evident when the measurement error of a dichotomous measurement is decomposed using the framework suggested by Bross (1954). Bross assumed that underlying the categorical measurement is a function of an individual's true classification and his or her propensity to be misclassified into other categories. For dichotomous variables, respondents may be either true positives ($\mu_i = 1$) or true negatives ($\mu_i = 0$). Bross then parsed the error (e_i) into two components: the false-negative error (misclassification of a true positive) and the false-positive error (misclassification of a true negative). The expected error for the ith unit is decomposed as

$$E(e_i \mid i) = -\mu_i\theta_i + (1-\mu_i)\phi_i \tag{8.6}$$

where $\theta_i = \Pr(y_i = 0 \mid \mu_i = 1)$ is the false-negative probability and $\phi_i = \Pr(y_i = 1 \mid \mu_i = 0)$ is the false-positive probability (see Section 3.1.1). This expression forms the basis for all classification error theory. As an example, the expression for measurement bias in (3.9) follows immediately from this expression. Likewise, the true score (or P_i) can be written as

$$\tau_i = \mu_i(1-\theta_i) + (1-\mu_i)\phi_i \tag{8.7}$$

Bross' simple model provides an important link between the earlier models and LC models. In the terminology of LCA, θ_i and ϕ_i are called *response probabilities* and μ_i is a latent class variable. Under the assumption that μ_i is the true classification, θ_i and ϕ_i can be interpreted as *classification error probabilities*. The population averages of these parameters were denoted by π, θ, and ϕ. In Chapter 3, (8.7) was used to convert the variance and bias formulas from the Census Bureau model into expressions that involve the LCA parameters π, θ, and ϕ. This formulation provides a valuable link between the early models and the LC models.

R can be easily rewritten in terms of π, θ, and ϕ by substituting for P_i in the Census Bureau formulation by its equivalent expression in (8.7). Doing so revealed a number of problems in using R to evaluate the error in categorical responses. As seen in (3.18), R is a very complex function of parameters π, θ, and ϕ. Figures 3.1 and 3.2 were used to demonstrate the potential for misinterpreting subgroup comparisons of R as indicators of relative data quality. They illustrate how two groups may have very different values of R and still have identical classification errors as a result of differing prevalence probabilities (i.e., π parameters). For many purposes, a better approach for assessing relative data quality is to compare the estimates of π, θ, and ϕ for subgroups directly.

Applications of LCA for classification error evaluation are essentially focused on estimating π, θ, and ϕ. Through (8.7), LCA is seen as an extension of the classical test theory models that formed the basis for reliability analysis. A common requirement for both LCA and reliability analysis is that at least two measurements are needed to estimate the model parameters. For LCA with two indicators, this involves writing the likelihood of the observed cross-classification table in terms of π, θ, and ϕ. Let A and B denote two parallel measurements of the same latent variable X. The parameters π, θ, and ϕ in LCA notation become π_1^X, $\pi_{2|1}^{A|X}$ and $\pi_{1|2}^{A|X}$, respectively, where, for parallel measurements, $\pi_{1|2}^{B|X} = \pi_{1|2}^{A|X}$ and $\pi_{2|1}^{B|X} = \pi_{2|1}^{A|X}$. Chapter 4 shows that this three-parameter model is not identifiable because of dependences in the ML equations arising from the model assumptions.

The Hui–Walter model (Hui and Walter 1980) provides one solution to this dilemma. The Hui–Walter model introduces a grouping variable G that has a nonzero interaction with X, but no three-way interactions; that is, the loglinear model terms XGA and XGB are both zero. For dichotomous variables, this produces a fully saturated model of six parameters. One advantage of the Hui–Walter model is that the assumption of parallel measurements is not required. A disadvantage of the approach is that it may be difficult to find a grouping variable that satisfies the model assumptions. Respondent gender seems to work quite well for some characteristics such as labor force data since males and females are often misclassified at the same rate, although they may have very different prevalence probabilities.

When three measurements are available—A, B, and C, say—the standard LC model $\{XA\ XB\ XC\}$ can be fit, which requires no additional restrictions or grouping variables. This model was introduced in Chapter 4 as a general methodology for modeling discrete latent phenomena. A critical assumption for the standard LC model is the assumption of local independence. *Local independence* essentially means that X, and only X, is required to explain the relationship among the indicator variables. In other words, after accounting for the influence of X on A, B, and C, the indicators are mutually independent. Mathematically, this assumption can be written as

$$\pi_{abc|x}^{ABC|X} = \pi_{a|x}^{A|X} \pi_{b|x}^{B|X} \pi_{c|x}^{C|X} \tag{8.8}$$

If (8.8) is violated, the standard LC model will produce biased parameter estimates. Section 5.2 considered this assumption in some detail, including the effects of local dependence on parameter estimates, how to detect it, and how to model its effects to produce unbiased estimates.

Although the assumption is simple to express, local independence is quite complex. Local independence is satisfied only when three conditions hold: univocality, homogeneity, and uncorrelated error (Section 5.2). *Univocality* essentially means that the indictor variables (A, B, C, D, etc.) that are assumed to be indicators of X are not indicators of a second latent variable, say, Y. If some measurements are indicators of X and others of Y, the indicators are bivocal and, consequently, locally dependent. One solution to this problem is to add a second latent variable, Y, to the model whenever such a model is identifiable or can be made such by appropriate parameter restrictions. Other solutions are discussed in detail in Section 5.2.

Homogeneity refers to the variation of the error probabilities within the population. For the elementary models described in Chapter 2, this variation was represented by the parameter $\gamma_{\theta\phi} = \pi\sigma_\theta^2 + (1-\pi)\sigma_\phi^2$ [equation (3.12)], which was treated as a nuisance parameter for reliability analysis. (*Unconditional*) homogeneity is equivalent to setting $\gamma_{\theta\phi}$ to 0; otherwise, the error probabilities are said to be heterogeneous (i.e., $\gamma_{\theta\phi} > 0$). Thus, the assumption that $\gamma_{\theta\phi} = 0$ is required for LCA as well as well as for reliability analysis. When the unconditional homogeneity assumption is violated, it may be possible to achieve *conditional* homogeneity through the addition of grouping variables. In other words, the assumption that $\gamma_{\theta\phi} = 0$ holds *within the cells formed by a grouping variable or by cross-classifying two or more grouping variables*. In that case, the indicators will be conditionally locally independent given these same grouping variables.

Uncorrelated error refers to the tendency for respondents to be misclassified in the same way by different indicators. For example, the errors in A and B are correlated if persons misclassified by A have a higher probability of being misclassified by B than do persons not misclassified by A. Correlated errors can be addressed by the addition of conditional probabilities (or loglinear model interaction terms) that allow the error probabilities for one indicator to depend on the presence or absence of an error on another indicator. The log-odds ratio check (LORC; see Section 5.2) was presented as one method for identifying locally dependent indicators that require this treatment.

8.1.3 Loglinear Models with Latent Variables

The next step in the evolution of LC modeling came when Leo Goodman (Goodman 1973b) and Shelby Haberman (Haberman 1979) introduced the loglinear parameterization of the latent class model. Reparameterizing the LC model as a loglinear model with latent variables greatly increased its generality for representing many complex relationships among the variables—relationships that could not be represented with the probabilistic parameterization. Likewise, the development of the EM algorithm was an important milestone for LCA. It paved the way for the development of generalized software for fitting these complex models, which enhanced the popularity of the models. The early software packages such at LAT and NEWTON were also essential developments for bringing LCA to the masses. Today, sophisticated software packages are available that aid the user in the modeling of local dependence as well as protecting them against the problems of local identifiability, boundary estimates, and the problems of testing model adequacy caused by data sparseness. More recent developments have made inroads into the model misspecification issues caused by complex survey sampling, although there is still much to do in this area.

Markov latent class analysis (MLCA) was a giant leap forward in this evolution. MLCA takes advantage of the built-in replication afforded by the usual panel survey design to estimate the LC model parameters. These models were introduced by Wiggins (1973) and extended and refined by Poulsen (1982) and Van de Pol and de Leeuw (1986). MLCA was a major breakthrough for survey error evaluation because it obviated the need for replicate measurements apart from those routinely collected in every panel survey with three or more waves. In Section 7.3.1, the standard Markov latent class model was viewed as a generalization of the standard LC model. In fact, any LC model can be expressed as an MLC model with latent transition probabilities constrained to 0 or 1. This relationship continues the linkage from the early Census Bureau model through to MLC models. It further illustrates the fact that LC and MLC models share many of the same estimation issues and model assumptions.

Because they can be applied to essentially any panel survey, MLC models provide enormous opportunities for exploring classification error for virtually any characteristic that is repeated at each wave. Many permutations on the standard MLC model are possible by imposing constraints on the structural or measurement components or both. These include mover–stayer models and second- and higher-order Markov models as well as models for the effects of local dependence, nonstationarity, and time-varying error probabilities. The richness of the modeling possibilities provides challenges to the modeler in forming the model that optimally balances parsimony, fit, and plausibility of the estimates. Another important challenge is data sparseness, which is even more critical for MLC than LC models because of the number of ways in which the data are partitioned by the MLC model structures, especially when grouping variables are involved.

8.2 CURRENT STATE OF THE ART

In the early twenty-first century, the use of LCA for error evaluations is still quite limited as only a few researchers in the field are exploiting these models in their work. One reason is that students of survey methodology find many LC model concepts somewhat opaque and proper application of the models to complex survey data challenging. Hopefully, this book will have a positive influence for expanding the use of these important methods. Clearly, LCA has tremendous potential as a tool for evaluating and exploring the sources of measurement error in surveys. Despite the many successful applications of LCA, skepticism that the methodology can identify real problems in survey questionnaires and methods still exists. This section provides a summary of what might be labeled *current best methods* for applying MLC models (which includes LC models as per the discussion above) to survey data for the purpose of evaluating classification error. We begin with a recounting of some of the criticisms surrounding the MLC methodology and then provide a modeling strategy that attempts to address many of these concerns.

8.2.1 Criticisms of LCA for Survey Error Evaluation

Skepticism regarding the validity of the MLC results is articulated very well in the comments by Vermunt (2004), Tucker (2004), and Miller and Polivka (2004) and the rejoinder by Biemer (2004c) to the results reported in Biemer (2004a). Biemer (2004a) presented evidence obtained through an MLCA of the 1992–1997 CPS that the 1994 CPS redesign actually increased the misclassification of unemployed persons. Biemer provided further evidence, again from MLCA, that modifications of the questions designed to distinguish between unemployed persons who are looking for work and those who are laid off caused the increase in misclassification after the 1994 redesign.

Vermunt (2004), Tucker (2004), and Miller and Polivka (2004) provided important commentaries on this work from various perspectives. Although all three commentaries acknowledged the compelling evidence for the author's conclusions, all three also pointed to one or more MLCA model assumptions that could have been violated in the analysis, including the first-order Markov assumption, time-homogeneous error probabilities, and ICE. If one or more of these assumptions were severely violated, then the conclusions of the paper would be rendered invalid. In a rejoinder, Biemer conducted a number of sensitivity analyses to determine the effects of violations of the MLC model assumptions on the final results. He concluded that the probable direction of the biases due to model misspecification would tend to further exaggerate the effects that he found, leaving his conclusions unchanged for the most part. He also provided additional evidence from traditional interview–reinterview analysis in support of his conclusions.

This summary is presented as an example of the dialog that the LC modeler is likely to encounter when presenting controversial evidence of error and bias in an important statistical data series. While the truth is almost never known, there are steps that the analyst can take to either support or refute the validity of LC model results. These include sensitivity analysis, triangulation, and an assessment of the plausibility of the results. Below, we discuss these approaches and others in the context of the common criticisms of LCA and MLCA.

It Doesn't Measure the Truth

A key assumption in LCA for survey error evaluations is that the latent variable X is the true value underlying the indicators (A, B, C, etc.). One's willingness to interpret the response probability estimates produced from an LCA as error probabilities depends on one's willingness to accept this assumption. Alexander (2001) states the issue quite succinctly as follows:

1. Even if we have in mind a particular meaning for X, in reality X includes the effects of all variables that have been omitted from the model. So even if the dependencies are due to unobserved variables, the X may not be what we think it is.
2. The latent class analysis is nothing but a way to interpret the dependencies among the observed variables. What we attribute to an unobserved X may be only a higher order interaction among the original variables, which we erroneously assumed to be zero.

This criticism was addressed to some extent in Section 4.2.1. There we argued that there can be no certainty about what is being measured in a survey unless the underlying true values are known. We distinguished between exploratory applications of LCA, like the Toby–Stouffer analysis, and applications where the indicators are specifically designed to measure a well-defined X. In the former analysis, the latent construct of interest was unobservable and consequently not directly measurable. Rather, the interpretation of X was determined by LCA. Certainly, Alexander's criticism applies to this situation.

By contrast, survey error evaluation deals with observable and measurable quantities such as labor force status, drug use, health conditions, crime victimizations, and behavioral characteristics. The survey questions were designed specifically to measure these characteristics but may do so inaccurately. A simple model that postulates what is being measured is the truth plus error seems reasonable. Such a model is the basis for all survey evaluation that seeks to estimate measurement bias. The difficulties in applying this model arises primarily from the assumptions made for the error term (e.g., local dependence in the standard LC model). When dealing with multiple indicators, bivocality can also be a problem as discussed in Section 5.2. However, the LC modeling framework provides a number of tools for dealing with this problem and other problems arising from model misspecification. In the end, however, Alexander's criticism can never be fully countered. The task of the analyst is to provide the most compelling case possible that the model assumptions hold and that the results are valid for the specified objectives of the evaluation.

LCA Makes Rather Strong Assumptions

It is difficult to argue against the veracity of this statement except to say that virtually all survey error models, including classical test theory and the US Census Bureau model, require rather strong assumptions. In addition to local independence, reliability analysis assumes parallel measurements. Error evaluations based on gold standard measurements must assume the criterion that such measurements are infallible—an assumption that is very difficult to test or substantiate. LCA can be viewed as an alternative modeling approach that replaces these assumptions with assumptions that are more plausible in many practical situations. In addition, many of the LCA assumptions can be formally tested, especially when more than two indicators are available.

Another advantage of LCA is the abundance of tools and methods available to the analyst for modeling critical assumptions such as local dependence and for dealing with many other data complexities. Indeed, when the assumptions associated with traditional analysis do not hold, LCA may be the only reasonable method for assessing the error in the original measurements. In fact, LCA assumptions are quite similar to those made for finite-mixture modeling and missing data imputation modeling, and these methods have received much wider acceptance among statisticians.

In a typical analysis, all model assumptions fail to some degree, and the results will be biased accordingly. The biases may be inconsequential for some evaluation objectives and critical for others. As an example, for identifying flawed questions, LCA model validity may be of secondary importance. The primary issue is whether the evaluation method is successful at identifying questions that are truly flawed and need repair. It may also be necessary to distinguish between questions with the high versus low levels of error. For this purpose, small biases in the LCA results due to model misspecification can be tolerated. For example, for the past-year marijuana use analysis (Section 4.4.6), LCA successfully identified two problems in the questionnaire that would have been difficult and/or costly to detect using other methods. Even if the validity of the model assumptions is unknown, the analysis still achieved its objectives. However, it is much riskier to publish official reports of past-year marijuana use prevalence that have been adjusted by these error probabilities. For estimating biases, a much higher standard of validity applies.

Markov LCA could also be criticized for its rather strong assumptions, including Markov transitions, time-invariant error probabilities, and ICE. The extent to which these assumptions can be relaxed depends on the data available to the modeler. As an example, with four panel waves, the first-order Markov assumption can be replaced by a second-order assumption. Similarly, if a third indicator is available, the Hui–Walter assumptions can also be relaxed to those of the standard LC model. This discussion highlights the need for new, improved tools and methods for identifying and rectifying model misspecification.

It Gives Poor Results with Sparse Data

Data sparseness can cause problems with model identification, model selection, parameter estimation, and categorical data analysis generally. Empirical results from a number of studies suggest that the estimation of LC model parameters is surprisingly robust to data sparseness, even in rather extreme situations (see Section 5.1.4). In addition, current software packages such as Latent GOLD, MPlus, and PANMARK provide robust model fit statistics that are based on pseudoreplication methods such as bootstrapping. A Bayesian stabilizing prior can be invoked in PROC LCA when sparseness is an issue for parameter estimation. Likewise, Latent GOLD and MPlus have employed Bayesian methods to address the problems of improper ML solutions.

Replicate Measurements are Difficult to Obtain

Most methods for survey error evaluation require at least two realizations of the same construct. These multiple measurements may come from separate interviewers (e.g., an interview followed some days later by a reinterview), embedded replicate measurements within the same interview, a survey combined with some other external source such as administrative records, and so on. Regardless of the methods for collecting the replicate measurements, these data often come at a cost to the investigator, a burden to the respondent and may pose many modeling challenges for the analyst.

As noted in Section 2.3, the reinterview survey is one of the more costly approaches for obtaining two measurements on a set of survey items. In addition, it can be challenging to maintain the underlying modeling assumptions for the interview–reinterview data. Most methods for analyzing these data require local independence, which can easily be violated by poor reinterview design. Likewise, reliability analysis requires parallel measurements, which also presents challenges. Still, reinterview surveys may be the only means for gathering replicate measurements on most or all of the items in a survey since in that situation embedded replicate measurements are not feasible. Reinterview surveys also excels in its ability to control local dependence. In addition, reinterview surveys with reconciliation can also provide information regarding the causes of the measurements errors [see, e.g., Morton et al. (2008)]. In that regard, it can be a very efficient method for investigating survey error. Unfortunately, repeatedly interviewing respondents for the same survey within a short timespan is not only costly but also burdensome for respondents. Two interviews (i.e., one reinterview) may be acceptable but that limits the LCA to Hui–Walter models.

Perhaps the most efficient method for obtaining replicate measurements that can be used for a limited number of items is embedded replication since it does not require recontacting the sample units. Further, if used sparingly, it should not be an appreciable burden to respondents. Its main drawback is the potential for correlated error among the indicators. To avoid this risk, alternate wordings of the questions designed to still measure the same construct can be used, but this incurs the risk of bivocality. Despite these problems, judicious LC modeling will compensate for potential local dependence issues in many cases. Therefore, we believe that replicate measurements are a highly effective and preferred method for investigating survey error.

Results are Easily Misused and/or Misinterpreted

Latent class analysis is a widely misunderstood methodology, particularly as it applies to survey error evaluation. In general applications, it may be viewed as the categorical variable analog to factor analysis. However, for survey error evaluation, it is perhaps better described as an extension of classical test theory or the US Census Bureau model to categorical data since much of the basic underlying theory is the same. LCA is a convenient methodology for modeling and estimating the error parameters associated with a categorical measurement process.

Regardless of the methodology, all survey error analysis is subject to abuse, misuse, and misinterpretation. This is primarily a result of the assumptions inherent in this methodology, failure of those assumptions to hold in many situations, and failure of the analyst to test the assumptions and respecify and refit the model as necessary. There are two risks associated with using LCA for exploring survey error: (1) being too accepting of the results of an improper analysis and (2) being too dismissive of the results of a valid analysis.

With regard to risk type 1, the risk of accepting invalid results varies inversely with the level of expertise of the LC modeler. This is because, as we have seen in the foregoing chapters, proper analysis requires a fair degree of knowledge of sampling theory, measurement error theory, categorical data analysis, and LC modeling. Lack of this knowledge and expertise can result in invalid analyses and biased results, especially in complex modeling situations. Even meeting these stringent requirements, an analyst's ability to produce valid results from LCA is often limited by the data available for modeling as previously noted.

With regard to risk type 2, numerous studies as well as the examples in this book confirm that LCA can produce informative results that are quite useful for understanding the sources, causes, and cures of survey error. Nevertheless, given the many pitfalls that can be encountered in a complex LCA, consumers of LCA results would be prudent to request evidence that the assumptions of the underlying models hold or that the key results are robust to whatever violations of the model assumptions seem likely for a given dataset. In light of this evidence, the results should be accepted as valid.

Another means of ameliorating risks of type 2 is triangulation, that is, the use of LCA in conjunction with other methods of analysis including reliability analysis and other traditional approaches. In fact, experience shows that it is often unwise to rely on any single method of error evaluation. Prudently, one should use multiple analysis methods since methods will differ somewhat in their sensitivity and specificity to various types of problem. Agreement of the results from multiple methods will engender confidence that the findings of the analysis are well founded. For example, for redesigning questionnaires, conflicting results from alternative methods can lead to speculation and new hypotheses regarding the underlying assumptions that are violated by the different methods. These conflicts often lead to a better understanding of the flaws in the questions themselves. Triangulation often yields much more knowledge about the underlying causes of the errors than if only one method was used [see, e.g., Biemer (2004c)].

8.2.2 General Strategy for Applying LC and MLC Models

As noted previously, the primary cause of model misspecification in the standard LC model is failure of the local independence assumption to hold. To address this failure, it is usually advantageous to treat each cause of local dependence (i.e., bivocality, heterogeneity, and correlated errors) separately with strategies that target the cause. The following general strategy for LC model fitting was developed from findings in the literature, empirical investigations, and personal experience.

The first step in the model-fitting process should be to inspect the indicators for bivocality. This is done for each indicator by substantively reviewing its wording and deciding whether it satisfies the definition of an indicator of X (see Section 4.2.1). When there are four or more indicators, one may consider adding a second latent variable to model bivocality. For two or three indicators, it is not possible to do this and maintain an identifiable model without further restrictions. In that case, one must rely on subsequent steps in the model-fitting process to compensate for local dependence.

The next step is to address the common problem of heterogeneity. This is done by adding grouping variables to the model. A beneficial byproduct of this approach is that it increases model degrees of freedom, which is important for addressing local dependence regardless of the source. One disadvantage is data sparseness, which, as we know, causes other problems. A general rule of thumb is that n/k should not be less than 5 where n is the sample size and k is the number of cells in the data table.

In determining what grouping variables should be added to the model, it is usually helpful to consider the structural and measurement components of the model separately (see Section 4.4.3). For the structural component, the usual grouping variables are respondent demographic and socioeconomic variables that help explain the variation of the latent variable(s) in the population. Such variables may also be quite effective at modeling heterogeneity in the measurement component. In addition, variables that describe the interview mode, setting, the respondent's demeanor and reactions to the interview, and other interviewer-supplied variables such as the condition of the housing unit, neighborhood, and so on may be important (see Figure 8.1 for other factors). The model selection procedures described in Chapter 4 should be applied to this pool of variables to identify the best model. The best model is one having the smallest BIC among models that fit the data well (e.g., models having a p value of at least 0.05).

Finally to address potentially correlated errors among the indicators, the LORC or an equivalent diagnostic approach can be used to identify intercorrelated pairs of indicators. This is an iterative procedure involving (1) running the LORC on all pairs of indicators, (2) determining the pair with the most significant interaction, (3) addressing the correlation in the model by the addition of an appropriate interaction term, and (4) repeating the first three steps until no further significant interaction is detected. With regard to step (3), it is usually preferable to add the three-way interaction containing the two correlated indicators and X rather than the simple two-way interaction (i.e., the direct effect between the indicators). For example, ABX is preferred over AB for indicator pair A and B. This is because the former is more easily interpreted than the latter. In addition, it is often the case that the degree of association between A and B depends on an individual's true state. The LORC procedure

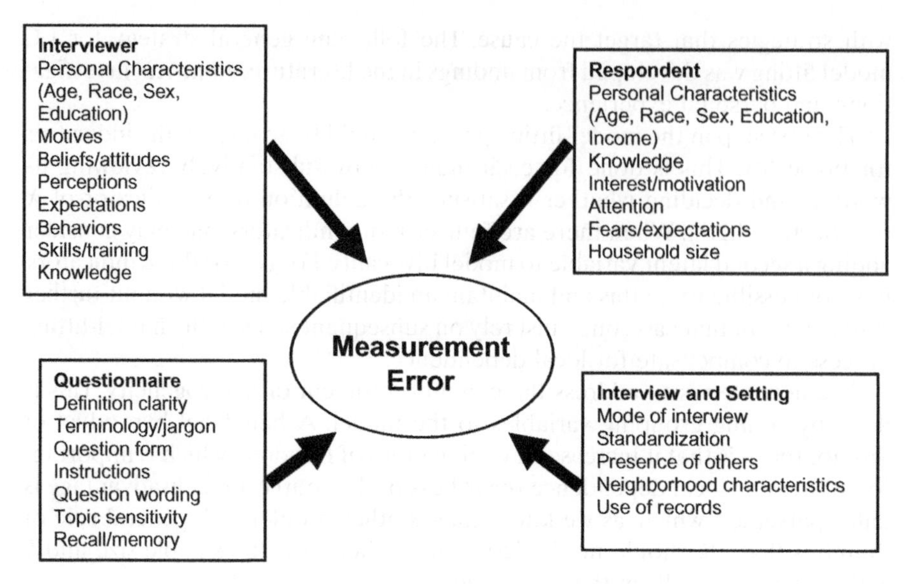

Figure 8.1 Factors related to measurement errors.

provides no guidance as to which interaction is preferred. Instead, the bivariate residual (BVR) statistic of Magidson and Vermunt (2004) can be used for this purpose.

In addition to model misspecification, the LC modeler should be aware of a number of important estimation issues that may afflict the estimation process. These include data sparseness, boundary issues, local maxima, and nonidentifiability. These issues are treated in some detail in Chapter 5. To briefly summarize, data sparseness may be due to low-frequency cells or structural zeros and can result in nonidentifiability, biased estimation, and invalid statistical tests. Sparseness may be particularly apparent in MLCA or for LCA when there are many grouping variables and/or indicators. Boundary estimates can occur for a number of reasons. The most problematic cause is due to improper solutions that occur when the MLE maximizing value of a parameter falls outside the parameter space. This type of boundary estimate can potentially bias other nonboundary parameters.

Local maxima, or the failure to converge to the global maximum of the likelihood, produces invalid parameter estimates. One potential solution for this problem is to select good starting values for a few key parameters. However, it is also prudent to run the model many times with random starting values to verify that a particular solution is optimal. This same approach can be used as an informal check on identifiability.

Finally, the modeling of complex survey data is still in its infancy. Although a number of strategies are available, the best choice still depends on the situation. For the ℓEM user, the only choice is to reweight and rescale the data and proceed as though the sample is selected by simple random sampling. In this case, the standard errors should be taken as approximations that may not be accurate for highly clustered samples with large weight variation. At this point in time, the safest approach appears to be the pseudo-maximum-likelihood (PML) approach, which is available in a few commercial software packages. However, as described in Section 5.3, it is still unclear whether these methods are appropriate when the primary focus of an analysis is on error probability estimation.

8.3 SOME IDEAS FOR FUTURE DIRECTIONS

As noted in the Preface, this book was written partly to stimulate interest in LC models among survey methodologists, analysts, and statisticians focused on evaluating, reducing, or adjusting for error in survey results. There are too few low-cost options available for survey error evaluation, particularly for categorical data. LCA, and especially MLCA, are important advancements in the field because they help fill this void. Moreover, research in LCA methodology is growing rapidly in the new millennium. In the future, our ability to extract valid information on misclassification from survey data will continue to be enhanced as this trend continues. In the course of writing this book, a number of issues that still plague the science of LC modeling were identified. This section summarizes some of these and suggests several new areas where progress in LCA is needed, especially for applications in survey error evaluation. These are listed in no particular order.

Measurement Error Theory

The problems of heterogeneity in LCA are addressed by appropriately modeling the measurement component of the LC model. Better modeling of the measurement component requires better theories about what influences measurement error and what grouping variables should be included in the model to capture that variation. Of course, only variables that are available to the analyst can be used in a model, and thus surveys should ensure that they are collected and included as part of the analysis dataset.

In addition to the usual demographic and socioeconomic variables, nontraditional so-called *paradata* or variables that relate to the data collection process, should be explored. For example, we know that interviewer characteristics (age, race, gender, experience, and attitude) can influence measurement error, especially in face-to-face surveys (Biemer and Lyberg 2003, Chapter 5). In addition, characteristics of respondents that are not traditionally collected in surveys can also affect survey error. These include attitude about the survey, motivation, and knowledge. Attributes of the questionnaire and the interview setting may be critical for some questions. Figure 8.1 lists these factors and others by source that could be explored in future analyses.

Latent Class Multilevel Models and Interviewer Effects

It is well known that in interviewer-administered surveys, the variance of estimates can be substantially increased by correlated interviewer error, also known as *between-interviewer variance* [see, e.g., Hansen et al. (1961), Kish (1962), Groves (1989, Chapter 8), and Biemer and Lyberg (2003, Chapter 5)]. The multiplicative interviewer effect on the variance of an estimated mean is similar to a design effect due to cluster sampling. Recall that the effect of sample clustering on the variance can be expressed as the design effect (deff) times the simple random sampling variance where

$$\text{deff} = 1 + \text{ICC}(m-1) \tag{8.9}$$

where ICC is the item-specific, intracluster correlation coefficient and m is the cluster size (Section 5.3.1). Now consider the interviewer's work assignment as the cluster, and let ρ_{int} denote the item-specific, intrainterviewer correlation defined by Kish (1962). Kish shows that the *interviewer design effect*, given by

$$\text{deff}_{\text{int}} = 1 + (m-1)\rho_{\text{int}} \tag{8.10}$$

[see also equation (1.7) in Chapter 1], where m is the average interviewer assignment size, has a similar effect on the variance as (8.9).[1] When responses to a particular question are more similar for respondents interviewed by the same interviewer than for respondents interviewed by different interviewers, ρ_{int} can become high, approaching a theoretical limit of 1. When responses to a particular question for respondents interviewed by the same interviewer are equivalent to responses obtained from a simple random sample of the total respondent pool, ρ_{int} approaches 0 and may even be negative. Near-zero intrainterviewer correlations can still have a substantial effect on the precision of an estimated mean. For example, assuming the assignment of $m = 41$ sample cases per interviewer, a value of ρ_{int} as small as 0.01 would still result in a 40% increase in the variance—or a 20% increase in the standard error—of an estimated mean, increasing confidence interval width and reducing effective sample sizes.

Traditionally, ANOVA models that treat the interviewer effects as random variables have been used to estimate ρ_{int} (Kish 1962; Groves 1989). More recently, multilevel models that treat the respondent as a level 1 unit and the interviewer as the level 2 unit have been employed [see, e.g., O'Muircheartaigh and Campanelli (1998), Hox et al. (1991), and Wiggins et al. (1992)]. One requirement of both of these approaches is that interviewer assignments be *interpenetrated*—that is, that each interviewer's assignment consists of a random sample of the same population. This condition seldom holds for field

[1]In complex surveys using interviewers for data collection, both the sample design deff and the interviewer deff_{int} are operating jointly on the variances in an approximately multiplicative fashion.

surveys (although they can be reasonably approximated in some cases for centralized telephone surveys). Especially for field surveys, interviewer interpenetration designs are fraught with logistical problems, and the costs for maintaining a rigorous interpenetration protocol are daunting. Since ρ_{int} is typically quite small, large samples are required to estimate it with acceptable precision. Further, deviations from interpenetration can cause substantial biass. Given these difficulties, interpenetration of face-to-face interviewer assignments is seldom attempted and, thus, very little is known about interviewer variance in modern field surveys.

On the other hand, it may be possible to relax the interpenetration requirement for surveys having repeated measurements. In that case, a *multilevel LC model* (Vermunt 2003), which have been implemented in Latent GOLD 4.0, could be employed. Such a model would be similar to the interpenetrated multilevel models but would assume a latent true value with error probabilities that would vary by interviewer according to a random-effects model. To our knowledge, such models have never been attempted; yet their potential for expanding our knowledge of interviewer effects on misclassification seems quite high.

Methods for Complex Survey Designs

In Section 5.3, five methods for dealing with complex sampling in LCA were described. There we noted that, for simple designs, the design variable approach seems to be the best choice since it accommodates the survey design while preserving the goodness-of-fit diagnostics. For more complex samples, the PML approach seems to be the best alternative, although weighting–scaling combined with replicate variance estimation and bootstrap fit diagnostics can provide results that are essentially equivalent to PML. The weighting and rescaling method with the $\sqrt{\text{deff}_{\text{all}}}$ correction to the standard error is a good choice for the novice user. The two-step approach also seems to be an important alternative, although more research is needed before it can be recommended. For example, how do unequal probability sampling and sample clustering affect the estimates of classification error? Is weighting the data an appropriate way of correcting the misclassification probability estimates for non-EPSEM sampling? When can the sampling weights be ignored in LCA? In general, there is still much research to be done for applying LCA to complex survey data, particularly for goodness-of-fit diagnostics such as BIC, AIC, and the chi-square tests for model adequacy.

Effects of Time-Varying Covariates on Error Probability Estimates

In panel surveys certain grouping variables can be set as fixed or varying in the MLC model. Does this choice impact the resulting classification error rates? Vermunt et al. (1999) suggest that the use of time-varying covariates removes error associated with a fixed covariate from the error rates and, therefore, makes them more accurate. However, it could also be argued that the time-invariant covariate is masking the true error rates and, thus, biasing the resulting model. More research is needed to determine how time-varying covariates affect the estimates of misclassification.

Effects on Estimates of Misclassification of Model Misspecification in the Structural Component

Survey methodologists are interested primarily in measurement error rather than the structural component of an LC or MLC model. A typical practice is to specify a fully saturated or to otherwise overspecify the structural component model rather than attempting to achieve submodel parsimony. Once specified, the structural component is essentially ignored in the model fitting process, where the focus turns to achieving the best specification for the measurement submodel. This begs the question "How important is the structural component if the focus of an analysis is on measurement error?" In particular, how do changes to the structural component affect estimates of the measurement parameters? For example, for an MLCA, the structural component might be a second- or higher-order Markov structure when a first-order Markov process with stationary transition probabilities is adequate. Research is needed to determine the degree to which an overspecified structural component affects the measurement component parameter estimates.

Within-Household Clustering

To what degree are classification error rates correlated within a PSU? While it is known that outcomes are usually positively correlated within a PSU, it is not known whether the error rates associated with those outcomes are correlated as well. This is important to determine because the LC assumes that errors are independent. One possible reason that they may not be is communication error. For example, in a survey where all family adults are interviewed, individuals may discuss their interview with other family members before all other household interviews are concluded. This could have the effect of reducing the measurement error if, for example, other household interviewees become better respondents as a result of this information. It could also adversely affect measurement error if, for example, other household interviewees are warned about sensitive information and advised not to be truthful. In either case, the errors may be correlated within the household, which is in violation of local independence. Since there may be thousands of households in the sample, simply adding a fixed effect for each household is not feasible. Alternatives for modeling types of correlated errors require further research.

Handling Imputed Data in LCA

Imputations for item and, in some cases, unit nonresponse is common place in surveys nowadays. Since they are model-based, imputations probably have an error structure that is very different from that of respondent reports, creating heterogeneity. If removed from the analysis, the cumulative missing data rate across many items could be quite large. If included in the analysis, imputations create heterogeneity unless appropriately modeled. How should imputations be handled in an analysis, and how can they be appropriately modeled?

Correlation Threshold for Bivocality

Consider two bivocal indicators, A and B, with underlying true values X and Y, respectively. If $\Pr(X = Y)$ is approximately 1, the indicators can be treated as univocal. However, as $\Pr(X = Y)$ becomes smaller, the assumption of univocality begins to fail, as does the assumption of local independence. What is the breaking point? In other words, how highly correlated should X and Y be in order for the standard LC model, which assumes local independence, to yield quality inference?

Correlation Threshold for Indicators

A similar question applies to univocal indicators A and B. If B is a very poor indicator of X (large misclassification error), the model may be nonidentifiable. What is the breaking point? In other words, how correlated should indicator variables be with their latent counterparts in order for the standard LC model to yield quality inference?

8.4 CONCLUSION

The field of survey error evaluation is central and essential to data quality improvement in surveys. With data quality being emphasized increasingly in survey work, error evaluations should continue to play an important role in survey methodology for many statistical organizations. The survey evaluation encompasses a wide range of methods and tools, but, for categorical data analysis, latent class analysis and its panel survey counterpart, Markov latent class analysis, are among the more powerful methods available today. Their generality and effectiveness for all sorts of applications is considerable, with many new extensions and applications still waiting to be discovered. Survey error analysts working with categorical data will benefit from mastering this methodology and applying it in their daily work. In addition, the field of LCA can move forward only if more researchers begin contributing to this methodology through the survey methods literature. Attracting more researchers to the field is a primary goal of this book.

The methods and modeling approaches described in this book will surely be considered outdated in time as the field continues to grow and mature. This could even happen soon should these developments continue at a rapid pace—a positive outcome in this author's view. However, as discussed in this chapter, many of the essential theories and concepts seem to have a very long "shelflife." This is evidence that the early ideas and theories on which contemporary LCA methodology was built are still sound, flexible, and supportive of the many new innovations that have come to light since the 1950s. Rather than replacing the classical modeling approaches, new approaches add opportunities for comparing the results from alternative modeling approaches (triangulation). Thus, the new and old methodologies combine to provide survey methodologists with an even richer toolset for exploring the root causes of survey error.

APPENDIX A

Two-Stage Sampling Formulas

The basic two-stage cluster sampling setup supposes a population comprising N clusters or primary-stage sampling units (PSUs), each containing M secondary-stage sampling units (SSUs). Let Y_{ij} denote the value of the characteristic of interest $\mathcal{Y}$ for the jth SSU in the ith PSU. Assume that Y_{ij} is measured without error, and let $\bar{Y}_i = E(Y_{ij} \mid i)$ denote the mean and $\sigma_i^2 = Var(Y_{ij} \mid i)$ denote the within-PSU variance for the ith PSU. Let $\bar{\bar{Y}} = E(Y_{ij})$ denote the population mean to be estimated:

$$\bar{\bar{Y}} = \frac{\sum_{i=1}^{N} \bar{Y}_i}{N} \tag{A.1}$$

Suppose that a sample of n PSUs is selected with SRS and, within each of these, m SSUs are selected also with SRS. Further, assume that $m \ll M$ so that the second-stage sampling fraction can be ignored in the variance formulas. Let $\bar{y}_i$ denote the sample proportion for primary i, $i = 1, \ldots, n$, and let $\bar{\bar{y}}$ denote the total sample proportion given by

$$\bar{\bar{y}} = \frac{1}{n} \sum_{i=1}^{n} \bar{y}_i \tag{A.2}$$

It is well known [see, e.g., Cochran 1977, p. 277] that $\bar{\bar{y}}$ is an unbiased estimator of $\bar{\bar{Y}}$ with variance

$$Var\,(\bar{\bar{y}}) = (1-f)\frac{S_1^2}{n} + \frac{S_2^2}{nm} \tag{A.3}$$

Latent Class Analysis of Survey Error By Paul P. Biemer

where

$$S_1^2 = \sum_{i=1}^{N} \frac{(\overline{Y}_i - \overline{\overline{Y}})^2}{N-1} \quad \text{and} \quad S_2^2 = \frac{\Sigma_{i=1}^{N} \sigma_i^2}{N} \tag{A.4}$$

is referred to as the *between-PSU variance component* and the *within-PSU variance component*, respectively, and f is the PSU sampling fraction, n/N. Here we have assumed that the within-PSU sampling fraction denoted by f_2 in Cochran's formulas is 0.

An unbiased estimator of this variance is

$$v(\overline{y}) = \frac{1-f}{n} s_1^2 + \frac{f}{nm} s_2^2 \tag{A.5}$$

where

$$s_1^2 = \frac{\Sigma_{i=1}^{n} (\overline{y}_i - \overline{\overline{y}})^2}{n-1} \qquad s_2^2 = \frac{\Sigma_{i=1}^{n} \Sigma_{j=1}^{m} (y_{ij} - \overline{y}_i)^2}{n(m-1)} \tag{A.6}$$

Note that no unbiased estimator of S_2^2 exists unless $m > 1$ because otherwise s_2^2 is undefined. In addition, s_1^2 and s_2^2 have expectations

$$E(s_1^2) = S_1^2 + \frac{1}{m} S_2^2 \quad \text{and} \quad E(s_2^2) = S_2^2 \tag{A.7}$$

(Cochran 1977, p. 278).

If $\mathcal{Y}$ is a dichotomous (i.e., 0 or 1) survey variable, then let P_i denote the proportion of 1s (or positives) in the ith PSU, $i = 1, \ldots, N$, and let P denote the population proportion to be estimated where $Q = 1 - P$. Further, let p_i denote the sample proportion in the ith PSU and let $p = \Sigma_{i=1}^{n} p_i / n$ with $q_i = 1 - p_i$ and $q = 1 - p$. Then substitute P_i and P everywhere in the preceding above equations for $\overline{Y}_i$ and $\overline{\overline{Y}}$, respectively, and p_i and p for $\overline{y}_i$ and $\overline{\overline{y}}$, respectively. Further, noting that $\sigma_i^2 = P_i Q_i$, we can rewrite (A.4) as

$$S_1^2 = \sum_{i=1}^{N} \frac{(P_i - P)^2}{N-1} \quad \text{and} \quad S_2^2 = \sum_{i=1}^{N} \frac{P_i Q_i}{N} \tag{A.8}$$

and (A.6) can be rewritten as

$$s_1^2 = \frac{\Sigma_{i=1}^{n} (p_i - p)^2}{n-1} \quad \text{and} \quad s_2^2 = \frac{m}{n(m-1)} \sum_{i=1}^{n} p_i q_i \tag{A.9}$$

APPENDIX B

Loglinear Modeling Essentials

This appendix reviews some of the basic ideas of loglinear and logit modeling and shows their equivalence when various modeling constraints are imposed. To simplify the exposition of ideas, the case where all variables are manifest is considered first before we generalize these results to latent variable models. The initial discussion will be confined to two dichotomous variables. Later these results will be generalized to three or more polytomous variables. For those familiar with analysis of variance (ANOVA) modeling, the transition to loglinear models begins with a discussion of the similarities and differences between ANOVA models and loglinear models.

B.1 LOGLINEAR VERSUS ANOVA MODELS: SIMILARITIES AND DIFFERENCES

Let A and B denote two manifest variables whose joint relationship we wish to model using a loglinear model. Let AB denote the cross-classification of A and B. The probabilistic parameterization of this relationship assumes that the cell frequencies $\{n_{ab}\}$ follow a multinomial distribution with parameters $\{\pi_{ab}^{AB}\}$. The loglinear model parameterization assumes that $\{n_{ab}\}$ follows a Poisson distribution with parameters $\{m_{ab}\}$, where $m_{ab} = E(n)\pi_{ab}^{AB}$. Here we use the expected value of n, denoted by $E(n)$, rather than n to emphasize that, under the Poisson assumption, the frequencies $\{n_{ab}\}$ are random variables and, hence, so is the total sample size, $n = \Sigma_a \Sigma_b n_{ab}$. Later in this appendix we will show how, by conditioning on n, the distribution of the $\{m_{ab}\}$ can be transformed into the corresponding multinomial distribution assumed in the probabilistic parameterization. Thus, the two parameterizations are essentially equivalent.

Let us first consider the likelihood of AB under the Poisson distribution assumption, according to which the probability of observing n_{ab} is

$$\Pr(n_{ab}) = \frac{\exp(-m_{ab})(m_{ab})^{n_{ab}}}{n_{ab}!} \tag{B.1}$$

where m_{ab} may be a function of additional parameters specified by a model for m_{ab}. Thus the loglikelihood kernel for the joint distribution of the AB cross-classification table is

$$\log(\mathcal{L}) = \sum_a \sum_b (n_{ab} \log m_{ab} - m_{ab}) \tag{B.2}$$

Differentiation of (B.2) with respect to the model parameters, setting to zero and solving for the parameters, will provide the maximum-likelihood estimates (MLEs) of the parameters. The form of the likelihood suggests that a natural way to express a relationship between $\{m_{ab}\}$ and the frequencies $\{n_{ab}\}$ is through the quantities $\log m_{ab}$ or equivalently, $\log \pi_{ab}^{AB}$.

Many aspects of loglinear analysis are analogous to those of other linear models such as ANOVA models. For readers who are more familiar with ANOVA models, some of these analogies will be demonstrated here for a two-way ANOVA. The left portion of Table B.1 shows data from a 2^2 factorial experiment for factors A and B and with a single experimental unit per cell. In an ANOVA, the observations y_{ij}, $i, j = 1,2$ are assumed to be measured on a continuous scale and the normal distribution is usually invoked for the observations. The observation y_{ij} is a response for unit (i, j), where (i, j) denotes the ith level of factor A and the jth level of factor B. A fully saturated ANOVA model for $Y_{ij} = E(y_{ij})$ is

$$Y_{ij} = \mu + A_i + B_j + AB_{ij} \tag{B.3}$$

where μ is the overall mean, A_i is the effect of A at level i, B_j is the effect of B at level j, and AB_{ij} is the interaction effect for the ith level of A and the jth level of B. When A and B each have only two levels, their interaction is the difference between the effects A_1 and A_2 at different levels of B. A zero interaction implies that the effect of one of the factors is the same across the levels of the other factor.

Table B.1 Data Tables for 2^2 Factorial and 2×2 Cross-Classification

2^2 Factorial Table[a]			2×2 Cross-Classification Table[b]		
—	$A = 1$	$A = 2$	—	$A = 1$	$A = 2$
$B = 1$	y_{11}	y_{12}	$B = 1$	n_{11}	n_{12}
$B = 2$	y_{21}	Y_{22}	$B = 2$	n_{21}	n_{22}

[a]Where y_{ij} is a continuous variable.
[b]Where n_{ij} is a count variable.

The right portion of Table B.1 is a cross-classification of a simple random sample of n observations, where $\{n_{ab}\}$ is the number of observations classified in cell (a,b) for $a = 1,2$ and $b = 1,2$ and where $E(n_{ab}) = m_{ab}$. Let $\ell_{ab} = \log m_{ab}$, the natural logarithm of the expected cell frequency. Then a fully saturated log-linear model for the AB table is given by

$$\ell_{ab} = u + u_a^A + u_b^B + u_{ab}^{AB} \tag{B.4}$$

Note the similarity of this model to (B.3). The interpretations of the effects and parameters are also quite similar. In (B.4), u denotes the overall mean of the $\{\log(m_{ab})\}$ (i.e., the mean of the log expected cell sizes), u_a^A denotes the effect associated with level a of A, u_b^B denotes the corresponding effect for B, and u_{ab}^{AB} denotes the interaction effect.

Table B.2 shows the form of the model effects in terms of the expected values of the observations for both models. Here, a bar over the letter with a subscript replaced by "+" indicates an average over that subscript. As examples, $\bar{Y}_{+1} = (Y_{11} + Y_{21})/2$, $\bar{\ell}_{++} = (\ell_{11} + \ell_{12} + \ell_{21} + \ell_{22})/4$, and so on. As shown in the table, the loglinear model effects (on the logarithm scale) can be obtained from the corresponding ANOVA effects after substituting ℓ_{ab} for Y_{ij} in each formula. Another important similarity that may not be apparent in the table

Table B.2 Comparison of Effects for the 2^2 Factorial ANOVA Model and the Loglinear Model for a 2 × 2 Cross-Classification

		Loglinear	
Effect	ANOVA	Logarithm Scale	Original Scale
Mean	$\bar{Y}_{++}$	$\bar{\ell}_{++}$	$(\pi_{11}^{AB}\pi_{12}^{AB}\pi_{21}^{AB}\pi_{22}^{AB})^{1/4}$
A	$A_1 = \bar{Y}_{1+} - \bar{Y}_{++}$	$u_1^A = \bar{\ell}_{1+} - \bar{\ell}_{++}$	$\left(\dfrac{\pi_{11}^{AB}\pi_{12}^{AB}}{\pi_{21}^{AB}\pi_{22}^{AB}}\right)^{1/4}$
	$A_2 = \bar{Y}_{2+} - \bar{Y}_{++}$	$u_2^A = \bar{\ell}_{2+} - \bar{\ell}_{++}$	$\left(\dfrac{\pi_{11}^{AB}\pi_{12}^{AB}}{\pi_{21}^{AB}\pi_{22}^{AB}}\right)^{-(1/4)}$
B	$B_1 = \bar{Y}_{+1} - \bar{Y}_{++}$	$u_1^B = \bar{\ell}_{+1} - \bar{\ell}_{++}$	$\left(\dfrac{\pi_{11}^{AB}\pi_{21}^{AB}}{\pi_{12}^{AB}\pi_{22}^{AB}}\right)^{1/4}$
	$B_2 = \bar{Y}_{+2} - \bar{Y}_{++}$	$u_2^B = \bar{\ell}_{+2} - \bar{\ell}_{++}$	$\left(\dfrac{\pi_{11}^{AB}\pi_{21}^{AB}}{\pi_{12}^{AB}\pi_{22}^{AB}}\right)^{-(1/4)}$
AB	$AB_{11} = Y_{11} - A_1 - B_1 - \mu$	$u_{11}^{AB} = \ell_{11} - u_1^A - u_1^B - u$	$\left(\dfrac{\pi_{11}^{AB}\pi_{22}^{AB}}{\pi_{12}^{AB}\pi_{21}^{AB}}\right)^{1/2}$
	$AB_{12} = Y_{12} - A_1 - B_2 - \mu$	$u_{12}^{AB} = \ell_{12} - u_1^A - u_2^B - u$	$\left(\dfrac{\pi_{11}^{AB}\pi_{22}^{AB}}{\pi_{12}^{AB}\pi_{21}^{AB}}\right)^{-(1/2)}$
	$AB_{21} = Y_{21} - A_2 - B_1 - \mu$	$u_{21}^{AB} = \ell_{21} - u_2^A - u_1^B - u$	$\left(\dfrac{\pi_{11}^{AB}\pi_{22}^{AB}}{\pi_{12}^{AB}\pi_{21}^{AB}}\right)^{-(1/2)}$
	$AB_{22} = Y_{22} - A_2 - B_2 - \mu$	$u_{22}^{AB} = \ell_{22} - u_2^A - u_2^B - u$	$\left(\dfrac{\pi_{11}^{AB}\pi_{22}^{AB}}{\pi_{12}^{AB}\pi_{21}^{AB}}\right)^{1/2}$

is that the sum of an effect over each subscript is zero for both models. For the ANOVA model, we obtain

$$\sum_i A_i = \sum_j B_j = \sum_i AB_{ij} = \sum_j AB_{ij} = 0 \tag{B.5}$$

while for the loglinear model, we have

$$\sum_a u_a^A = \sum_b u_b^B = \sum_a u_{ab}^{AB} = \sum_b u_{ab}^{AB} = 0 \tag{B.6}$$

Both models require these constraints as a conditional of estimability of the model parameters.

Table B.2 lists the definitions of the loglinear model effects for both the logscale and the original scale. As an example, the effect u_a^A is defined on the log scale as

$$u_1^A = \ell_{1+} - \ell_{++} \tag{B.7}$$

which is equivalent to $\sqrt[4]{\pi_{11}^{AB}\pi_{12}^{AB} / \pi_{21}^{AB}\pi_{22}^{AB}}$ in the original scale (i.e., after taking the exponential of the ℓ or log expression). To see this, note that

$$\begin{aligned}\exp(\ell_{1+} - \ell_{++}) &= \exp\left(\frac{\log m_{11} + \log m_{12}}{2} - \frac{\log m_{11} + \log m_{12} + \log m_{21} + \log m_{22}}{4}\right) \\ &= \exp\left(\frac{\log m_{11} + \log m_{12} - \log m_{21} - \log m_{22}}{4}\right) \\ &= \left(\frac{m_{11}m_{12}}{m_{21}m_{22}}\right)^{1/4} = \left(\frac{\pi_{11}^{AB}\pi_{12}^{AB}}{\pi_{21}^{AB}\pi_{22}^{AB}}\right)^{1/4}\end{aligned} \tag{B.8}$$

The ANOVA model analogy can help in the interpretation of the log-linear effects. In the ANOVA model, the effect of the first level of A is

$$2A_1 = \tfrac{1}{2}[(Y_{11} - Y_{21}) + (Y_{12} - Y_{22})] \tag{B.9}$$

Further, because $A_2 = -A_1$ [from (B.5)], it follows that $A_1 - A_2$ is the mean difference in responses at $A = 1$ and $A = 2$ taken over the levels of B. By analogy, the loglinear model form is

$$2u_1^A = \tfrac{1}{2}[(\ell_{11} - \ell_{21}) + (\ell_{12} - \ell_{22})] \tag{B.10}$$

or, on the original scale

$$\begin{aligned}\exp(2u_1^A) &= \exp\left(\frac{1}{2}[(\ell_{11}-\ell_{21})+(\ell_{12}-\ell_{22})]\right)\\ &= [\exp(\ell_{11}-\ell_{21})]^{1/2}[\exp(\ell_{12}-\ell_{22})]^{1/2}\\ &= \left[\frac{\exp(\ell_{11})}{\exp(\ell_{21})}\frac{\exp(\ell_{12})}{\exp(\ell_{21})}\right]^{1/2}\\ &= \left(\frac{\pi_{11}^{AB}\pi_{12}^{AB}}{\pi_{21}^{AB}\pi_{22}^{AB}}\right)^{1/2}\end{aligned} \tag{B.11}$$

and, hence

$$\begin{aligned}\exp(u_1^A - u_2^A) &= \frac{[(\pi_{11}^{AB}\pi_{12}^{AB})]/(\pi_{21}^{AB}\pi_{22}^{AB})]^{1/4}}{[(\pi_{11}^{AB}\pi_{12}^{AB})/(\pi_{21}^{AB}\pi_{22}^{AB})]^{1/4}}\\ &= \left(\frac{\pi_{11}^{AB}\pi_{12}^{AB}}{\pi_{21}^{AB}\pi_{22}^{AB}}\right)^{1/2}\\ &= \left[\frac{(\pi_{11}^{AB}/\pi_1^B)(\pi_{12}^{AB}/\pi_2^B)}{(\pi_{21}^{AB}/\pi_1^B)(\pi_{22}^{AB}/\pi_2^B)}\right]^{1/2}\\ &\left(\frac{\pi_{1|1}^{A|B}\ \pi_{1|2}^{A|B}}{\pi_{2|1}^{A|B}\ \pi_{2|2}^{A|B}}\right)^{1/2}\end{aligned} \tag{B.12}$$

which can be interpreted as the mean odds of being in category 1 of A taken over the categories of B; here *mean* refers to the geometric mean defined as $\bar{x} = \sqrt[k]{x_1 x_2 \cdots x_k}$. A similar interpretation applies to the B effect given by $\exp(u_1^B - u_2^B)$.

For the ANOVA model, the interaction term can be rewritten as

$$2A_{11} = \tfrac{1}{2}[(Y_{11}-Y_{21})-(Y_{12}-Y_{22})] \tag{B.13}$$

where $Y_{11}-Y_{21}$ is the effect of A at the low level of B and $Y_{12}-Y_{22}$ is the effect of A at the high level of B. If these two effects differ, there is an interaction between A and B. Thus, the interaction effect defined as $A_{11}-A_{12}=A_{21}-A_{22}=2A_{11}$ is the average difference between these two differences. Likewise, for the loglinear model using the results in Table B.2, we obtain

$$\begin{aligned}\exp(u_{11}^{AB}-u_{12}^{AB}) &= \left(\frac{\pi_{11}^{AB}\pi_{22}^{AB}}{\pi_{12}^{AB}\pi_{21}^{AB}}\right)^{1/4}\left(\frac{\pi_{11}^{AB}\pi_{22}^{AB}}{\pi_{12}^{AB}\pi_{21}^{AB}}\right)^{1/4}\\ &= \left(\frac{\pi_{1|1}^{A|B}\pi_{2|1}^{A|B}}{\pi_{1|2}^{A|B}\pi_{2|2}^{A|B}}\right)^{1/2}\end{aligned} \tag{B.14}$$

in the original scale. In this expression, $\pi_{1|b}^{A|B} / \pi_{2|b}^{A|B}$ is the odds of being classified in $A = 1$ for the same category of $B = b$. The ratio of these odds is analogous to the difference in odds on the log scale. Therefore, like the ANOVA interaction, the loglinear interaction term is essentially a mean difference in the effect of A and different levels of B. When A and B are independent, the interaction of A and B is zero for loglinear models as it is for the ANOVA model. For example, $\pi_{a|b}^{AB} = \pi_a^A$ for all a,b, which implies that (B.14) is 1 and $u_{11}^{AB} - u_{12}^{AB} = 0$. In fact, it can be shown that $u_{ab}^{AB} = 0$ for all a,b.

Illustration of the Calculations

An illustration will help illuminate these loglinear modeling concepts. Table B.3 shows the results of cross-classifying a simple random sample of 830 cases by classifiers A and B. The left portion of the table shows the original frequencies $\{n_{ab}\}$ and right portion of the table shows the log-transformed frequencies, $\{\log(n_{ab})\}$. In the case of a saturated model, MLEs of effects for the model in (B.4) can be obtained by replacing m_{ab} in the formulas in Table B.2 by their estimates n_{ab}. Thus, wc have

$$\ell_{++} = \frac{5.35 + 5.19 + 5.30 + 5.48}{4} = 5.33$$

$$\ell_{1+} = \frac{5.35 + 5.19}{2} = 5.27,\ \ell_{2+} = \frac{5.30 + 5.48}{2} = 5.39$$

$$\ell_{+1} = \frac{5.35 + 5.30}{2} = 5.32,\ \ell_{+2} = \frac{5.19 + 5.48}{2} = 5.34$$

$$u_1^A = -0.0597, u_2^A = 0.0597, u_1^B = -0.0070, u_2^B = 0.0070$$

$$u_{11}^{AB} = 0.0841, u_{12}^{AB} = -0.0841, u_{21}^{AB} = -0.0841, u_{22}^{AB} = 0.0841$$

The parameter estimates can be used to obtain model estimates of the four cell frequencies via (B.4). For example, an estimate of m_{11} is exp(–5.33 + 0.0597 – 0.0070 + 0.0841) = exp(5.34) = 210. The other cells can be estimated similarly. In the case of a fully saturated model, the estimated cell frequencies will

Table B.3 An Illustrative *AB* Table

Original Scale Frequencies			Log-Transformed Frequencies		
—	$A = 1$	$A = 2$	—	$A = 1$	$A = 2$
$B = 1$	210	200	$B = 1$	5.35	5.30
$B = 2$	180	240	$B = 2$	5.19	5.48

exactly equal the observed frequencies. For unsaturated models, this may not true. In that case, a comparison of the model estimates and observed frequencies can provide information on model adequacy, as we shall see subsequently.

B.2 MODELING CELL AND OTHER CONDITIONAL PROBABILITIES

The loglinear parameter estimates can also be used to estimate cell probabilities and other conditional probabilities. To illustrate for the saturated loglinear model, noting that

$$m_{ab} = \exp(u + u_a^A + u_b^B + u_{ab}^{AB}) \tag{B.15}$$

the cell (a,b) probability is

$$\pi_{ab}^{AB} = \frac{\exp(u + u_a^A + u_b^B + u_{ab}^{AB})}{\Sigma_a \Sigma_b \exp(u + u_a^A + u_b^B + u_{ab}^{AB})} \tag{B.16}$$

Note that, in this expression, the u term cancels out of the numerator and denominator. Using the estimates of the loglinear parameters from above, $\Sigma_a \Sigma_b \exp(u_a^A + u_b^B + u_{ab}^{AB}) = 4.02$, and thus

$$\pi_{11}^{AB} = \frac{\exp(-0.060 - 0.0070 + 0.084)}{4.02} = 0.25$$

$$\pi_{12}^{AB} = \frac{\exp(-0.060 + 0.0070 - 0.084)}{4.02} = 0.22$$

$$\pi_{21}^{AB} = \frac{\exp(0.060 - 0.0070 - 0.084)}{4.02} = 0.24$$

$$\pi_{22}^{AB} = \frac{\exp(0.060 + 0.0070 + 0.084)}{4.02} = 0.29$$

As will be the case for saturated loglinear models, these estimates exactly agree with the direct estimates of π_{ab}^{AB} computed from the data table as n_{ab}/n.

In classification error modeling, interest is on estimating probabilities rather than frequencies. If the cell frequencies $\{n_{ab}\}$ are Poisson random variables, the joint distribution of the cell probabilities, $p_{ab} = n_{ab}/n$ is multinomial with parameters n and $\{\pi_{ab}^{AB}\}$. To see this, note that

$$
\begin{aligned}
\Pr(p_{11}, p_{12}, p_{21}, p_{22}) &= \Pr(n_{11}, n_{12}, n_{21}, n_{22} \mid \Sigma_a \Sigma_b m_{ab} = n) \\
&= \frac{\Pi_a \Pi_b \Pr(n_{ab} \mid \Sigma_a \Sigma_b m_{ab} = n)}{\Pr(\Sigma_a \Sigma_b m_{ab} = n)} \\
&= \frac{\Pi_a \Pi_b \dfrac{\exp(-m_{ab})(m_{ab})^{n_{ab}}}{n_{ab}!}}{\dfrac{\exp(-\Sigma_a \Sigma_b m_{ab})(\Sigma_a \Sigma_b m_{ab})^n}{n!}} \qquad \text{(B.17)} \\
&= n! \prod_a \prod_b \frac{(\pi_{ab}^{AB})^{n_{ab}}}{n_{ab}!}
\end{aligned}
$$

which is a multinomial density with parameters n and $\{\pi_{ab}^{AB}\}$. A similar result can be shown for conditional probabilities. Consider the distribution of $\{n_{ab}\}$ holding the marginals n_{a+} constant, which is equivalent to the distribution of $\{p_{b|a}\}$. The distribution of $p_{b|a}$ is binomial with parameters n and $\{\pi_{b|a}^{B|A}\}$ for $a = 1,2$. The joint distribution of $p_{b|a}$, $a = 1$, 2 is called the *product multinomial.*

The conditional estimates $\pi_{b|a}^{B|A}$ (or $\pi_{a|b}^{A|B}$) can also be computed from the loglinear model parameters. For example

$$
\begin{aligned}
\pi_{b|a}^{B|A} &= \frac{\exp(u + u_a^A + u_b^B + u_{ab}^{AB})}{\Sigma_a \exp(u + u_a^A + u_b^B + u_{ab}^{AB})} \\
&= \frac{\exp(u_b^B + u_{ab}^{AB})}{\Sigma_a \exp(u_b^B + u_{ab}^{AB})}
\end{aligned} \qquad \text{(B.18)}
$$

We can also model $\pi_{b|a}^{B|A}$ directly using the *logit* function defined as $\text{logit}(\pi) = \log[\pi/(1 - \pi)]$. We write

$$
\begin{aligned}
\log\left(\frac{\pi_{1|a}^{B|A}}{1 - \pi_{1|a}^{B|A}}\right) &= \log\left(\frac{\pi_{1|a}^{B|A}}{\pi_{2|a}^{B|A}}\right) \\
&= \log \pi_{1|a}^{B|A} - \log \pi_{2|a}^{B|A} \\
&= \log\left(\frac{\exp(u_1^B + u_{a1}^{AB})}{\Sigma_b \exp(u_b^B + u_{ab}^{AB})}\right) - \log\left(\frac{\exp(u_2^B + u_{a2}^{AB})}{\Sigma_b \exp(u_b^B + u_{ab}^{AB})}\right) \qquad \text{(B.19)} \\
&= u_1^B - u_2^B + u_{a1}^{AB} - u_{a2}^{AB} \\
&= \lambda + \lambda_a^A
\end{aligned}
$$

where $\lambda = u_1^B - u_2^B$ and $\lambda_a^A = u_{a1}^{AB} - u_{a2}^{AB}$. The loglinear modeling constraints imply that

$$
\lambda_1^A + \lambda_2^A = u_{11}^{AB} - u_{12}^{AB} + u_{21}^{AB} - u_{22}^{AB} = 0 \qquad \text{(B.20)}
$$

The model in (B.20) is a saturated logit model for $\pi_{b|a}^{B|A}$. Estimates of $\pi_{1|a}^{B|A}$ can be obtained from the estimates of λ and λ_a^A using the inverse transformation

$$\pi_{1|a}^{B|A} = \frac{\exp(\lambda + \lambda_a^A)}{1 + \exp(\lambda + \lambda_a^A)} \quad \text{(B.21)}$$

Because the exponential function is strictly positive, $\pi_{b|a}^{B|A}$ will always be in the interval [0,1] regardless of the values of λ and λ_a^A.

B.3 GENERALIZATION TO THREE VARIABLES

The results of the previous section can easily be generalized to three dichotomous variables. For example, let A, B, and C denote three dichotomous variables, and consider the following nonsaturated loglinear model for the expected cell frequencies $\{m_{abcd}\}$:

$$\log m_{abc} = u + u_a^A + u_b^B + u_c^C + u_{ab}^{AB} + u_{bc}^{BC} \quad \text{(B.22)}$$

The saturated model contains two additional terms: u_{ac}^{AC} and u_{abc}^{ABC}; however, the present model assumes that both of these terms are zero. As we did for two variables, we must impose zero-equality constraints to obtain an estimable model:

$$\begin{aligned} &\sum_a u_a^A = \sum_b u_b^B = \sum_a u_c^C = 0 \\ &\sum_a u_{ab}^{AB} = \sum_b u_{ab}^{AB} = \sum_b u_{bc}^{BC} = \sum_a u_{ac}^{AC} = \sum_c u_{ac}^{AC} = 0 \end{aligned} \quad \text{(B.23)}$$

For the saturated model, these additional constraints would be needed:

$$\sum_a u_{ac}^{AC} = \sum_c u_{ac}^{AC} = \sum_a u_{abc}^{ABC} = \sum_b u_{abc}^{ABC} = \sum_c u_{abc}^{ABC} = 0 \quad \text{(B.24)}$$

The model in (B.22) is referred to as a *hierarchical loglinear model* in that whenever an interaction effect is included in the model, all lower-order effects containing the variables in the higher-order effect must also be included. So, for example, if the AB interaction is included, then so must be the effects for A alone and B alone. Likewise, the AC interaction must not be in the model unless both the A and C main effects are also included. In this book, only hierarchical loglinear and logit models have been considered.

Because a hierarchical model can be specified by knowing only the highest-order interaction in which a particular variable appears, a shorthand notation has been developed. In this notation, (B.22) can be specified as $\{AB\ BC\}$

because the lower-order terms are automatically in the model. Likewise, $\{ABC\}$ is the saturated model and $\{A\ B\ C\}$ is the three-way independence model containing just the u terms for A, B, and C.

The interpretation of the loglinear model effects will be discussed in terms of the saturated model, $\{ABC\}$ which is analogous to a 2^3 factorial ANOVA model. There are, of course, eight parameters of this model because the ABC table contains 8 degrees of freedom. These are the overall mean, u; the three main effects, u_1^A, u_1^B, and u_1^C; the three first-order interactions, u_{11}^{AB}, u_{11}^{AC}, and u_{11}^{BC}; and one second-order interaction, u_{111}^{ABC}. These effects can be expressed as simple contrasts of the eight cells of the ABC table. Extending the notation of the previous section, let $\ell_{abc} = \log(m_{abc})$. The main effect for A is computed by subtracting the sum of the log counts at the low level of A from the sum of log counts at the high level of A ignoring the other factors as follows:

$$\begin{aligned} 8u_1^A &= -(\ell_{111} + \ell_{112} + \ell_{121} + \ell_{122}) + (\ell_{211} + \ell_{212} + \ell_{221} + \ell_{222}) \\ &= -\overline{\ell}_{1++} + \overline{\ell}_{2++} \end{aligned} \qquad \text{(B.25)}$$

This reflects the difference in moving from category 1 to category 2 of A irrespective of the other factors. The B and C main effects are computed similarly:

$$8u_1^B = -\ell_{+1+} + \ell_{+2+} \quad \text{and} \quad 8u_1^C = -\ell_{++1} + \ell_{++2} \qquad \text{(B.26)}$$

The AB interaction is computed as the difference in the effect of A going from the low level to the high level of B ignoring C:

$$8u_{11}^{AB} = (\overline{\ell}_{11+} - \overline{\ell}_{21+}) - (\overline{\ell}_{12+} - \overline{\ell}_{22+}) \qquad \text{(B.27)}$$

Therefore, there is no AB interaction if the average effect of A is the same effect at both levels of B. Likewise, we have

$$8u_{11}^{AC} = (\overline{\ell}_{1+1} - \overline{\ell}_{2+1}) - (\overline{\ell}_{1+2} - \overline{\ell}_{2+2}) \quad \text{and} \quad 8u_{11}^{BC} = (\overline{\ell}_{+11} - \overline{\ell}_{+21}) - (\overline{\ell}_{+12} - \overline{\ell}_{+22}) \qquad \text{(B.28)}$$

Finally, the ABC interaction is the interaction of AB at the low level of C minus the interaction of AB at the high level of C:

$$8u_{111}^{ABC} = [(\ell_{111} - \ell_{211}) - (\ell_{121} - \ell_{221})] - [(\ell_{112} - \ell_{212}) - (\ell_{122} - \ell_{222})] \qquad \text{(B.29)}$$

The ABC interaction is zero if the combination of A and B has the same effect on the log count at both levels of C. These equations are summarized in Table B.4.

Note that, as in ANOVA, the interaction coefficients can be obtained by multiplying the columns of main effects for variables contained in the interac-

Table B.4 Coefficients for Effects in the ABC Cross-Classification for Dichotomous Variables

	$8u$	$8u_1^A$	$8u_1^B$	$8u_1^C$	$8u_{11}^{AB}$	$8u_{11}^{AC}$	$8u_{11}^{BC}$	$8u_{111}^{ABC}$
ℓ_{111}	+	−	−	−	+	+	+	−
ℓ_{112}	+	−	−	+	+	−	−	+
ℓ_{121}	+	−	+	−	−	+	−	+
ℓ_{122}	+	−	+	+	−	−	+	−
ℓ_{211}	+	+	−	−	−	−	+	+
ℓ_{212}	+	+	−	+	−	+	−	−
ℓ_{221}	+	+	+	−	+	−	−	−
ℓ_{222}	+	+	+	+	+	+	+	+

tion. For example, the column of coefficients for u_{11}^{BC} is equal to the product of the coefficients for u_1^B and u_1^C.

Interpreting the effects on the original scale [$\exp(u_1^A)$, $\exp(u_1^B)$, etc.] requires the use of odds ratios and ratios of odds ratios. For example, the three-way interaction can be expressed as

$$\exp(8u_{111}^{ABC}) = \frac{(m_{111}/m_{211})/(m_{121}/m_{221})}{(m_{112}/m_{212})/(m_{122}/m_{222})} = \frac{R_1}{R_2} \tag{B.30}$$

where R_c is the ratio of odds of being classified in the first category of A going from the low level of B to the high level of B when $C = c$. If $R_1 = R_2$, then $u_{111}^{ABC} = 0$. The greater the extent to which R_1/R_2 departs from 1, the larger will be the three-way interaction.

The conditional probability $\pi_{clab}^{C|AB}$ can be expressed as a function of the parameters of the model in (B.22) as follows:

$$\begin{aligned}\pi_{clab}^{C|AB} &= \frac{\exp(m_{abc})}{\sum_c \exp(m_{abc})}\\ &= \frac{\exp(u+u_a^A+u_b^B+u_c^C+u_{ab}^{AB}+u_{bc}^{BC})}{\sum_c \exp(u+u_a^A+u_b^B+u_c^C+u_{ab}^{AB}+u_{bc}^{BC})}\\ &= \frac{\exp(u_c^C+u_{bc}^{BC})}{\sum_c \exp(u_c^C+u_{bc}^{BC})}\end{aligned} \tag{B.31}$$

Because C is dichotomous, the logit can also be written as

$$\begin{aligned}\log\left(\frac{\pi_{1lab}^{C|AB}}{\pi_{2lab}^{C|AB}}\right) &= (u_1^C - u_2^C) + (u_{a1}^{AC} - u_{a2}^{AC}) + (u_{b1}^{BC} - u_{b2}^{BC})\\ &= \lambda + \lambda_a^A + \lambda_b^B\end{aligned} \tag{B.32}$$

For logit models, we will also consider only hierarchical models. For example, (B.32) can be succinctly specified as $\pi_{c|ab}^{C|AB} = \{A \ B\}$.

Finally, these results can be easily extended to polytomous variables. Suppose now that A, B, and C have K_A, K_B, and K_C categories, respectively. For the saturated model in (B.22), the same constraints are required where now the summations extend to all categories of the variables. The parameters of the saturated model, $\{ABC\}$, are also defined similarly as follows:

$$
\begin{aligned}
u &= \bar{\ell}_{+++} \\
u_a^A &= \bar{\ell}_{a++} - \bar{\ell}_{+++} \\
u_b^B &= \bar{\ell}_{+b+} - \bar{\ell}_{+++} \\
u_c^C &= \bar{\ell}_{++c} - \bar{\ell}_{+++} \\
u_{ab}^{AB} &= \bar{\ell}_{ab+} - \bar{\ell}_{a++} - \bar{\ell}_{+b+} + \bar{\ell}_{+++} \\
u_{ac}^{AC} &= \bar{\ell}_{a+c} - \bar{\ell}_{a++} - \bar{\ell}_{++c} + \bar{\ell}_{+++} \\
u_{bc}^{BC} &= \bar{\ell}_{+bc} - \bar{\ell}_{+b+} - \bar{\ell}_{++c} + \bar{\ell}_{+++} \\
u_{abc}^{ABC} &= \ell_{abc} - \bar{\ell}_{ab+} - \bar{\ell}_{a+c} - \bar{\ell}_{+bc} + \bar{\ell}_{a++} + \bar{\ell}_{+b+} + \bar{\ell}_{++c} - \bar{\ell}_{+++}
\end{aligned}
$$

It can be easily shown that, for dichotomous variables, these equations are equivalent to those given in Table B.4.

The results for logit models can also be easily generalized. Suppose that $K_C = 2$ while K_A, K_B are both larger than 2. The saturated logit model

$$
\log\left(\frac{\pi_{1|ab}^{C|AB}}{\pi_{2|ab}^{C|AB}}\right) = \lambda + \lambda_a^A + \lambda_b^B + \lambda_{ab}^{AB} \tag{B.33}
$$

can be derived as in (B.19) with similar zero equality constraints on all indexed parameters. The model contains $K_A - 1$ λ^A parameters, $K_B - 1$ λ^B parameters and $(K_A - 1)(K_B - 1)$ λ^{AB} parameters.

B.4 ESTIMATION OF LOGLINEAR AND LOGIT MODELS

The likelihood given in (B.2) for two variables can be easily generalized for any number of variables. For three variables, it may be written as

$$
\log(\mathcal{L}) = \sum_a \sum_b \sum_c (n_{abc} \log m_{abc} - m_{abc}) \tag{B.34}
$$

where the expected frequencies m_{abc} are functions of the unknown u parameters. The parameters are estimated by maximum-likelihood (ML) estimation. First, the likelihood equations are obtained by differentiating the likelihood kernel with respect to each parameter and setting each partial derivative to 0. The maximum-likelihood estimates (MLEs) are obtained by solving the likelihood equations simultaneously. To illustrate the basic idea, we rewrite the model as

$$\log m_{abc} = \sum_a \sum_b \sum_c \beta_j x_{abcj} \tag{B.35}$$

or, in matrix form, as

$$\log \mathbf{m} = \mathbf{X}\boldsymbol{\beta} \tag{B.36}$$

where β_j are the various u parameters, $\boldsymbol{\beta}$ is the column vector $[\beta_j]$, x_{abcj} are known constants, and $\mathbf{X}$ is the design matrix $[x_{abcj}]$.

For example, for the model $\{AB\ BC\}$ where A, B, and C are all dichotomous, $\boldsymbol{\beta}' = [u, u_a^A, u_b^B, u_c^C, u_{ab}^{AB}, u_{bc}^{BC}]$. The first column of $\mathbf{X}$ corresponds to $\beta_1 = u$ and is therefore all 1s. The second column corresponds to u_a^A, the third to u_b^B, and so on. Note that, because of the model constraints, the values in each column of $\mathbf{X}$, which are either 1 or −1, sum to 0. The rows of $\mathbf{X}$ correspond to the eight cells of the table. For instance, the row of $\mathbf{X}$ corresponding to cell 121 is

$$\begin{aligned} \log m_{121} &= u + u_1^A + u_2^B + u_1^C + u_{12}^{AB} + u_{21}^{BC} \\ &= u + u_1^A - u_1^B + u_1^C - u_{12}^{AB} - u_{21}^{BC} \end{aligned} \tag{B.37}$$

which produces the row vector [1 1 −1 1 −1 −1].

It can be shown [see, e.g., Agresti (1990)] that the likelihood equations for (B.36) are

$$\mathbf{X}'\mathbf{n} = \mathbf{X}'\hat{\mathbf{m}} \tag{B.38}$$

where $\mathbf{n}$ and $\hat{\mathbf{m}}$ are column vectors of observed and estimated expected cell counts, respectively. For example, for the model $\{AB\ BC\}$, it can be shown that the likelihood equations are

$$\begin{aligned} n_{+++} &= \hat{m}_{+++} \\ n_{a++} &= \hat{m}_{a++} \\ n_{+b+} &= \hat{m}_{+b+} \\ n_{++c} &= \hat{m}_{++c} \\ n_{ab+} &= \hat{m}_{ab+} \\ n_{+bc} &= \hat{m}_{+bc} \end{aligned} \tag{B.39}$$

Bishop et al. (1975, pp. 74–75) show how to obtain direct estimates of the parameters in these equations. In general, however, iterative methods such as IPF can be used to solve the equations (Bishop et al. 1975, p. 830).

Note that only the last two equations are important because, if these are satisfied, the first four equations will be as well. Thus, the MLEs for the model $\{AB\ BC\}$ are obtained by equating table margins for AB and BC to their expected values and solving for the parameters. The margins n_{ab+} and n_{bc+} are called ***minimally sufficient statistics***. This result generalizes to all hierarchical loglinear models. It can further be shown that including an effect in the loglinear model forces the estimated and observed marginal totals corresponding to that effect to be equal. Bishop et al. (1975, pp. 74–75) show how to obtain direct estimates of the parameters for the equations in (B.39). In general, however, iterative methods such as IPF can be used to solve the equations (Bishop et al. 1975, p. 830).

A well-known result is that loglinear models can be easily transformed into logit models by including particular effects in the loglinear model. For example, the logit model in which C is the dependent variable and A and C are the independent variables is

$$\pi_{c|ac} = \frac{\exp(u_c^C + u_{ac}^{AC} + u_{bc}^{BC})}{\Sigma_c \exp(u_c^C + u_{ac}^{AC} + u_{bc}^{BC})} = \frac{m_{abc}}{m_{ab+}} \tag{B.40}$$

where it is assumed that $m_{ab+} = n_{ab+}$. It can be shown that the likelihood equations corresponding to (B.40) are the same as those for the loglinear model having the same effects (i.e., C, AC, BC) conditional on m_{ab+}. This is just the loglinear model $\{AB\ AC\ BC\}$. In general, the parameters of any hierarchical logit model can be estimated by a loglinear model having all the u parameters of the logit model plus the logit model's conditioning effects and all its lower-order interactions and main effects. A proof of this result can be found in Vermunt (1997a, pp. 306–308).

References

Abowd, J. M., and Zellner, A. (1985), Estimating gross labor-force flows, *Journal of Business and Economic Statistics*, *3*, 254–283.

Achenbach, T. M. (1991a), *Manual for the Child Behavior Checklist 2–3 and 1991 Profile*, Department of Psychiatry, University of Vermont, Burlington.

Achenbach, T. M. (1991b), *Manual for the Child Behavior Checklist 4–18 and 1991 profile*, Department of Psychiatry, University of Vermont, Burlington.

Agresti, A. (1990), *Categorical Data Analysis*, 1st ed., Wiley, New York.

Agresti, A. (2002), *Categorical Data Analysis*, 2nd ed., Wiley, New York.

Agresti, A., and Yang, M. (1987), An empirical investigation of some effects of sparseness in contingency tables, *Computational Statistics & Data Analysis*, *5*, 9–21.

Aitken, M., Anderson, D., and Hinde, J. (1981), Statistical modeling of data on teaching styles, *Journal of the Royal Statistical Society, Series A*, *144*, 419–461.

Alexander, C. A. (2001), "Formal discussion," *Proceedings of the 2001 Federal Committee on Survey Methodology Conference*, Session XI-B, Washington, DC.

Alho, J., Mulry, M., Wurdeman, K., and Kim, J. (1993), Estimating heterogeneity in the probabilities of enumeration for dual-system estimation, *Journal of the American Statistical Association*, *88*, 1130–1136.

Allison, P. D. (2001), *Missing Data*, Sage University Papers Series on Quantitative Applications in the Social Sciences, Series 07-136, Thousand Oaks, CA.

Alwin, D. F. (2007), *Margins of Error: A Study of Reliability in Survey Measurement*, Wiley, Hoboken, NJ.

American National Election Studies (ANES) (2008), *The ANES Guide to Public Opinion and Electoral Behavior*, University of Michigan, Center for Political Studies, Ann Arbor, MI (retrievable from `www.electionstudies.org`).

Andersen, R., Kasper, J., Frankel, M., and Associates (1979), *Total Survey Error: Applications to Improve Health Surveys*, Jossey-Bass, San Francisco.

Bailar, B. A. (1968), Recent research on reinterview procedures, *Journal of the American Statistical Association*, *63*, 41–63.

Baker, S. G. (1992), A simple method for computing the observed information matrix when using the EM algorithm with categorical data, *Journal of Computational and Graphical Statistics*, *1*, 63–76.

Barkume, A. J., and Horvath, F. W. (1995), Using gross flows to explore movements in the labor force, *Monthly Labor Review*, April 28–35.

Bartholomew, D. J. (1982), *Stochastic Models for Social Processes*, 3rd ed., Wiley, New York.

Bartholomew, D. J., and Knott, M. (1999), *Latent Variable Models and Factor Analysis*, Arnold, London.

Bartolucci, F., and Forcina, A. (2001), Analysis of capture-recapture data with a Rasch-type model allowing for conditional dependence and multidimensionality, *Biometrics*, *57*, 714–719.

Bartolucci, F., and Pennoni, F. (2007), A class of latent Markov models for capture–recapture data allowing for time, heterogeneity and behavior effects, *Biometrics*, *63*, 568–578.

Bassi, F. (1997), Identification of latent class Markov models with multiple indicators and correlated measurement errors, *Journal of the Royal Statistical Society, Series A*, *3*, 201–211.

Bell, W. (1993), Using information from demographic analysis in post-enumeration survey estimation, *Journal of the American Statistical Association*, *88*, 1106–1118.

Bensmail, H., Celeux, G., Raftery, A. E., and Robert, C. (1997), Inference in model-based cluster analysis, *Statistics and Computing*, *7*, 1–10.

Berzofsky, M. (2009), Survey classification error analysis: Critical assumptions and model robustness, paper presented at the 2009 Classification Error Society Conference.

Biemer, P. P. (1988), Measuring data quality, in R. Groves, P. P. Biemer, L. Lyberg, J. Massey, W. Nicholls, and J. Waksberg, eds., *Telephone Survey Methodology*, Wiley, New York, pp. 273–282.

Biemer, P. P. (2000), *An Application of Markov Latent Class Analysis for Evaluating Reporting Error in Consumer Expenditure Survey Screening Questions*, RTI Technical Report for the US Bureau of Labor Statistics, RTI International, Research Triangle Park, NC.

Biemer, P. (2001), Nonresponse bias and measurement bias in a comparison of face to face and telephone interviewing, *Journal of Official Statistics*, *17*(2), 295–320.

Biemer, P. (2004a), An analysis of classification error for the revised Current Population Survey employment questions, *Survey Methodology*, *30*(2), 127–140.

Biemer, P. (2004b), Simple response variance: Then and now, *Journal of Official Statistics*, *20*, 417–439.

Biemer, P. P. (2004c), Response from the author to comments on "An analysis of classification error for the revised Current Population Survey employment questions." *Survey Methodology*, *30*(2), 154–158.

Biemer, P. (2009), Measurement errors in surveys, in D. Pfeffermann and C. R. Rao, eds., *Handbook of Statistics*, Vol. *29*, *Sample Surveys: Design, Methods and Applications*, Elsevier, Oxford, UK, Chapter 12, pp. 277–312.

Biemer, P. P. (2010), Overview of design issues: Total survey error, in P. Marsden and J. Wright, eds., *Handbook of Survey Research, Second Edition*, Bingley, Emerald Group Publishing, UK. Chapter 2.

Biemer, P., and Bushery, J. (2001), On the validity of Markov latent class analysis for estimating classification error in labor force data, *Survey Methodology*, *26*(2), 136–152.

Biemer, P., and Christ, S. (2008), Weighting survey data, in J. Hox, E. De Leeuw, and D. Dillman, eds., *International Handbook on Survey Methodology*, Lawrence Erlbaum Associates, Mahwah, NJ.

Biemer, P., and Davis, M. (1991), *Measurement of Census Erroneous Enumerations—Clerical Error Made in the Assignment of Enumeration Status*, Evaluation Project P10, Internal US Bureau of the Census Report, July.

Biemer, P. P., and Forsman, G. (1992), On the quality of reinterview data with applications to the Current Population Survey, *Journal of the American Statistical Association*, *87*(420), 915–923.

Biemer, P. P., and Lyberg, L. E. (2003), *Introduction to Survey Quality*, Wiley, Hoboken, NJ.

Biemer, P. P., and Stokes, S. L. (1991), Approaches to modeling measurement error, in P. P. Biemer, R. Groves, L. Lyberg, N. Mathiowetz, and S. Sudman, eds., *Measurement Errors in Surveys*, Wiley, New York.

Biemer, P., and Trewin, D. (1997), A review of measurement error effects on the analysis of survey data, in L. Lyberg, P. Biemer, M. Collins, E. De Leeuw, C. Dippo, N. Schwarz, and D. Trewin, eds., *Survey Measurement and Process Quality*, Wiley, New York.

Biemer, P. P., and Tucker, C. (2001), Estimation and correction for underreporting errors in expenditure data: A Markov latent class modeling approach, paper presented at the International Statistical Institute, Seoul, South Korea.

Biemer, P., and Wiesen, C. (2002), Latent class analysis of embedded repeated measurements: An application to the National Household Survey on Drug Abuse, *Journal of the Royal Statistical Society, Series A*, *165*(1), 97–119.

Biemer, P. P., and Woltman, H. (2002), Estimating reliability and bias from reinterviews with application to the 1998 dress rehearsal race question, paper presented at the 2002 Federal Committee on Survey Methodology Conference.

Biemer, P. P., Groves, R. M., Lyberg, L., Mathiowetz, N. A., and Sudman, S, eds., (1991), *Measurement Errors in Surveys*, Wiley, New York.

Biemer, P. P., Woltman, H., Raglin, D., and Hill, J. (2001), Enumeration accuracy in a population census: An evaluation using latent class analysis, *Journal of Official Statistics*, *17*(1), 129–149.

Biemer, P., Brown, G., and Wiesen, C. (2002), Triple system estimation with erroneous enumerations in the administrative records list, paper presented at the Joint Statistical Meetings, Survey Research Methods Section.

Biemer, P., Christ, S., and Wiesen, C. (2009), A general approach for estimating scale score reliability for panel survey data, *Psychological Methods*, *4*(14), 400–412.

Binder, D. A. (1983), On the variances of asymptotically normal estimators from complex surveys, *International Statistical Review*, *51*, 279–292.

Bishop, Y. M. M., Fienberg, S. E., and Holland, P. W. (1975), *Discrete Multivariate Analysis*, MIT Press, Cambridge, MA.

Blumen, I., Kogan, M., and McCarthy, P. J. (1966), Probability models for mobility, in P. F. Lazarsfeld and N. W. Henry, eds., *Readings in Mathematical Social Science*, MIT Press, Cambridge, MA, pp. 318–334.

Bolesta, M. S. (1998), *Comparison of Standard Errors within a Latent Class Framework Using Resampling and Newton Techniques*, doctoral dissertation, University of Maryland, College Park.

Bollen, K. (1989), *Structural Equations with Latent Variables*, Wiley-Interscience, New York.

Brooks, C., and Bailar, B. (1978), *An Error Profile: Employment as Measured by the Current Population Survey*, Statistical Working Paper 3, US Office for Management and Budget, Washington, DC.

Bross, I. (1954), Misclassification in 2 × 2 tables, *Biometrics*, *10*(4), 478–486.

Brown, G., and Biemer, P. (2004), Estimating erroneous enumerations in the decennial census using four lists, paper presented at the Joint Statistical Meetings, Survey Research Methods Section.

Brown, J. J., Diamond, I. D., Chambers, R. L., Buckner, L. J., and Teague, A. D. (1999), A methodological strategy for a one-number census in the UK, *Journal of the Royal Statistical Association, Series A*, *162*(Part 2), 247–287.

Burnham, K. P., and Anderson, D. R. (2002), *Model Selection and Multimodel Inference: A Practical-Theoretic Approach*, 2nd ed., Springer-Verlag.

Centers for Disease Control and Prevention (2009), *2008 National Health Interview Survey (NHIS) Public Use Data Release: NHIS Survey Description*, Centers for Disease Control, US Department of Health and Human Services, Hyattsville, MD.

Chakrabarty, R., and Torres, G. (1996), *American Housing Survey: A Quality Profile*, US Department of Housing and Urban Development and US Department of Commerce, Washington, DC.

Chao, A., and Tsay, P. K. (1998), A sample coverage approach to multiple-system estimation with application to the census undercount, *Journal of the American Statistical Association*, *93*, 283–293.

Chromy, J., and Meyers, L. (2001), Variance models applicable to the NHSDA, paper presented at the Annual Meeting of the American Statistical Association.

Chua, T. C., and Fuller, W. A. (1987), A model for multinomial response error applied to labor flows, *Journal of the American Statistical Association*, *82*, 46–51.

Cicchetti, D. V. (1994), Guidelines, criteria, and rules of thumb for evaluating normed and standardized assessment instruments in psychology, *Psychological Assessment*, *6*, 284–290.

Clogg, C. C. (1979), Some latent structure models for the analysis of Likert-type data, *Social Science Research*, *8*, 287–301.

Clogg, C., and Eliason, S. (1987), Some common problems in log-linear analysis, *Sociological Methods and Research*, *16*, 8–14.

Cochran, W. G. (1968), Errors of measurement in statistics, *Technometrics*, *10*(4) 637–666.

Cochran, W. G. (1977), *Sampling Techniques*, Wiley, New York.

Cohen, J. (1960), A coefficient of agreement for nominal scales, *Educational and Psychological Measurements*, *20*, 37–46.

Collins, L., and Lanza, S. (2010), *Latent Class and Latent Transition Analysis*, Wiley, Hoboken, NJ.

Collins, L., and Tracy, A. (1997), Estimation in complex latent transition models with extreme data sparseness, *Kwantitatieve Methoden*, *55*, 57–71.

Collins, L. M., Fidler, P. F., Wugalter, S. E., and Long, L. D. (1993), Goodness-of-fit testing for latent class models, *Multivariate Behavioral Research*, *28*(3), 375–389.

Collins, L., Fidler, P., and Wugalter, S. (1996), Some practical issues related to estimation of latent class and latent transition parameters, in A. von Eye and C. Clogg, eds., *Analysis of Categorical Variables in Developmental Research*, Academic Press, Inc., San Diego, CA, pp. 133–146.

Converse, P. E. (1964), The nature of belief systems in mass publics, in D. E. Apter, ed., *Ideology and Discontent*, Free Press, New York, pp. 206–261.

Cressie, N., and Read, T. R. C. (1984), Multinomial goodness-of-fit tests, *Journal of the Royal Statistical Society Society*, *B*, *46*, 440–464.

Cronbach, L. J. (1951), Coefficient alpha and the internal structure of tests, *Psychometrika*, *16*, 297–334.

Croon, M. (1990), Latent class analysis with ordered latent classes, *British Journal of Mathematical and Statistical Psychology*, *43*, 171–192.

Croon, M. A. (1991), Investigating Mokken scalability of dichotomous items by means of ordinal latent class analysis, *British Journal of Mathematical & Statistical Psychology*, *44*, 315–331.

Croon, M. A. (2002), Ordering the classes, in J. A. Hagenaars and A. L. McCutcheon, eds., *Applied Latent Class Analysis*, Cambridge University Press, Cambridge, UK, pp. 137–162.

Dalenius, T. (1985), Relevant official statistics, *Journal of Official Statistics*, *1*(1), 21–33.

Darroch, J. N., Fienberg, S. E., Glonek, G. F. V., and Junker, B. W. (1993), A three-sample multiple recapture approach to census population estimation with heterogeneous catchability, *Journal of the American Statistical Association*, *88*, 1137–1148.

Dayton, C. M., and MacReady, G. B. (1980), A scaling model with response errors and intrinsically unscalable respondents, *Psychometrika*, *45*, 343–356.

De Leeuw, E., and van der Zouwen, J. (1988), Data quality in telephone surveys and face to face surveys: A comparative meta-analysis, in R. Groves, P. P. Biemer, L. Lyberg, J. Massey, W. Nicholls, and J. Waksberg, eds., *Telephone Survey Methodology*, Wiley, New York, pp. 273–282.

De Menezes, L. M. (1999), On fitting latent class models for binary data: The estimation of standard errors, *British Journal of Mathematical and Statistical Psychology*, *52*, 149–168.

Deming, W. E. (1944), On errors in surveys, *American Sociological Review*, *9*(4), 359–369.

Dempster, A. P., Laird, N. M., and Rubin, D. B. (1977), Maximum likelihood from incomplete data via the EM algorithm, *Journal of the Royal Statistical Society*, *B*, 1–38.

De Waal, T., and Haziza, D. (2009), Statistical editing and imputation, in C. R. Rao and D. Pfeffermann, eds., *Handbook of Statistics*, Vol. *29A*, *Sample Surveys: Design, Methods and Applications*, Elsevier, Amsterdam.

Dillman, D. (2007), *Mail and Internet Surveys: The Tailored Design Method*, 2nd ed., Wiley, Hoboken, NJ.

Dillman, D., Smyth, J., and Christian, L. (2008), *Internet, Mail, and Mixed-Mode Surveys: The Tailored Design Method*, 3rd ed., Wiley, Hoboken, NJ.

Dohrmann, S., Han, D., and Mohadjer, L. (2007), Improving coverage of residential address lists in multistage area samples, paper presented at the American Statistical Association, Survey Research Methods Section, Salt Lake City, UT.

Eckler, A. (1972), *The Bureau of the Census*, Praeger Publishers, New York.

Efron, B. (1979), Bootstrap methods: Another look at the jackknife, *The Annals of Statistics*, *7*(1), 1–26.

Efron, B., and Hinkley, D. (1978), Assessing the accuracy of the maximum likelihood estimator: Observed versus expected Fisher information, *Biometrika*, *65*, 457–487.

Efron, B., and Tibshirani, R. J. (1993), *An Introduction to Bootstrap*, Chapman & Hall, London.

Energy Information Administration (1996), *Residential Energy Consumption Survey Quality Profile*, US Department of Energy, Washington, DC.

Espeland, M. A., and Handelman, S. L. (1989), Using latent class models to characterize and assess relative error in discrete measurements, *Biometrics*, *45*(2), 585–599.

Eurostat (2003), Item 4.2—methodological documents—definition of quality statistics, Working Group on Assessment of Quality in Statistics, 6th Meeting, Luxembourg, available at http://epp.eurostat.ec.europa.eu/portal/page/portal/quality/documents/ess%20quality%20definition.pdf.

Everitt, B. S., and Hand, D. J. (1981), *Finite Mixture Models*, Chapman & Hall, New York.

Fay, R. E. (1986), Causal models for patterns of nonresponse, *Journal of the American Statistical Association*, *81*, 354–365.

Fay, R. E. (2002), *Evidence of Additional Erroneous Enumerations from the Person Duplication Study*, Executive Steering Committee for ACE Policy II, Report 9, revised March 2002, Washington, DC, available at http://www.census.gov/dmd/www/pdf/report9revised.pdf, US Census Bureau.

Fecso, R., and Pafford, B. (1988), Response errors in establishment surveys with an example from an agribusiness survey, paper presented at the Joint Statistical Meetings, Survey Research Methods Section, American Statistical Association, pp. 315–320.

Feder, M. (2007), Variance estimation of the survey-weighted kappa measure of agreement, paper presented at the American Statistical Association, Survey Research Methods Section, Anaheim, CA, pp. 3002–3007.

Feiser, M., and Lin, Y. (1999), A goodness of fit test for the latent class model when the expected frequencies are small, in M. E. Sobel and M. P. Becker, eds., *Sociological Methods*, Blackwell, Oxford, pp. 81–112.

Fellegi, I. (1964), Response variance and its estimation, *Journal of the American Statistical Association*, *59*, 1016–1041.

Fellegi, I. (1996), Characteristics of an effective statistical system, *International Statistical Review, 64*(2).

Fellegi, I., and Sunter, A. (1974), Balance between different sources of survey errors—some canadian experiences, *Sankhya, Series C, 36*, 119–142.

Fienberg, S. E. (1992), Bibliography on capture-recapture modeling with application to census undercount adjustment, *Survey Methodology, 18*, 143–154.

Fienberg, S. E., Hersh, P., Rinaldo, A., and Zhou, Y. (2007), *Maximum Likelihood Estimation in Latent Class Models for Contingency Table Data, Technical Report*, Carnegie Mellon University, available at `http://www.stat.cmu.edu/tr/`.

Fienberg, S. E., Johnson, M. S., and Junker, B. W. (1999), Classical multilevel and Bayesian approaches to population size estimation using multiple lists, *Journal of the Royal Statistical Society, A, 162*, 383–405.

Fleiss, J. L. (1981), *Statistical Methods for Rates and Proportions*, 2nd ed., Wiley, New York.

Fletcher, R. (1987), *Practical Methods of Optimization*, 2nd ed., Wiley, Chichester, UK.

Formann, A. K. (1985), Constrained latent class models: Theory and applications, *British Journal of Mathematical and Statistical Psychology, 38*, 87–111.

Forsman, G., and Schreiner, I. (1991), The design and analysis of reinterview: An overview, in P. Biemer, R. Groves, L. Lyberg, N. Mathiowetz, and S. Sudman, eds., *Measurement Errors in Surveys*, Wiley, New York, pp. 279–302.

Forsyth, B., and Lessler, J. (1991), Cognitive laboratory methods: A taxonomy, in P. Biemer, R. Groves, L. Lyberg, N. Mathiowetz, and S. Sudman, eds., *Measurement Errors in Surveys*, Wiley, New York, pp. 393–418.

Fuller, W. (1987), *Measurement Error Models*, New York: John Wiley & Sons.

Fuller, W., and Chua, T. C. (1985), Gross change estimation in the presence of response error, paper presented at the Conference on Gross Flows in Labor Force Statistics, Washington, DC.

Galindo-Garre, F., and Vermunt, J. K. (2006), Avoiding boundary estimates in latent class analysis by Bayesian posterior mode estimation, *Behaviormetrika, 33*, 43–59.

Garrett, E. S., and Zeger, S. L. (2000), Latent class model diagnosis, *Biometrics, 56*(4), 1055–1067.

Goodman, L. A. (1961), Statistical methods for the Mover-Stayer model, *Journal of the American Statistical Association, 56*(296), 841–868.

Goodman, L. A. (1973a), Causal analysis of data from panel studies and other kinds of surveys, *American Journal of Sociology, 78*, 1135–1191.

Goodman, L. A. (1973b), The analysis of multidimensional contingency tables when some variables are posterior to others: A modified path analysis approach, *Biometrika, 60*, 179–192.

Goodman, L. A. (1974a), Exploratory latent structure analysis using both identifiable and unidentifiable models, *Biometrika, 61*, 215–231.

Goodman, L. A. (1974b), The analysis of systems of qualitative variables when some of the variables are unobservable. Part I: A modified latent structure approach, *American Journal of Sociology, 79*, 1179–1259.

Goodman, L. A. (2002), Latent class analysis: The empirical study of latent types, latent variables, and latent structures, in J. A. Hagenaars and A. L. McCutcheon, eds., *Applied Latent Class Analysis*, Cambridge University Press, Cambridge, UK, pp. 3–55.

Groves, R. (1989), *Survey Errors and Survey Costs*, Wiley, New York.

Groves, R., and Couper, M. (1998), *Household Survey Nonresponse*, Wiley, New York.

Groves, R., Fowler, F., Couper, M., Lepkowski, J., Singer, E., and Tourangeau, R. (2009), *Survey Methodology*, 2nd ed., Wiley, Hoboken, NJ.

Guggenmoos-Holzmann, I. (1996), The meaning of kappa: Probabilistic concepts of reliability and validity revisited, *Journal of Clinical Epidemiology*, *49*(7), 775–782.

Guggenmoos-Holzmann, I., and Vonk, R. (1998), Kappa-like indices of observer agreement viewed from a latent class perspective, *Statistics in Medicine*, *17*, 797–812.

Haberman, S. J. (1977), Log linear models and frequency tables with small expected cell counts, *Annals of Statistics*, *5*, 1148–1169.

Haberman, S. J. (1979), *Analysis of Qualitative Data, Vol. 2, New Developments*, Academic Press, New York.

Habibullah, M., and Katti, S. K. (1991), A modified steepest descent method with applications to maximizing likelihood functions, *Annals of the Institute of Statistical Mathematics*, *43*(2), 391–404.

Haertel, E. (1984a), An application of latent class models to assessment data, *Applied Psychological Measurement*, *8*, 333–346.

Haertel, E. (1984b), Detection of a skill dichotomy using standardized achievement test items, *Journal of Educational Measurement*, *21*, 59–72.

Haertel, E. (1989), Using restricted latent class models to map the skill structure of achievement items, *Journal of Educational Measurement*, *26*, 301–321.

Hagenaars, J. A. (1988a), LCAG—loglinear modeling with latent variables: A modified LISREL approach, in W. E. Saris and I. N. Gallhofer, eds., *Sociometric Research, Vol. 2, Data Analysis*, Macmillan, London, pp. 111–130.

Hagenaars, J. A. (1988b), Latent structure models with direct effects between indicators: Local dependence models, *Sociological Methods and Research*, *16*, 379–405.

Hagenaars, J. A. (1990), *Categorical Longitudinal Data Log-Linear Panel, Trend, and Cohort*, Sage Publications, Newbury Park, CA.

Hagenaars, J. A., and McCutcheon, A. L. (2002), *Applied Latent Class Analysis*, Cambridge University Press, New York.

Hájek, Jaroslav (1981), *Sampling from a Finite Population*, Marcel Dekker, New York.

Hansen, M., Hurwitz, W., Marks, E., and Mauldin, W. (1951), Response errors in surveys, *Journal of the American Statistical Association*, *46*, 147–190.

Hansen, M., Hurwitz, W., and Bershad, M. (1961), Measurement errors in censuses and surveys, *Bulletin of the International Statistical Institute*, *38*(2), 359–374.

Hansen, M., Hurwitz, W. N., and Pritzker, L. (1964), The estimation and interpretation of gross differences and the simple response variance, in C. R. Rao, ed., *Contributions to Statistics* (presented to P. C. Mahalanobis on the occasion of his 70th birthday), Statistical Publishing Society, Calcutta.

Hansen, M. H., Madow, W. G., and Tepping, B. J. (1983), An evaluation of model-dependent and probability-sampling inferences in sample surveys (with discussion), *Journal of the American Statistical Association*, *78*, 776–807.

Harrison, L. (1997), The validity of self-reported drug use in survey research: An overview and critique of research methods, in L. Harrison and A. Hughes, eds., *NIDA Research Monograph 97–4147*, National Institute of Drug Abuse, Vol. *167*, pp. 17–36.

Hartley, H. O. (1958), Maximum likelihood estimation from incomplete data, *Biometrics*, *14*, 174–194.

Hauck, W., Jr., and Donner, A. (1977), Wald's test as applied to hypotheses in logit analysis, *Journal of the American Statistical Association*, *72*(360), 851–853.

Heinen, T. (1996), *Latent Class and Discrete Latent Trait Models*, Sage Publications, Thousand Oaks, CA.

Hess, J., Singer, E., and Bushery, J. (1999), Predicting test-retest reliability from behavior coding, *International Journal of Public Opinion Research*, *11*(4), 346–360.

Hill, M. (1991), *The Panel Study of Income Dynamics: A User's Guide*, Vol. *2*, Sage, Publications, retrieved from `http://www.sagepub.com/book.aspx?pid=2286`.

Hogan, H., and Wolter K. (1988), Measuring accuracy in a post-enumeration survey, *Survey Methodology*, *14*, 99–116.

Hox, J. J., de Leeuw, E. D., and Kreft, G. G. (1991), The effect of interviewer and respondent characteristics on the quality of survey data: A multilevel model, in P. P. Biemer, R. M. Groves, L. E. Lyberg, N. A. Mathiowetz, and S. Sudman, eds., *Measurement Errors in Surveys*, Wiley, New York.

Hui, S. L., and Walter, S. D. (1980), Estimating the error rates of diagnostic tests, *Biometrics*, *36*, 167–171.

Iannacchione, V., Staab, J., and Redden, D. (2003), Evaluating the use of residential mailing addresses in a metropolitan household survey, *Public Opinion Quarterly*, *67*(2), 202–210.

Iannacchione, V., McMichael, J., Chromy, J., Cunningham, D., Morton, K., Czajka, J., and Curry, R. (2007), Comparing the coverage of a household sampling frame based on mailing addresses to a frame based on field enumeration, paper presented at the American Statistical Association, Survey Research Methods Section, Salt Lake City, UT.

Jabine, T., King, K., and Petroni, R. (1990), *Quality Profile for the Survey of Income and Program Participation (SIPP)*, US Bureau of the Census, Washington, DC.

Jamshidian, M., and Jennrich, R. (2000), Standard errors for EM estimation, *Journal of the Royal Statistical Society, Series B*, *62*(2), 257–270.

Jay, G., Belli, R., and Lepkowski, J. (1994), Quality of last doctor visit reports: A comparison of medical records and survey data, paper presented at the American Statistical Association, Section on Survey Research Methods.

Juran, J., and Gryna, F. (1980), *Quality Planning and Analysis*, 2nd ed., McGraw-Hill, New York.

Kalton, G., Winglee, M., Krawchuk, S., and Levine, D. (2000), *Quality Profile for SASS: Rounds 1–3: 1987–1995*, US Department of Education, National Center for Education Statistics (NCES 2000-308), Washington, DC.

Kasprzyk, D., and Kalton, G. (2001), Quality profiles in U.S. statistical agencies, paper presented at the International Conference on Quality in Official Statistics, Stockholm, May.

Kish, L. (1962), Studies of interviewer variance for attitudinal variables, *Journal of the American Statistical Association*, *57*, 92–115.

Kish, L. (1965), *Survey Sampling*, Wiley, New York.

Körmendi, E. (1988), The quality of income information in telephone and face to face surveys, in R. Groves, P. Biemer, L. Lyberg, J. Massey, W. Nicholls, II, and J. Waksberg, eds., *Telephone Survey Methodology*, Wiley, New York, pp. 341–375.

Korn, E., and Graubard, B. (1999), *Analysis of Health Surveys*, Wiley, New York.

Kreuter, F., Yan, T., and Tourangeau, R. (2008), Good item or bad—can latent class analysis tell? The utility of latent class analysis for the evaluation of survey questions, *Journal of the Royal Statistical Society, Series A: Statistics in Society*, *171*, 723–738.

Krosnick, J. (1991), Response strategies for coping with the cognitive demands of attitude measures in surveys, *Applied Cognitive Psychology*, *5*, 213–236.

Langeheine, R. (1988), Manifest and latent Markov chain models for categorical panel data, *Journal of Educational Statistics*, *13*, 299–312.

Langeheine, R., and Rost, J. (1988), *Latent Trait and Latent Class Models*. Plenum Press, New York.

Langeheine, R. and van de Pol, F. (1990), A unifying framework for Markov modeling in discrete space and discrete time. *Sociological Methods and Research*, *18*, 416–441.

Langeheine, R., and Van de Pol, F. (1993), Multiple indicator Markov models, in R. Steyer, K. F. Wender, and K. F. Widaman, eds., *Proceedings of the 7th European Meeting of the Psychometric Society in Trier*, Fischer Stuttgart, pp. 248–252.

Langeheine, R., and Van de Pol, F. (1994), Discrete-time mixed Markov latent class models, in A. Dale and R. B. Davies, eds., *Analyzing Social and Political Change: A Casebook of Methods*, Sage Publications, London, pp. 170–197.

Langeheine, R., Pannekoek, J., and Van de Pol, F. (1996), Bootstrapping goodness-of-fit measures in categorical data analysis, *Sociological Methods and Research*, *24*, 492–516.

Lanza, S. T., Collins, L., Lemmon, D., and Schafer, J. L. (2007), PROC LCA: A SAS procedure for latent class analysis, *Structural Equation Modeling: A Multidisciplinary Journal*, *14*(4), 671–694.

Lanza, S. T., Collins, L. M., Schafer, J. L., and Flaherty, B. P. (2005), Using data augmentation to obtain standard errors and conduct hypothesis tests in latent class and latent transition analysis, *Psychological Methods*, *10*, 84–100.

Laska, E. M. (2002), The use of capture-recapture methods in public health, *Bulletin of the World Health Organization*, *80*(11), 845.

Lazarsfeld, P. F. (1950a), The logical and mathematical foundations of latent structure analysis, in S. A. Stouffer et al., eds., *Measurement and Prediction*, Princeton University Press, Princeton, NJ, Chapter 10, pp. 362–412.

Lazarsfeld, P. F. (1950b), Some latent structures, in S. A. Stouffer et al., eds., *Measurement and Prediction*, Princeton University Press, Princeton, NJ, Chapter 11, pp. 362–412.

Lazarsfeld, P. F., and Henry, N. W. (1968), *Latent Structure Analysis*, Houghton-Mifflin, Boston.

Lessler, J., Forsyth, B., and Hubbard, M. (1992), Cognitive evaluation of the questionnaire, in C. Turner, J. Lessler, and J. Gfroerer, eds., *Survey Measurement of Drug Use: Methodological Studies*, National Institute on Drug Use, Rockville, MD, pp. 13–52.

Levy, P., and Lemeshow, S. (2008), *Sampling of Populations: Methods and Applications*, 4th ed., Wiley, Hoboken, NJ.

Lin, T. S., and Dayton, C. M. (1997), Model-selection information criteria for non-nested latent class models, *Journal of Educational and Behavioral Statistics*, *22*, 249–264.

Little, R. J. A., and Rubin, D. B. (1987), *Statistical Analysis with Missing Data*, Wiley, New York.

Lohr, S. (1999), *Sampling: Design and Analysis*, Duxbury Press, Pacific Grove, CA.

Lord, F., and Novick, M. (1968), *Statistical Theories of Mental Test Scores*. Addison-Wesley, Reading, MA.

Louis, T. A. (1982), Finding the observed information matrix when using the EM algorithm, *Journal of the Royal Statistical Society, Series B*, *44*, 226–233.

Lyberg, L., Felme, S., and Olsson, L. (1977), *Kvalitetsskydd av data (Data quality protection)*, Liber (in Swedish).

Magidson, J. and Vermunt, J. (2004), Latent class analysis, in D. Kaplan, ed. *The Sage Handbook of Quantitative Methodology for the Social Sciences*, Sage Publishers, Thousand Oaks, CA, Chapter 10.

Marks, E., Seltzer, W., and Krotki, K. (1974), *Population Growth Estimation: A Handbook of Vital Statistics Measurement*, The Population Council, New York.

Marquis, K. (1978), Inferring health interview response bias from imperfect record checks, paper presented at the American Statistical Association, Section on Survey Research Methods.

Marquis, K., and Moore, J. (1990), Measurement errors in the Survey of Income and Program Participation (SIPP) program reports, paper presented at the US Bureau of the Census Annual Research Conference.

McCutcheon, A. L. (1987), *Latent Class Analysis, Quantitative Applications in the Social Sciences Series*, Vol. *64*, Sage Publications, Thousand Oaks, CA.

McLachlan, G., and Krishnan, T. (1997), *The EM Algorithm and Extensions*, Wiley, New York.

McLachlan, G., and Peel, D. (2000), *Finite Mixture Models*, Wiley, New York.

Meng, X. L., and Rubin, D. B. (1993), Maximum likelihood estimation via the ECM algorithm: A general framework, *Biometrika*, *80*, 267–278.

Mieczkowski, T. (1991), The accuracy of self-reported drug use: An analysis and evaluation of new data, in R. Weisheit, ed., *Drugs, Crime, and Criminal Justice System*, Anderson, Cincinnati, pp. 275–302.

Miller, S. M. and Polivka, A.E. (2004), Comment on "An analysis of classification error for the revised Current Population Survey employment questions," *Survey Methodology*, *30*(2), 145–150.

Mooihaart, A., and van der Heijden, P. G. M. (1992), The EM algorithm for latent class analysis with equality constraints, *Psychometrika*, *57*(2), 261–269.

Moore, J. C. (1988), Self/proxy response status and survey response quality, *Journal of Official Statistics*, *4*(2), 155–122.

Moré, J. J. (1978), The Levenberg-Marquardt algorithm: Implementation and theory, in G. A. Watson, ed., *Lecture Notes in Mathematics 630: Numerical Analysis*, Berlin: Springer-Verlag, Berlin, pp. 105–116.

Morton, J. E., Mullin, P., and Biemer, P. (2008), Using reinterview and reconciliation methods to design and evaluate survey questions, *Survey Research Methods*, *2*(2), 75–82.

Mosteller, F., and Tukey, J. W. (1968), Data analysis, including statistics, in G. Lindzey and E. Aronson, eds., *Handbook of Social Psychology*, Vol. *2*, Addison-Wesley, Reading, MA.

Muthén, L. K., and Muthén, B. O. (1998–2005), *Mplus*, Muthén & Muthén, Los Angeles.

National Health Interview Survey (2008), Design of the National Health Interview Survey, available from `http://www.cdc.gov/nchs/nhis/about_nhis.htm`.

Neyman, J. (1934), On the two different aspects of the representative method: The method of stratified sampling and the method of purposive selection, *Journal of the Royal Statistical Society*, *97*, 558–606.

Noreen, E. W. (1989), *Computer Intensive Methods for Testing Hypotheses: An Introduction*, Wiley, New York.

Nunnally, J., and Bernstein, I. (1994), *Psychometric Theory*, 3rd ed., McGraw-Hill, New York.

O'Muircheartaigh, C. (1991), Simple response variance: estimation and determinants, in P. P. Biemer, R. M., Groves, L. Lyberg, N. A. Mathiowetz, and S. Sudman, eds., *Measurement Errors in Surveys*, Wiley, New York, pp. 551–574.

O'Muircheartaigh, C. and Campanelli, P. (1998), The relative impact of interviewer effects and sample design effects on survey precision, *Journal of the Royal Statistical Society*, *A*, *161*(1), 63–77.

O'Muircheartaigh, C., English, E., and Eckman, S. (2007), Predicting the relative quality of alternative sampling frames, paper presented at the American Statistical Association, Survey Research Methods Section, Anaheim, CA.

Patterson, B., Dayton, C. M., and Graubard, B. (2002), Latent class analysis of complex survey data: Application to dietary data, *Journal of the American Statistical Association*, *97*, 721–729.

Pfeffermann, D. (1993), The role of sampling weights when modeling survey data, *International Statistical Review*, *61*(2), 317–337.

Pledger, S. (2000), Unified maximum likelihood estimates for closed capture-recapture models using mixtures, *Biometrics*, *56*, 434–442.

Pollock, K. H., Nichols, J. D., Brownie, C., and Hines, J. E. (1990), *Statistical Inference for Capture-Recapture Experiments*, Wildlife Society Monograph 107.

Poterba, J., and Summers, L. (1995), Unemployment benefits and labor market transitions: A multinomial logit model with errors in classification, *Review of Economics and Statistics*, *77*, 207–216.

Poulsen, C. A. (1982), *Latent Structure Analysis with Choice Modeling Applications*, Aarhus School of Business Administration and Economics, Arhus, Denmark.

Qu, Y., Tan, M., and Kuther, M. H. (1996), Random effects models in latent class analysis for evaluating accuracy of diagnostics test, *Biometrics*, *53*(3), 797–810.

Rabe-Hesketh, S., Skrondal, A., and Pickles, A. (2004), *GLLAMM Manual*, UC Berkeley Division of Biostatistics Working Paper Series, October.

Raglin, D., Griffin, D., and Kromar, R. (1998), *1996 Census Test Evaluations*, Internal Census Bureau Report, available from the US Census Bureau.

Rai, S. N., and Matthews, D. E. (1993), Improving the EM algorithm, *Biometrics*, *49*, 587–591.

Ramsey, J. B. (1969), Tests for specification errors in classical least-squares regression analysis, *Journal of the Royal Statistical Society, Series B*, *31*, 350–371.

Robert, C. P., and Casella, G. (2004), *Monte Carlo Statistical Methods*, 2nd ed., Springer-Verlag, New York.

Rost, J., and Langeheine, R., eds. (1997), *Applications of Latent Trait and Latent Class Models in the Social Sciences*, Waxmann Münster, New York.

Rubin, D. (1987), *Multiple Imputation for Nonresponse in Surveys*, Wiley, New York.

Schafer, J. L. (1997), *Analysis of Incomplete Multivariate Data*, Chapman & Hall/CRC Press, London.

Scheuren, F. (1999), *What Is a Survey?* Web booklet available at `http://www.whatisasurvey.info/`.

Schwarz, N., and Sudman, S., eds. (1994), *Autobiographical Memory and the Validity of Retrospective Reports*, Springer-Verlag, New York.

Seber, G. A. F. (1982), *The Estimation of Animal Abundance and Related Parameters*, 2nd ed., Blackburn Press, Caldwell, NJ.

Sekar, C. C., and Deming, W. E. (1949), On a method of estimating birth and death rates and extent of registration, *Journal of the American Statistical Association*, *44*, 101–115.

Sepulveda, R., Vicente-Villardon, J. L., and Galindo, M. P. (2008), The Biplot as a diagnostic tool of local dependence in latent class models: A medical application, *Statistics in Medicine*, *27*, 1855–1869.

Shores, R., and Sands, R. (2003), Correlation bias in the Accuracy and Coverage Evaluation Survey Revision II, paper presented at the American Statistical Association, on CD-ROM.

Sinclair, M. (1994), *Evaluating Reinterview Survey Methods for Measuring Response Errors*, unpublished doctoral dissertation, George Washington University, Washington, DC.

Sinclair, M., and Gastwirth, J. (1996), On procedures for evaluating the effectiveness of reinterview survey methods: application to labor force data, *Journal of the American Statistical Association*, *91*, 961–969.

Sinclair, M. D., and Gastwirth, J. L. (1998), Estimates of the errors in classification in the labor force survey and their effect on the reported unemployment rate, *Survey Methodology*, *24*, 157–169.

Sinclair, M. D., and Gastwirth, J. L. (2000), Properties of the Hui and Walter and related methods for estimating prevalence rates and error rates of diagnostic testing procedures, *Drug Information Journal*, *34*, 605–615.

Singh, A. C., and Rao, J. N. K. (1995), On the adjustment of gross flow estimates for classification error with application to data from the Canadian Labour Force Survey, *Journal of the American Statistical Association*, *90*(430), 478–488.

Skinner, C. J. (1989), Introduction to part A, in C. Skinner, D. Holt, and T. M. F. Smith, eds., *Analysis of Complex Surveys*, Wiley, Chichester, UK, Chapter 2.

Skinner, C., and Vallet, L. (in press), Fitting log-linear models to contingency tables from surveys with complex sampling designs: An investigation of the Clogg-Eliason approach, *Sociological Methods and Research*.

Skrondal, A., and Rabe-Hesketh, S. (2004), *Generalized Latent Variable Modeling: Multilevel, Longitudinal, and Structural Equation Models*, Chapman & Hall/CRC, Boca Raton, FL.

Smith, T. M. F. (1990), Comment on Rao and Bellhouse: Foundations of survey-based estimation and analysis, *Survey Methodology*, *16*, 26–29.

Stapleton, L. (2008), Analysis of data from complex surveys, in J. Hox, E. de Leeuw, and D. Dillman, eds., *International Handbook of Survey Methodology*, Psychology Press, Taylor & Francis, NY, pp. 342–369.

Statistics Canada (2001), *Coverage, Census Technical Report 2*, Statistics Canada Catulog no. 92-394-XIE, available at `http://www12.statcan.ca/english/census01/Products/Reference/tech_rep/coverage/offline%20documents/92-394-XIE.pdf`.

Statistics Canada (2006), *Quality*, available at `http://www.statcan.ca/english/edu/power/ch3/quality/quality.htm`.

Stouffer, S. A., and Toby, J. (1951), Role conflict and personality, *American Journal of Sociology*, *56*, 395–406.

Taylor, M. C. (1983), The black-and-white model of attitude stability: A latent class examination of opinion and nonopinion in the American public, *American Journal of Sociology*, *89*, 373–401.

Thurstone, L. L. (1947), *Multiple-Factor Analysis*, University of Chicago Press.

Tourangeau, R., Rips, L. J., and Rasinski, K. (2000), *The Psychology of Survey Response*, Cambridge University Press, Cambridge, UK.

Trewin, D. (2006), *Measuring Net Undercount in the 2006 Population Census Australia*, Australian Bureau of Statistics Information Paper, Catalog No. 2940.0.55.001, Canberra, Australia.

Tucker, C. (2004), Comment on "An analysis of classification error for the revised Current Population Survey employment questions," *Survey Methodology*, *30*(2), 151–153.

Tucker, C., Biemer, P., and Meekins, B. (2003), Latent class modeling and estimation of error in consumer expenditure reports, *Proceedings of the the American Statistical Association, Survey Research Methods Section*, San Francisco.

Tucker, C., Biemer, P., Meekins, B., and Shields, J. (2004), Estimating the level of underreporting of expenditures among expenditure reporters: A micro-level latent class analysis, *Proceedings of the American Statistical Association, Survey Research Methods Section*, Toronto.

Tucker, C., Meekins, B., and Biemer, P. (2005), Estimating the level of underreporting of expenditures among expenditure reporters: A micro-level latent class analysis, *Proceedings of the American Statistical Association, Survey Research Methods Section*, Minneapolis, MN.

Turner, C., Lessler, J., and Devore, J. (1992), Effects of mode of administration and wording of reporting of drug use, in C. Turner, J. Lessler, and J. Gfroerer, eds., *Survey Measurement of Drug Use: Methodological Studies*, US Department of Health and Human Services, Washington, DC.

US Census Bureau (1985), *Evaluating Censuses of Population and Housing*, STD-ISP-TR-5, US Government Printing Office, Washington, DC.

US Census Bureau (2006), *Current Population Survey: Design and Methodology, Technical Paper 66*, available at `http://www.census.gov/prod/2006pubs/tp-66.pdf`.

Uebersax, J. (1999), Probit latent class analysis with dichotomous or ordered category measures: Conditional independence/dependence models, *Applied Psychological Measurement*, *23*(4), 283–297.

Uebersax, S. L. (2000), *CONDEP Program*, available at `http://ourworld.compuserve.com/homoepages/jsuebersax/condep.htm`.

Vacek, P. M. (1985), The effect of conditional dependence on the evaluation of diagnostic tests, *Biometrics*, *41*, 959–968.

Van de Pol, R., and De Leeuw, J. (1986), A latent Markov model to correct for measurement error, *Sociological Methods and Research*, *15*, 118–141.

Van de Pol, F., and Langeheine, R. (1990), Mixed Markov latent class models, in C. C. Clogg, ed., *Sociological Methodology*, Blackwell, Oxford, pp. 213–247.

Van de Pol, F., and Langeheine, R. (1997), Separating change and measurement error in panel surveys with an application to labor market data, in L. L. Lyberg, P. Biemer, M. Collins, E. De Leeuw, C. Dippo, N. Schwarz, and D. Trewin, eds., *Survey Measurement and Process Quality*, Wiley, New York.

Van de Pol, E., Langeheine, R., and De Jong, W. (1991), *PANMARK User Manual: Panel Analysis Using MARKov Chains*, Netherlands Central Bureau of Statistics, Voorburg.

Van de Pol, F., Langeheine, R., and De Jong, W. (1998), *PANMARK User Manual, Version 3*, Netherlands Central Bureau of Statistics, Voorburg.

Vermunt, J. K. (1996), *Log-Linear Event History Analysis: A General Approach with Missing Data, Latent Variables, and Unobserved Heterogeneity*, Tilburg University Press, Tilburg, The Netherlands.

Vermunt, J. (1997a), *Log-Linear Models for Event Histories*, Sage Publications, Thousand Oaks, CA.

Vermunt, J. (1997b), *ℓEM: A General Program for the Analysis of Categorical Data* (user's manual), Department of Methodology and Statistics, Tilburg University, Tilburg, The Netherlands.

Vermunt, J. K. (2002), Comments on "Latent class analysis of complex sample survey data," *Journal of the American Statistical Association*, *97*, 736–737.

Vermunt, J. K. (2003), Multilevel latent class models, *Sociological Methodology*, *33*, 213–239.

Vermunt, J. (2004), Comment on "An analysis of classification error for the revised Current Population Survey employment questions," *Survey Methodology*, *30*(2), pp. 141–144.

Vermunt, J. (2007), Latent class analysis with sample weights: A maximum likelihood approach, *Sociological Methods and Research*, *36*(1), 87–111.

Vermunt, J. K., Langeheine, R., and Bockenholt, U. (1999), Discrete-time discrete-state latent Markov models with time-constant and time-varying covariates, *Journal of Educational and Behavioral Statistics*, *24*(2), 179–207.

Vermunt, J. K., and Magidson, J. (2001), Latent class factor and cluster models, bi-plots and related graphical displays, *Sociological Methodology*, *31*, 223–264.

Vermunt, J., and Magidson, J. (2005a), *Latent GOLD 4.0 User's Guide*, Statistical Innovations, Inc., Belmont, MA.

Vermunt, J. K., and Magidson, J. (2005b), *Technical Guide to Latent Gold 4.0: Basic and Advanced*, Statistical Innovations, Belmont, MA.

Vermunt, J. K., and Magidson, J. (2007), Latent class analysis with sampling weights: A maximum-likelihood approach, *Sociological Methods Research*, *36*(87), 87–111.

Visher, C., and McFadden, K. (1991), *A Comparison of Urinalysis Technologies for Drug Testing in Criminal Justice*, US Department of Justice, National Institute of Justice, Washington, DC.

Wang, Y., and Zhang, N. L. (2006), Severity of local maxima for the EM algorithm: Experiences with hierarchical latent class models, paper presented at the Third European Workshop on Probabilistic Graphical Model (PGH-06).

Wedel, M., ter Hofstede, F., and Steenkamp, J. E. M. (1998), Mixture model analysis of complex surveys, *Journal of Classification*, *15*, 225–244.

White, G. C., and Burnham, K. P. (1999), Program MARK: Survival estimation from populations of marked animals, *Bird Study 46 Supplement*, 120–138.

Wiggins, L. M. (1973), *Panel Analysis, Latent Probability Models for Attitude and Behavior Processing*, Elsevier SPC, Amsterdam.

Wiggins, R. D., Longford, N. T., and O'Muircheartaigh, C. A. (1992), A variance components approach to interviewer effects, in A.Westlake, R. Banks, C. Payne, and T. Orchard, eds., *Survey and Statistical Computing*, North-Holland, Amsterdam, pp. 243–254.

Willis, G. (2005), *Cognitive Interviewing: A Tool for Improving Questionnaire Design*, Sage Publications, Thousand Oaks, CA.

Wolter, K. M. (1986), Some coverage error models for census data, *Journal of the American Statistical Association*, *81*, 157–162.

Wolter, K. (2007), *Introduction to Variance Estimation*, 2nd ed., Springer-Verlag, New York.

Zaslavsky, A., and Wolfgang, G. (1993), Triple-system modeling of census, post-enumeration survey, and administrative-list data, *Journal of Business and Economic Statistics*, *11*, 279–288.

Index

WILEY SERIES IN SURVEY METHODOLOGY

Established in Part by WALTER A. SHEWHART AND SAMUEL S. WILKS

The ***Wiley Series in Survey Methodology*** covers topics of current research and practical interests in survey methodology and sampling. While the emphasis is on application, theoretical discussion is encouraged when it supports a broader understanding of the subject matter.

The authors are leading academics and researchers in survey methodology and sampling. The readership includes professionals in, and students of, the fields of applied statistics, biostatistics, public policy, and government and corporate enterprises.

ALWIN · Margins of Error: A Study of Reliability in Survey Measurement
BETHLEHEM · Applied Survey Methods: A Statistical Perspective
BETHLEHEM, COBBEN, and SCHOUTEN · Handbook of Nonresponse in Household Surveys
BIEMER · Latent Class Analysis of Survey Error
*BIEMER, GROVES, LYBERG, MATHIOWETZ, and SUDMAN · Measurement Errors in Surveys
BIEMER and LYBERG · Introduction to Survey Quality
BIEMER · Latent Class Analysis of Survey Error
BRADBURN, SUDMAN, and WANSINK · Asking Questions: The Definitive Guide to Questionnaire Design—For Market Research, Political Polls, and Social Health Questionnaires, *Revised Edition*
BRAVERMAN and SLATER · Advances in Survey Research: New Directions for Evaluation, No. 70
CHAMBERS and SKINNER (editors) · Analysis of Survey Data
COCHRAN · Sampling Techniques, *Third Edition*
CONRAD and SCHOBER · Envisioning the Survey Interview of the Future
COUPER, BAKER, BETHLEHEM, CLARK, MARTIN, NICHOLLS, and O'REILLY (editors) · Computer Assisted Survey Information Collection
COX, BINDER, CHINNAPPA, CHRISTIANSON, COLLEDGE, and KOTT (editors) · Business Survey Methods
*DEMING · Sample Design in Business Research
DILLMAN · Mail and Internet Surveys: The Tailored Design Method
FULLER · Sampling Statistics
GROVES and COUPER · Nonresponse in Household Interview Surveys
GROVES · Survey Errors and Survey Costs
GROVES, DILLMAN, ELTINGE, and LITTLE · Survey Nonresponse
GROVES, BIEMER, LYBERG, MASSEY, NICHOLLS, and WAKSBERG · Telephone Survey Methodology
GROVES, FOWLER, COUPER, LEPKOWSKI, SINGER, and TOURANGEAU · Survey Methodology, *Second Edition*
*HANSEN, HURWITZ, and MADOW · Sample Survey Methods and Theory, Volume 1: Methods and Applications
*HANSEN, HURWITZ, and MADOW · Sample Survey Methods and Theory, Volume II: Theory

*Now available in a lower priced paperback edition in the Wiley Classics Library.

HARKNESS, BRAUN, EDWARDS, JOHNSON, LYBERG, MOHLER, PENNELL, and SMITH (editors) · Survey Methods in Multinational, Multiregional, and Multicultural Contexts
HARKNESS, VAN DE VIJVER, and MOHLER (editors) · Cross-Cultural Survey Methods
KALTON and HEERINGA · Leslie Kish Selected Papers
KISH · Statistical Design for Research
*KISH · Survey Sampling
KORN and GRAUBARD · Analysis of Health Surveys
LEPKOWSKI, TUCKER, BRICK, DE LEEUW, JAPEC, LAVRAKAS, LINK, and SANGSTER (editors) · Advances in Telephone Survey Methodology
LESSLER and KALSBEEK · Nonsampling Error in Surveys
LEVY and LEMESHOW · Sampling of Populations: Methods and Applications, *Fourth Edition*
LUMLEY · Complex Surveys: A Guide to Analysis Using R
LYBERG, BIEMER, COLLINS, de LEEUW, DIPPO, SCHWARZ, TREWIN (editors) · Survey Measurement and Process Quality
MAYNARD, HOUTKOOP-STEENSTRA, SCHAEFFER, and VAN DER ZOUWEN · Standardization and Tacit Knowledge: Interaction and Practice in the Survey Interview
PORTER (editor) · Overcoming Survey Research Problems: New Directions for Institutional Research, No. 121
PRESSER, ROTHGEB, COUPER, LESSLER, MARTIN, MARTIN, and SINGER (editors) · Methods for Testing and Evaluating Survey Questionnaires
RAO · Small Area Estimation
REA and PARKER · Designing and Conducting Survey Research: A Comprehensive Guide, *Third Edition*
SARIS and GALLHOFER · Design, Evaluation, and Analysis of Questionnaires for Survey Research
SÄRNDAL and LUNDSTRÖM · Estimation in Surveys with Nonresponse
SCHWARZ and SUDMAN (editors) · Answering Questions: Methodology for Determining Cognitive and Communicative Processes in Survey Research
SIRKEN, HERRMANN, SCHECHTER, SCHWARZ, TANUR, and TOURANGEAU (editors) · Cognition and Survey Research
SUDMAN, BRADBURN, and SCHWARZ · Thinking about Answers: The Application of Cognitive Processes to Survey Methodology
UMBACH (editor) · Survey Research Emerging Issues: New Directions for Institutional Research No. 127
VALLIANT, DORFMAN, and ROYALL · Finite Population Sampling and Inference: A Prediction Approach

Printed and bound by CPI Group (UK) Ltd, Croydon, CR0 4YY
16/04/2025
14658520-0002